Lecture Notes in Physics

Volume 1022

The series Lecture Notes in Physics (LNP), founded in 1969, reports new developments in physics research and teaching - quickly and informally, but with a high quality and the explicit aim to summarize and communicate current knowledge in an accessible way. Books published in this series are conceived as bridging material between advanced graduate textbooks and the forefront of research and to serve three purposes:

- to be a compact and modern up-to-date source of reference on a well-defined topic;
- to serve as an accessible introduction to the field to postgraduate students and non-specialist researchers from related areas;
- to be a source of advanced teaching material for specialized seminars, courses and schools.

Both monographs and multi-author volumes will be considered for publication. Edited volumes should however consist of a very limited number of contributions only. Proceedings will not be considered for LNP.

Volumes published in LNP are disseminated both in print and in electronic formats, the electronic archive being available at springerlink.com. The series content is indexed, abstracted and referenced by many abstracting and information services, bibliographic networks, subscription agencies, library networks, and consortia.

Proposals should be sent to a member of the Editorial Board, or directly to the responsible editor at Springer:

Dr Lisa Scalone
lisa.scalone@springernature.com

Betti Hartmann · Jutta Kunz
Editors

Gravity, Cosmology, and Astrophysics

A Journey of Exploration and Discovery with Female Pioneers

Foreword by Jocelyn Bell Burnell

Springer

Editors
Betti Hartmann
Department of Mathematics
University College London
London, UK

Jutta Kunz
Department of Physics
Carl von Ossietzky University of Oldenburg
Oldenburg, Germany

ISSN 0075-8450 ISSN 1616-6361 (electronic)
Lecture Notes in Physics
ISBN 978-3-031-42095-5 ISBN 978-3-031-42096-2 (eBook)
https://doi.org/10.1007/978-3-031-42096-2

This eBook was published Open Access with funding support from the Sponsoring Consortium for Open Access Publishing in Particle Physics (SCOAP3).

This Springer imprint is published by the registered company Springer Nature Switzerland AG
The registered company address is: Gewerbestrasse 11, 6330 Cham, Switzerland

Paper in this product is recyclable.

We dedicate this book to all scientists from underrepresented groups.

Foreword

Maybe every generation says or feels like this, but this seems a time of multiple breakthroughs in astrophysics and cosmology. Of course, new, better equipment delivers new, better data. Some of it is as expected, but some startling, contradicting assumptions we have, unwittingly perhaps, made. The opening of new windows, such as the detection of gravitational radiation, may confirm what some of us (a few of us!) already believed to be there and observable, but in turn reveal unexpected things. These unexpected things may be physical entities, or may be new angles on old theories, or even confirmation of what some had come to suspect that our beloved theories were in need of an extension or serious modification.

Learning new things can be tough, but for practising scientists it is accepted as an essential part of life. Note, I do not say of 'normal life', as with hindsight this is often seen to be a turning point, a place where mentally spinning on the spot opens new avenues, new lines of enquiry, and new understanding. Many of us can likely point to such instances where a window, or a gulf, opened.

This collection of essays will stimulate, reveal, provoke thought, and maybe even surprise and amaze. All these are the starting points of new ideas, new theories, and new understanding. May it also provoke out-of-the-box thinking, and new branches bearing new fruit.

April 2023

Jocelyn Bell Burnell
University of Oxford
Oxford, UK

Preface

During the COVID-19 pandemic, online seminars and conferences became increasingly popular. Using this opportunity, we established an online *London-Oldenburg Relativity Seminar* series that runs to this day on a weekly basis.

From the beginning, our aim has been two-fold: (a) providing advanced undergraduate and postgraduate students with introductory talks on recent and exciting developments in gravity as well as (b) making the work of women in this area of research more visible. Ever since the seminar series started we have been successful in recruiting excellent female speakers. From this sprang the idea to collect written versions of these often very pedagogical and insightful seminars and make them available to a larger audience.

This book is authored solely by women and highlights some of the highly topical contributions of female scientists to research in Cosmology, Astrophysics, and Gravity.

London, UK
Oldenburg, Germany
July 2023

Betti Hartmann
Jutta Kunz

The original version of this book was previously published without open access. A correction to this book is available at https://doi.org/10.1007/978-3-031-42096-2_13

Acknowledgements

BH would like to thank all the members of her close family for the continuous support and help they have provided over the past decades, especially her parents, Elisabeth and Gerhard, who never had the opportunity to pursue higher education, but took the risk to allow her to. Also, thanks to all her great friends who so often provide her with words of encouragement, support, and wisdom—she could not do without them. Thanks to all the researchers she have had the pleasure to collaborate with over the past decades. Also, sincere thanks to all the scientists that contributed to this book and hence made it possible.

JK would like to deeply thank her family, her husband Gerhard and her daughters Esther and Sandra, for their love, support, and understanding in all these years. She would also like to express her sincere thanks to all her teachers and colleagues for their backing, when doing physics was not so common among young women, and when there were few role models around. Moreover, she'd very much like to thank all her collaborators and students over the years: It was wonderful for her to work with them and make together progress in science, she very much enjoyed their company along this exciting path.

BH and JK would like to thank Jens-Christian Drawer for helping with the Latex typesetting of this book.

Contents

Extended Gravity Models

Cosmology

Contributors

Bahareh Azad Institute for Physics, University of Oldenburg, Oldenburg, Germany

Roopkatha Banerjee Department of Computational and Data Sciences, Indian Institute of Science, Bangalore, India

Nele Callebaut Institute for Theoretical Physics, University of Cologne, Köln, Germany

Carla Cederbaum Universität Tübingen, Fachbereich Mathematik, Tübingen, Germany

Celia Escamilla-Rivera Instituto de Ciencias Nucleares, Universidad Nacional Autónoma de México, Circuito Exterior C.U., México, México

Melanie Graf Universität Tübingen, Fachbereich Mathematik, Tübingen, Germany;
Fachbereich Mathematik, Universität Hamburg, Hamburg, Germany

Ruth Gregory Department of Physics, King's College London, The Strand, London, UK

Merce Guerrero Physics Department, Autonomous University of Barcelona, Barcelona, Spain

Fech Scen Khoo Institute for Physics, University of Oldenburg, Oldenburg, Germany

Sayantani Lahiri Center of Applied Space Technology and Microgravity (ZARM), Universität Bremen, Bremen, Germany

Petya Nedkova Faculty of Physics, Sofia University, Sofia, Bulgaria

Laura L. Parrilla Instituto de Ciencias Nucleares, Universidad Nacional Autónoma de México, Circuito Exterior C.U., México, México

Suparna Roychowdhury Department of Physics, St. Xavier's College, Kolkata, India

Lara Sousa Centro de Astrofísica da Universidade do Porto, Porto, Portugal; Instituto de Astrofísica e Ciências do Espaço, Porto, Portugal; Departamento de Física e Astronomia, Faculdade de Ciências, Universidade do Porto, Porto, Portugal

Aneta Wojnar Departamento de Física Teórica, Universidad Complutense de Madrid, Madrid, Spain

Acronyms

ADM	Arnowitt-Deser-Miser
AdS	Anti-de Sitter space-time
BBN	Big Bang Nucleosynthesis
BH	Black hole
BTZ	Bañados-Teitelboim-Zanelli
CDL	Coleman-de Luccia
CFT	Conformal Field Theory
CMB	Cosmic Microwave background
CMC	Constant mean curvature
CRTBP	Circular restricted three-body problem
dS	de Sitter space-time
EHT	Event Horizon Telescope
EIH	Einstein-Infeld-Hoffmann
EOM	Equation of motion
EOS	Equation of state
FHP	Fodor-Hoenselaers-Perjés
FLI	Fast Lyapunov Indicators
FLRW	Friedman-Lemâitre-Robertson-Walker
FV	False vacuum
GALI	General Alignment Index
GR	General Relativity
GRMHD	General Relativistic Magneto-Hydrodynamics
GW	Gravitational Wave
HR	Hertzsprung-Russell
IR	Infrared
ISCO	Innermost stable circular orbit
JLMS	Jafferis-Lewkowycz-Maldacena-Suh
JNW	Janis-Newman-Winicour
KAM	Kolmogorov-Arnold-Moser
LCE	Lyapunov Characteristic exponent
LE	Lane-Emden
LIGO	Laser Interferometer Gravitational-Wave Observatory
LISA	Laser Interferometer Space Antenna
LM	Lewkowycz-Maldacena

M87	Messier 87
MG	Modified gravity
MIS	Müller-Israel-Stewart
MLEE	Modified Lane-Emden equation
MLE	Maximal Lyapunov Exponent
ODE	Ordinary differential equation
PDE	Partial differential equation
PMS	Pre-main sequence
PN	Post-Newtonian
RMS	Root-mean-squared
RT	Ryu-Takayanagi
QFT	Quantum Field Theory
QNM	Quasi-normal mode
RTBP	Restricted three-body problem
SALI	Small Alignement Index
SM	Standard Model
SMBH	Supermassive black hole
STCMC	Space-time constant mean curvature
STT	Scalar-tensor theory
TFD	Thermofield double
TV	True vacuum
UCO	Ultra-compact object
UV	Ultraviolet
VOS	Velocity-dependent one-scale
WD	White dwarf

Astrophysics and Observations

Quasinormal Modes of Static Ellis-Bronnikov Wormholes

Bahareh Azad

Abstract

Quasinormal modes represent characteristic features of compact objects, since they entail their reaction to perturbations. While black holes and neutron stars in general relativity have been extensively studied and their quasinormal modes are well-known, the complete set of quasinormal modes for wormholes was only recently obtained. Here we discuss how to determine these quasinormal modes of Ellis-Bronnikov wormholes. These wormholes are static spherically symmetric solutions of general relativity coupled to a phantom scalar field. We obtain the axial and polar quasinormal modes and show, that for the massless wormhole a threefold degeneracy of the modes arises, i.e., there is isospectrality of the modes.

1 Introduction

The theory of wormholes, also known as Einstein-Rosen bridges, is a fascinating concept in theoretical physics that stems from the theory of general relativity. Originally introduced by Einstein and Rosen [29], wormholes represent hypothetical structures in spacetime that could provide shortcuts or connections between distant regions, allowing for space travel.

In the classical formulation, wormholes were considered non-traversable and encumbered by event horizons [54]. However, subsequent investigations have explored the possibility of traversable wormholes by incorporating exotic matter or non-standard scalar fields. Ellis [30,31] and Bronnikov [21] demonstrated that wormhole solutions could be obtained by employing phantom fields, which are scalar fields with negative energy densities. These fields violate the energy conditions of classical general relativity, allowing for the existence of traversable wormholes.

B. Azad (✉)
University of Oldenburg, D-26111 Oldenburg, Germany
e-mail: bahareh.azad@uni-oldenburg.de

B. Hartmann and J. Kunz (eds.), *Gravity, Cosmology, and Astrophysics*,
Lecture Notes in Physics 1022, https://doi.org/10.1007/978-3-031-42096-2_1

Alternative theories of gravity, such as Einstein-scalar-Gauss-Bonnet theories [7,41,42], have also been explored as frameworks that can accommodate wormholes without the need for exotic matter. In these theories, the violation of the energy conditions is achieved through the gravitational degrees of freedom alone. Furthermore, quantum degrees of freedom, including Dirac particles [17–19,51] and 3-form fields as well [13,20], have been investigated as potential sources for violating the energy conditions and enabling the formation of traversable wormholes.

The theoretical interest in wormholes extends beyond their existence and properties. Detecting and characterizing these structures is a topic of great interest. Various detection methods have been proposed, including the observation of gravitational lensing effects [4,12,25,52,55,56,59,61,64–67], the search for wormhole shadows [12,20,36,37,57,58,62], and the study of accretion disks and associated radiation [11,27,28,38,39,53,70], etc.

An alternative fascinating option to identify wormholes and differentiate them from black holes may emerge through investigations of how a scalar field scatters. Lately, mathematical formulations have been derived for the transmission and reflection amplitudes of the relevant effective potential, as well as the absorption cross section of wormholes. These calculations have demonstrated that, whether the field is massless or has mass, an observer positioned at infinity can readily distinguish between a wormhole and a Schwarzschild black hole by analyzing the scattering information of the scalar field [9].

But the detecting of gravitational waves in Ligo-Virgo opened a new window to study compact objects. Currently, gravitational wave astrophysics [1–3] is giving us a new potential approach to observe wormholes through their damping modes called quasinormal modes (see e.g. [14,44,47]).

Quasinormal modes are characteristic modes of freely oscillating spacetime. When a compact object such as a black hole, a neutron star, or a wormhole undergoes oscillations, the system is open, and gravitational waves gradually lose energy and decay over time. Hence, these modes are referred to as quasinormal modes. They are described by a complex frequency $\omega = \omega_R + i\omega_I$, where ω_R is the real part and ω_I is the imaginary part. The real part determines the frequency of the oscillations, while the imaginary part represents the damping rate of the oscillations [44,47].

The investigation of quasinormal modes of wormholes has been previously explored in various contexts [6,16,24,35,40,43,45,46,48–50,69]. There is an interesting aspect when we look at the inverse problem, which corresponds to determining the shape of a wormhole from the knowledge of its quasinormal modes [49,69]. (However in black holes a family of effective potentials produces the same quasinormal mode spectrum [23].) Quasinormal modes have been utilized to reconstruct the metric near the throat in the case of symmetric Ellis-Bronnikov wormholes, providing a specific example [49]. The quasinormal modes of the general family of Ellis-Bronnikov wormholes have been investigated completely now [10,16,43,45]. A systematic analysis of the scalar, axial, and radial perturbations has been performed first [16], and a complete set of modes of the polar perturbations has been provided recently [10].

In this chapter, we derive a complete set of quasinormal modes for Ellis-Bronnikov wormholes. Section 2 introduces the concept of Ellis-Bronnikov wormholes. In Sect. 3 we examine scalar perturbations as a toy model and explain the theoretical and numerical calculations for determining the quasinormal modes. Then, in Sect. 4, we apply the same method to study the full set of metric and scalar perturbations and explore the spectrum of modes. Finally, we conclude in Sect. 5.

2 The Ellis-Bronnikov Wormhole Spacetime

2.1 Introduction

The solution to the vacuum Einstein equations in the context of the theory of general relativity for the static spherically symmetric scenario [29] leads to the formation of the Einstein-Rosen bridge. It denotes a non-traversable wormhole. Morris and Thorne [54] provided a comprehensive discussion on the indispensability of a phantom field or, more generally, some form of unconventional matter that violates the energy conditions in classical general relativity. Furthermore, they explored the idea of using wormholes for quick interstellar travel, see also [5,68]. Ellis [30,31] and Bronnikov [21] independently derived wormhole solutions that can be traversed. However, the attainment of such traversable wormholes necessitated the utilization of an unconventional scalar field known as a phantom field. In the text we call them Ellis-Bronnikov or for simplicity just E-B wormholes.

2.2 The Ellis-Bronnikov Wormhole Metric

With a massless minimally coupled phantom field ϕ to the Einstein-Hilbert action we consider the action

$$S = \frac{1}{16\pi G}\int d^4x\sqrt{-g}\Big(R + 2\nabla_\mu\phi\nabla^\mu\phi\Big). \tag{1}$$

The coupled set of equations of motion is obtained by varying the action,

$$R_{\mu\nu} = -2\,\partial_\mu\phi\,\partial_\nu\phi, \tag{2}$$

$$\nabla_\mu\nabla^\mu\phi = 0. \tag{3}$$

Static and spherically symmetric E-B wormholes can be expressed as

$$\phi = \phi^{(b)}(r) = \frac{Q}{r_0}\left[\tan^{-1}\left(\frac{r}{r_0}\right) - \frac{\pi}{2}\right], \tag{4}$$

$$ds^2 = g^{(b)}_{\mu\nu}dx^\mu dx^\nu = -e^f dt^2 + \frac{1}{e^f}\Big[dr^2 + \left(r^2 + r_0^2\right)\left(d\theta^2 + \sin^2\theta d\varphi^2\right)\Big], \tag{5}$$

with

$$f = \frac{C}{r_0}\left[\tan^{-1}\left(\frac{r}{r_0}\right) - \frac{\pi}{2}\right], \tag{6}$$

and Q, r_0, and C are constants that are discussed below and that satisfy

$$4Q^2 = C^2 + 4r_0^2. \tag{7}$$

The superscript (b) indicates, that these are the background solutions to be employed in the perturbation expansions.

If we look at the asymptotic behaviour, when $r \to +\infty$, the metric function f goes to zero, $f \to 0$, and the metric approaches asymptotically Minkowski spacetime. On the other hand, when $r \to -\infty$, we need a coordinate transformation to approach Minkowski spacetime,

$$\bar{t} = e^{-\frac{C\pi}{2r_0}} t\,, \quad \bar{r} = e^{\frac{C\pi}{2r_0}} r\,, \quad \bar{r}_0 = e^{\frac{C\pi}{2r_0}} r_0\,. \tag{8}$$

In the spacetime the two asymptotically flat regions are connected by a throat. where the circumferential radius $R(r)$,

$$R^2(r) = e^{-f}(r^2 + r_0^2)\,, \tag{9}$$

assumes its minimal value.

The constant Q represents the charge of the phantom field, and the constant C determines the mass of the wormhole, obtained in the asymptotically flat region as r tends to infinity. It is defined by

$$C = 2M\,. \tag{10}$$

The constant C represents also a measure of the symmetry of the wormhole. For $C = 0$, the wormhole is massless and it is symmetric with respect to the throat, when considering reflection at $r = 0$, i.e., the transformation $r \to -r$. On the other hand, when $C \neq 0$, the wormhole acquires mass and loses its symmetry. In this case, the throat of the wormhole is situated either in the region $r < 0$ or in the region $r > 0$. Moreover, there exists a symmetry relationship between solutions when C is positive and when it is negative,

$$f(r, C) = f(-r, -C) - \frac{\pi C}{r_0}, \tag{11}$$

$$\phi(r, C) = -\phi(-r, -C) - \frac{\pi Q}{r_0}\,. \tag{12}$$

The above solutions are now employed as the background solutions in perturbation theory. Perturbation theory is a mathematical technique used to study the behavior of small deviations from a known solution in general relativity. In general relativity quasinormal modes are the characteristic modes of freely oscillating black holes, wormholes or other compact objects.

To calculate the quasinormal modes using perturbation theory, we first start with a known metric solution, such as the Schwarzschild metric or E-B metric. We then perturb the metric by adding small fluctuations to it. The quasinormal modes can be determined by studying the asymptotic behavior of the perturbations.

3 Scalar Perturbations

3.1 Theoretical Set-Up

Before going to the full set of perturbations of the E-B wormhole, it is instructive to consider a simpler problem. Therefore we here study only the perturbations ψ of the phantom field ϕ

$$\phi = \phi^{(b)} + \psi, \tag{13}$$

where the superscript (b) stands for the background field, on the E-B wormhole background. Thus we assume that the metric is not perturbed. The ψ is then viewed as a "test" scalar field whose amplitude is small, so that we can ignore its backreaction on the metric.

The variation of the scalar field equation leads to

$$\delta\Box\phi = \Box\psi = 0. \tag{14}$$

So we need to calculate

$$\Box\psi \equiv \nabla_\mu \nabla^\mu \psi = 0 \tag{15}$$

where $\psi = \psi(t, r, \theta, \phi)$ is small. Also ∇^μ is the covariant derivative operator and $\Box$ is the d'Alembertian operator, i.e.,

$$\nabla_\mu \nabla^\mu \psi = \frac{1}{\sqrt{-g}}\partial_\mu \left(\sqrt{-g}\partial^\mu \psi\right) = 0. \tag{16}$$

Since the background is spherically symmetric we employ a spherical harmonic decomposition ($m = 0$) of the perturbation ψ of the scalar field, yielding

$$\psi(t, r) = \sum_l \int d\omega e^{-i\omega t} u_l(r)\, P_l(\cos\theta)\,. \tag{17}$$

For given values of l and ω the Laplacian of the scalar field perturbation then becomes

$$u'' = -\frac{2r}{r^2 + r_0^2}u' + \left[\frac{l(l+1)}{r^2 + r_0^2} - \omega^2 e^{-2f}\right]u. \tag{18}$$

Fig. 1 Effective potential $V_l^{scalar}(r)$ for scalar perturbations versus radial coordinate r for $C = 0, 0.1, 0.25, 0.5$ and 1 ($l = 0$, $r_0 = 1$)

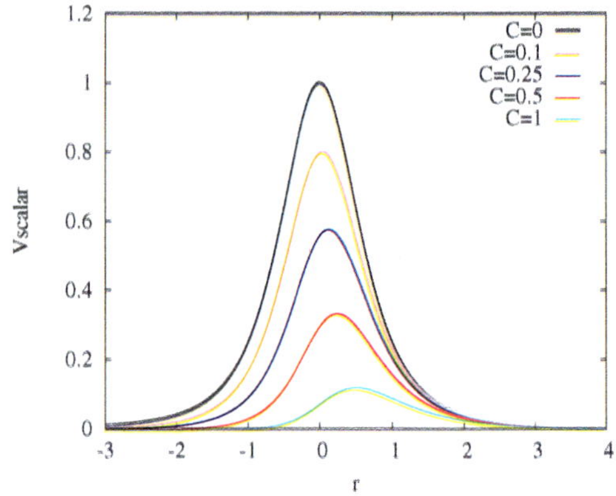

By defining

$$u(r) = A(r)Z(r^*) \tag{19}$$

with

$$A(r) = \frac{e^{f/2}}{\sqrt{r^2 + r_0^2}} \tag{20}$$

and employing the tortoise coordinate r^*

$$\frac{dr^*}{dr} = e^{-f} \tag{21}$$

we rewrite Eq. (18) as a second order Schrödinger-like equation,

$$\frac{d^2 Z}{dr^{*2}} + \left(\omega^2 - V^{scalar}(r)\right) Z = 0, \tag{22}$$

with the effective potential

$$V_l^{scalar}(r) = \mathrm{e}^{2f} \left(\frac{l\,(l+1)+1}{r_0^2 + r^2} - \frac{(2r - C)^2}{4\left(r_0^2 + r^2\right)^2} \right). \tag{23}$$

In Fig. 1 we exhibit the effective potential $V_l^{scalar}(r)$ for $C = 0, 0.1, 0.25, 0.5$ and 1 ($r_0 = 1, l = 0$).

3.2 Numerical Method

To obtain the damped quasinormal modes for the scalar perturbations, we use the direct integration method. We solve Eq. (22), the second order Schrödinger-like equation, with the scalar perturbation potential (23) (see [16]). We have to impose that we do not have any incoming waves from infinity. Therefore the modes are purely outgoing at infinity $r^* \to \pm\infty$

$$Z \simeq e^{\pm i\omega r^*}, \quad r^* \to \pm\infty. \tag{24}$$

To numerically determine the quasinormal modes of the scalar perturbations, we divide the space at a specific value r_c into two regions. Beyond r_c ($r > r_c$), we describe the asymptotic behavior of the perturbation function $Z(r)$ as r approaches positive infinity, following the approach presented in [22] as follows,

$$r > r_c, \quad Z^+(r) = e^{i\omega r_*} Z_P(r). \tag{25}$$

In this case, we obtain the leading terms in the asymptotic expansion of $Z_P(r)$

$$Z_P(r) = a_0\left(1 + \frac{il(l+1)}{2\omega}\frac{1}{r} + \left(\frac{-iC(l^2+l-1)}{4\omega} - \frac{l(l+2)(l^2-1)}{8\omega^2}\right)\frac{1}{r^2} + \cdots\right). \tag{26}$$

Here, a_0 represents an arbitrary amplitude. In the region r_c ($r < r_c$), we describe the asymptotic behavior as r approaches negative infinity as

$$r < r_c, \quad Z^-(r) = e^{-i\omega r_*} Z_N(r), \tag{27}$$

and the leading terms in the asymptotic expansion of $Z_N(r)$ are

$$Z_N(r) = a_0\left(1 - \frac{il(l+1)}{2\omega e^{C\pi/r_0}}\frac{1}{r} + \left(\frac{-iC(l^2+l-1)}{4\omega e^{C\pi/r_0}} - \frac{l(l+2)(l^2-1)}{8\omega^2 e^{2C\pi/r_0}}\right)\frac{1}{r^2} + \cdots\right). \tag{28}$$

For a given value of ω, we generate solutions for the functions $Z_N(r)$ and $Z_P(r)$ that satisfy the expansions near infinity, while ensuring $Z_P(r_c) = Z_N(r_c) = 1$. The quasinormal modes are determined when the matching condition is satisfied,

$$\left.\frac{1}{Z^-}\frac{dZ^-}{dr}\right|_{r=r_c} - \left.\frac{1}{Z^+}\frac{dZ^+}{dr}\right|_{r=r_c} = 0. \tag{29}$$

To numerically integrate the equations with the corresponding boundary conditions, we employ the package Colsys [8], a collocation method for systems of ordinary differential equations with error estimation and adaptive mesh selection.

3.3 Spectrum

Let us now examine the spectrum of quasinormal modes for the scalar perturbations. The quasinormal modes corresponding to the scalar phantom field have been obtained in Ref. [16]. For every value of l, we observe a unique set of fundamental modes. To put it simply, these modes represent stable disturbances that gradually diminish

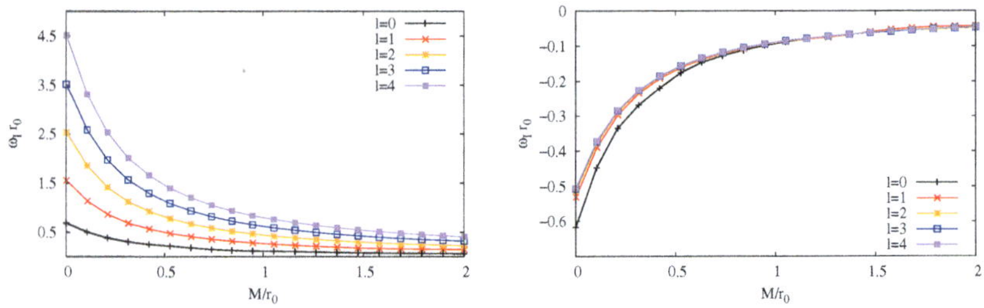

Fig. 2 Scalar quasinormal modes: dimensionless frequency $\omega_R r_0$ (left) and dimensionless decay rate $\omega_I r_0$ (right) versus dimensionless mass M/r_0 ($r_0 = 1$) for multipole numbers $l = 0, 1, 2, 3$ and 4

over time in a consistent manner, following an exponential trend. Figure 2 presents the spectrum of quasinormal modes for the scalar perturbations.

Inspecting the figure, we note that as the mass of the wormhole increases, the frequency decreases while the imaginary component grows. Additionally, when we increase the value of l, the frequency increases, although the imaginary part remains relatively stable without significant changes.

4 Metric Perturbations

Now we examine the scenario of the full set of perturbations including those of the spacetime. As we will see, the method for analyzing this issue bears resemblance to the one discussed above for the scalar field only. However, it becomes more intricate because of the tensorial nature of the metric. Employing the metric from Eq. (5), we decompose the metric and the phantom field as follows

$$g_{\mu\nu} = g_{\mu\nu}^{(b)} + h_{\mu\nu}, \tag{30}$$

$$\phi = \phi^{(b)} + \psi, \tag{31}$$

where the superscript (b) stands again for the background, and the perturbations are assumed to be small. The variation of the Einstein equations with the phantom field source term is

$$\delta R_{\mu\nu} = -2\delta(\partial_\mu\phi\,\partial_\nu\phi)\,, \tag{32}$$

where

$$\delta R_{\mu\nu} = \frac{1}{2}\left(\nabla_\rho\nabla_\mu\, h^\rho_\nu + \nabla_\rho\nabla_\nu\, h^\rho_\mu - \nabla_\nu\nabla_\mu\, h - \Box h_{\mu\nu}\right)\,. \tag{33}$$

From the variation of the scalar field equation we obtain

$$\delta\Box\phi = \Box\psi + \frac{1}{2}\nabla_\lambda h\;\partial^\lambda\phi^{(b)} - \nabla_\mu\left(h^{\mu\nu}\partial_\nu\phi^{(b)}\right) = 0. \tag{34}$$

We shall analyze and separate Eqs. (32) and (34) in polar coordinates. Our initial step is to decompose the perturbations into products of four functions, where each function depends on only one coordinate. To accomplish this, we employ the widely used technique of spherical harmonics (which is well-known for scalars and vectors) also for tensors. For a comprehensive and thorough explanation of spherical harmonics, please refer to the detailed information provided in [32]. As the background metric is spherically symmetric such a decomposition will then allow us to solve for the various multipole numbers separately. For the metric perturbations, we consider a very general perturbation matrix. Since the background is spherically symmetric it is sufficient to consider only axisymmetric modes of the perturbations ($m = 0$). The non axisymmetric modes can be calculated from the axisymmetric modes as there is no preferences in choosing the axes. In fact, considering the axisymmetric modes we see that the equations for the perturbations are separable in all four variables t, r, θ and φ.

Specializing to $m = 0$, we use the Regge-Wheeler gauge for simplicity [60]. The Regge-Wheeler gauge, also known as the Regge-Wheeler coordinate system, is a specific choice of coordinates. It was introduced by Tullio Regge and John Archibald Wheeler in their study of the stability properties of the Schwarzschild black hole. In summary the Regge-Wheeler gauge is a coordinate choice that simplifies the analysis of the perturbations of the metric. It is helpful when separating the perturbations into even and odd modes (with respect to parity transformations), making it easier to analyze their behavior and to determine their frequencies. In the coming sections we study the perturbations in their canonical form for odd waves and even waves of multipole number l and projection $m = 0$.

4.1 Axial Perturbations

4.1.1 Theoretical Set-Up

Employing the Regge-Wheeler gauge [60] the canonical form for an odd wave, i.e., an axial perturbation, of multipole number l and projection $m = 0$ of the metric is given by

$$h^{axial}_{\mu\nu}(t, r, \theta, \phi) = \sum_l \int d\omega e^{-i\omega t} \begin{bmatrix} 0 & 0 & 0 & h_0 S(\theta) \\ 0 & 0 & 0 & h_1 S(\theta) \\ 0 & 0 & 0 & 0 \\ h_0 S(\theta) & h_1 S(\theta) & 0 & 0 \end{bmatrix}, \tag{35}$$

where $S(\theta)$ is

$$\begin{aligned} S(\theta) &= e^{im\varphi} \sin(\theta) \partial_\theta P_l(\cos(\theta)) \\ &= -(l+1)\Big[\cos(\theta) P(l+1, \cos\theta) - P(l+1, \cos\theta)\Big]. \end{aligned} \tag{36}$$

From the Einstein equations we obtain one second order equation ($\delta R_{t\varphi}$) and two first order equations ($\delta R_{r\varphi}$ and $\delta R_{\theta\varphi}$) which are not trivial, the rest of the equations are trivially zero. From the two first order equations we obtain the equations for the first order derivatives of perturbation functions

$$\frac{dh_0(r)}{dr} = \frac{C - 2r}{r^2 + r_0^2} h_0(r) + i\Big(\frac{(l-1)(l+2)e^{2f}}{\omega(r^2 + r_0^2)} - \omega\Big) h_1(r), \tag{37}$$

$$\frac{dh_1(r)}{dr} = -i\omega e^{2f} h_0(r) - \frac{C}{r^2 + r_0^2} h_1(r), \tag{38}$$

which are consistent with the second order equation. We then reduce the system to a second order equation for $h_1(r)$, and by defining

$$h_1(r) = Ze^{-\frac{3f}{2}}(r^2 + r_0^2)^{1/2} \tag{39}$$

and employing the tortoise coordinate r^*

$$\frac{dr^*}{dr} = e^{-f} \tag{40}$$

we obtain a Schrödinger-like master equation

$$\frac{d^2 Z}{dr^{*2}} + \left(\omega^2 - V(r)\right) Z = 0 \tag{41}$$

with the effective potential

$$V^{axial}(r) = \mathrm{e}^{2f(r)} \left(\frac{l\,(l+1) - 3}{r_0^2 + r^2} + \frac{3\,(2r - C)^2}{4\left(r_0^2 + r^2\right)^2} \right). \tag{42}$$

In Fig. 3 we exhibit the effective potential $V_l^{axial}(r)$ for $C = 0, 0.1, 0.25, 0.5$ and 1 ($r_0 = 1, l = 2$).

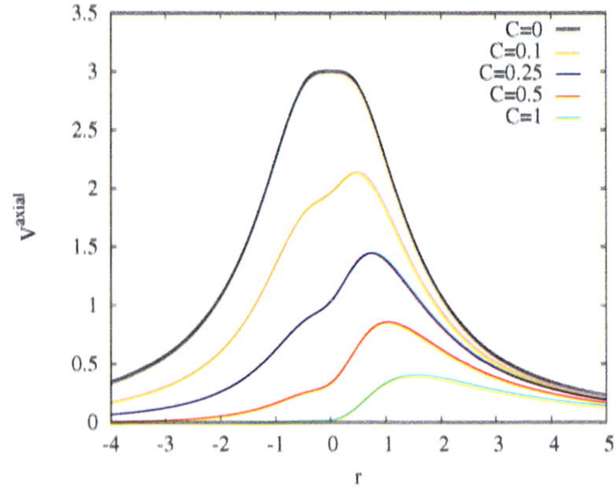

Fig. 3 Effective potential $V_l^{axial}(r)$ for axial perturbation versus radial coordinate r for $C = 0, 0.1, 0.25, 0.5$ and 1 ($l = 2$, $r_0 = 1$)

4.1.2 Numerical Method

After simplification, we are left with a single second-order equation. To calculate the damped quasinormal modes for the axial perturbations, we employ the same approach we used previously for the scalar perturbations, as described in Sect. 3.2. The only difference lies in the modification of the function, with asymptotic expansions

$$Z_P(r) = a_0\left(1 + \frac{il(l+1)}{2\omega}\frac{1}{r} + \left(\frac{-iC(l^2+l+3)}{4\omega} - \frac{l(l+2)(l^2-1)}{8\omega^2}\right)\frac{1}{r^2} + \cdots\right) \tag{43}$$

and

$$Z_N(r) = a_0\left(1 - \frac{il(l+1)}{2\omega e^{C\pi/r_0}}\frac{1}{r} + \left(\frac{-iC(l^2+l+3)}{4\omega e^{C\pi/r_0}} - \frac{l(l+2)(l^2-1)}{8\omega^2 e^{2C\pi/r_0}}\right)\frac{1}{r^2} + \cdots\right). \tag{44}$$

Again the quasinormal modes are determined when the matching condition below is satisfied

$$\left.\frac{1}{Z^-}\frac{dZ^-}{dr}\right|_{r=r_c} - \left.\frac{1}{Z^+}\frac{dZ^+}{dr}\right|_{r=r_c} = 0. \tag{45}$$

4.1.3 Spectrum

Let us shift our focus now to the spectrum of quasinormal modes for the axial perturbations. The quasinormal modes corresponding to the axial perturbations have been calculated in Ref. [16]. Similar to the previous case of scalar perturbations, these modes represent consistently stable disturbances that gradually fade away over time, following an exponential damping pattern. The spectrum of the fundamental quasinormal modes for the axial perturbations is given in Fig. 4. As seen in the figure, we observe that as the mass increases, both the frequency and the imaginary component of the mode also increase. Furthermore, when we increase the value of l, the frequencies also experience an increase whereas the imaginary parts decrease.

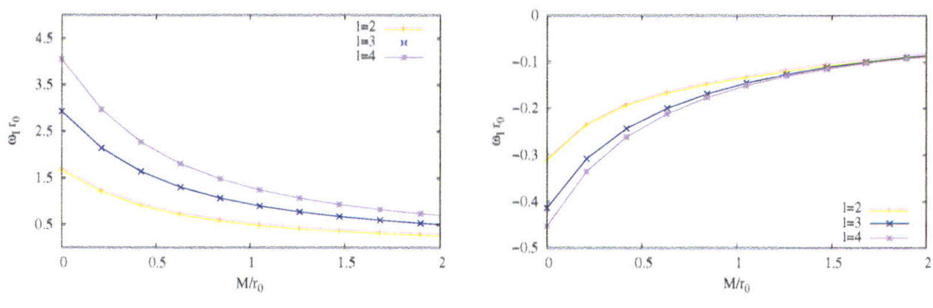

Fig. 4 Axial quasinormal modes: dimensionless frequency $\omega_R r_0$ (left) and dimensionless decay rate $\omega_I r_0$ (right) versus dimensionless mass M/r_0 ($r_0 = 1$) for multipole numbers $l = 2, 3, 4$

4.2 Polar Perturbations

Utilizing the Regge-Wheeler gauge [60] the canonical form for an even wave, i.e., a polar perturbation, of multipole number l and projection $m = 0$ of the metric is given by

$$h^{pol}_{\mu\nu}(t, r, \theta, \phi) = \sum_l \int d\omega e^{-i\omega t} P_l(\cos\theta) \times$$
$$\begin{bmatrix} e^{f(r)} H_{0l}(r) & H_{1l}(r) & 0 & 0 \\ H_{1l}(r) & e^{-f(r)} H_{2l}(r) & 0 & 0 \\ 0 & 0 & e^{-f(r)}(r^2 + r_0^2) K_l(r) & 0 \\ 0 & 0 & 0 & e^{-f(r)}(r^2 + r_0^2)\sin^2\theta K_l(r) \end{bmatrix}. \tag{46}$$

Again, we use a spherical harmonic decomposition ($m = 0$) of the perturbation ψ for the scalar field

$$\psi(t, r) = \sum_l \int d\omega e^{-i\omega t} u_l(r)\, P_l(\cos\theta)\,. \tag{47}$$

The Laplacian of the phantom field perturbation then becomes

$$\Box\psi = \left(\frac{u_l(r)}{e^{f(r)}}\omega^2 + e^{f(r)}\left(\frac{2r}{r^2 + a^2}\partial_r u_l(r) + \partial_r^2\, u_l(r) - \frac{l(l+1)}{r^2 + r_0^2} u_l(r)\right)\right) \times$$
$$e^{-i\omega t} P_l(\cos\theta)\,. \tag{48}$$

We simplify the notation by omitting the index l for the perturbation functions of the metric and the scalar field. The scalar, vector, and tensor spherical harmonics are defined for $l \geq 0$, $l \geq 1$, and $l \geq 2$, respectively. In the coming subsections, we will separately examine the monopole ($l = 0$) and dipole ($l = 1$) cases. We here start with the most general case which is $l \geq 2$. In this scenario, we must analyze the cases of the wormhole having mass and the cases where it is massless. For each part we first study the theoretical set-up and then we address the numerical method.

4.2.1 Theoretical Set-Up for $l \geq 2$ and $C \neq 0$

We study Eq. (32) and we obtain seven nontrivial equations δR_{tt}, δR_{tr}, $\delta R_{t\theta}$ ($\delta R_{t\varphi}$), δR_{rr}, $\delta R_{r\theta}$ ($\delta R_{r\varphi}$), $\delta R_{\theta\theta}$ and $\delta R_{\varphi\varphi}$. Replacing K'', which is extracted from $\delta R_{\theta\theta}$, in $\delta R_{\varphi\varphi}$ gives us the equality.

$$H_2 = H_0\,. \tag{49}$$

Then we eliminate H_2 in the remaining six equations and obtain

$$\delta R_{tt} = \Bigg[-\frac{e^{2f}}{2} H_0'' - \frac{e^{2f}\,(C + 2r)}{2\left(r^2 + r_0^2\right)} H_0' + \left(\frac{l(l+1)e^{2f}}{2(r^2 + r_0^2)} + \frac{\omega^2}{2}\right) H_0,$$

$$
\begin{aligned}
&+ i\omega e^{2f}\left(-H_1' + \frac{C-4r}{2(r^2+r_0^2)}H_1\right) + \frac{Ce^{2f}}{2(r^2+r_0^2)}K' + \omega^2 K\Bigg]\times \\
&e^{-i\omega t}P_l(\cos\theta) = 0\,, \qquad (50)\\
\delta R_{tr} &= \Bigg[-i\omega\left(\frac{C-2r}{2(r^2+r_0^2)}H_0 + \frac{C-r}{r^2+r_0^2}K - K'\right) + \frac{l(l+1)e^{2f}}{2(r^2+r_0^2)}H_1\Bigg]\times \\
&e^{-i\omega t}P_l(\cos\theta) \\
&= 2i\omega\, e^{-i\omega t}\, u(r)\, P_l(\cos\theta)\, \partial_r\phi^{(b)}\,, \qquad (51)\\
\delta R_{t\theta} &= \frac{1}{2}\Bigg[i\omega(H_0+K) + e^{2f}\left(\frac{C}{r^2+r_0^2}H_1 + H_1'\right)\Bigg]e^{-i\omega t}\,\partial_\theta\, P_l(\cos\theta) = 0\,, \qquad (52)\\
\delta R_{rr} &= \Bigg[\frac{1}{2}\left(\left(\frac{l(l+1)}{r^2+r_0^2} - \omega^2 e^{2f}\right)H_0 + \frac{C+2r}{r^2+r_0^2}H_0' + H_0''\right) \\
&+ i\omega e^{f}\left(\frac{C}{2(r^2+r_0^2)}H_1 + H_1'\right) + \frac{C-4r}{2(r^2+r_0^2)}K' - K''\Bigg]e^{-i\omega t}P_l(\cos\theta) \\
&= -4\, e^{-i\omega t}\,\partial_r u(r)\, P_l(\cos\theta)\,\partial_r\phi^{(b)}\,, \qquad (53)\\
\delta R_{r\theta} &= \frac{1}{2}\left(\frac{C}{r^2+r_0^2}H_0 + H_0' + \omega e^{-f}H_1 - K'\right)e^{-i\omega t}\partial_\theta P_l(\cos\theta) \\
&= -2\, e^{-i\omega t}\, u(r)\,\partial_\theta P_l(\cos\theta)\,\partial_r\phi^{(b)}\,, \qquad (54)
\end{aligned}
$$

and

$$
\begin{aligned}
\delta R_{\theta\theta} = \Bigg[&H_0 - \frac{C-2r}{2}H_0' + \frac{i\omega}{2}(C-2r)e^{-f}H_1 \\
&+ \frac{1}{2}\left((l-1)(l+2) - \omega^2(r^2+r_0^2)e^{-2f}\right)K \\
&+ \frac{1}{2}(C-4r)K' - \frac{1}{2}(r^2+r_0^2)K''\Bigg]e^{-i\omega t}P_l(\cos\theta) = 0\,. \qquad (55)
\end{aligned}
$$

The Eqs. (51), (52) and (54) are of first order. We calculate $K'(r)$ from Eq. (51) (δR_{tr}), $H_1'(r)$ from Eq. (52) ($\delta R_{t\theta}$), and $H_0'(r)$ from Eq. (54) ($\delta R_{r\theta}$) after inserting $K'(r)$ in the last two equations. This leads to

$$
\begin{aligned}
K' = \frac{1}{r^2+r_0^2}\Bigg[&\frac{C-2r}{2}H_0 + \frac{il(l+1)}{2\omega}e^{f}H_1 + (C-r)K \\
&+ \sqrt{C^2+4r_0^2}\,u(r)\Bigg]\,, \qquad (56)
\end{aligned}
$$

$$
H_1' = -i\omega e^{-f}(H_0+K) - \frac{C}{r^2+r_0^2}H_1\,, \qquad (57)
$$

$$H_0' = \frac{1}{r^2+r_0^2}\left[\frac{2r-3C}{2}H_0 + \frac{i}{2}\left(\frac{l(l+1)}{\omega}e^f - 2\omega(r^2+r_0^2)e^{-f}\right)H_1 + (C-r)K - \sqrt{C^2+4r_0^2}\,u(r)\right]. \tag{58}$$

Inserting $K''(r)$ that we already extracted from $\delta R_{\theta\theta}$, $H_0''(r)$ that we extract from δR_{tt} and the first derivatives of the metric functions together in δR_{rr} results in the *algebraic relation*

$$\begin{aligned}
&\sqrt{C^2+4r_0^2}\left[\frac{2u'}{r^2+r_0^2} - \frac{u\,(C-4r)}{\left(r^2+r_0^2\right)^2}\right] + \left[\frac{2\,(l-1)\,(l+2)}{r^2+r_0^2} - \frac{3C\,(C-2r)}{\left(r^2+r_0^2\right)^2}\right]H_0 \\
&+ i\left[\frac{2\omega\,(C-2r)}{r^2+r_0^2}\,e^{-f} + \frac{l\,(l+1)\,C}{\omega\left(r^2+r_0^2\right)^2}\,e^f\right]H_1 \\
&+ 2\left[2\omega^2\,e^{-2f} - \frac{(l-1)(l+2)}{r^2+r_0^2} + \frac{C(C-r)}{(r^2+r_0^2)^2}\right]K = 0.
\end{aligned} \tag{59}$$

Considering that we solve the *algebraic relation* for u', we end up with four first-order equations for the four variables. We then obtain the three metric perturbation functions H_0, H_1, and K in terms of u and its derivatives.This means that if we would set $u(r) = 0$ (we turn off the perturbation of the scalar field), the metric perturbations would also vanish, indicating that the metric perturbations are coupled with the phantom field perturbation. So in general we cannot decouple the phantom field perturbations from the perturbations of the metric.

4.2.2 Numerical Method for $l \geq 2$ and $C \neq 0$

As before we need to study the perturbation functions at infinity, and we obtain the series expansions assuming asymptotic flatness. Employing the tortoise coordinate r^*

$$\frac{dr^*}{dr} = e^{-f} \tag{60}$$

the expansion for $r \to +\infty$ becomes

$$u(r) = e^{i\omega r^*}\left(\frac{u_1}{r} + \frac{u_2}{r^2} + \frac{u_3}{r^3} + \cdots\right), \tag{61}$$

$$H_0(r) = e^{i\omega r^*}\left(i\omega K_0 r + \frac{(3iC\omega - (l{+}2)(l{-}1))}{2}K_0 + \cdots\right), \tag{62}$$

$$H_1(r) = e^{i\omega r^*}\left(-i\omega K_0 r - \frac{(3iC\omega - (l{+}2)(l{-}1))}{2}K_0 + \cdots\right), \tag{63}$$

$$K(r) = e^{i\omega r^*}\left(K_0 - \frac{\omega^2(C^2+4r_0^2) + 6iC\omega - l^4 - 2l^3 + l^2 + 2l}{8\omega^2 r^2}K_0 + \cdots\right), \tag{64}$$

where $u_1 = A_s^+$ and $K_0 = A_g^+$ are free amplitudes, and u_2 and u_3 are given in terms of u_1

$$u_2 = \frac{1}{4\omega}\Big(\Big(\sqrt{C^2+4r_0^2}K_0 - 2u_1\Big)C\omega + 2il(l+1)u_1\Big), \tag{65}$$

$$u_3 = \frac{1}{8\omega^2}\Big(\big(\omega^2(C^2-4r_0^2) - 2i\omega C(2l^2+2l-1) - l^4 - 2l^3 + l^2 + 2l\big)u_1 + i\omega C\sqrt{C^2+4r_0^2}(l+2)(l-1)K_0\Big). \tag{66}$$

For $r \to -\infty$ the expansion becomes

$$\bar{u}(r) = e^{-i\omega r^*}\Big(\frac{\bar{u}_1}{r} + \frac{\bar{u}_2}{r^2} + \frac{\bar{u}_3}{r^3} + \cdots\Big), \tag{67}$$

$$\bar{H}_0(r) = e^{-i\omega r^*}\Big(-i\omega r e^{C\pi/r_0}\bar{K}_0 - \frac{3iC\omega e^{C\pi/r_0} + (l+2)(l-1)}{2}\bar{K}_0 + \cdots\Big), \tag{68}$$

$$\bar{H}_1(r) = e^{-i\omega r^*}\Big(-i\omega r e^{C\pi/r_0}\bar{K}_0 - \frac{3iC\omega e^{C\pi/r_0} + (l+2)(l-1)}{2}\bar{K}_0 + \cdots\Big), \tag{69}$$

$$\bar{K}(r) = e^{-i\omega r^*}\Big(\bar{K}_0 - \frac{\omega^2(C^2+4r_0^2) - 6iC\omega e^{-C\pi/r_0}}{8\omega^2 r^2}\bar{K}_0 - \frac{(l^4+2l^3-l^2-2l)e^{-2C\pi/r_0}}{8\omega^2 r^2}\bar{K}_0 + \cdots\Big), \tag{70}$$

where again $\bar{u}_1 = A_s^-$ and $\bar{K}_0 = A_g^-$ are free amplitudes and $\bar{u}_2$ and $\bar{u}_3$ are given in terms of $\bar{u}_1$

$$\bar{u}_2 = \frac{1}{4\omega}\Big(\Big(\sqrt{C^2+4r_0^2}\,\bar{K}_0 - 2\bar{u}_1\Big)C\omega - 2il(l+1)e^{-C\pi/r_0}\bar{u}_1\Big), \tag{71}$$

$$\bar{u}_3 = \frac{1}{8\omega^2}\Big(\Big(\omega^2(C^2-4r_0^2) + 2i\omega e^{-C\pi/r_0}C(2l^2+2l-1) - (l^4+2l^3-l^2-2l)e^{-2C\pi/r_0}\Big)\bar{u}_1 + i\omega e^{-C\pi/r_0}C\sqrt{C^2+4r_0^2}(l+2)(l-1)\bar{K}_0\Big). \tag{72}$$

To calculate the quasinormal modes with $l \geq 2$ numerically we rewrite the system of Eqs. (56)–(59) as a second order ordinary differential equation (ODE) for u, which is coupled to the two first order ODEs for the metric functions H_1 and K (the other perturbation functions are then calculated algebraically in terms of these three functions). We express this set of equations as follows,

$$\frac{d}{dr}\mathbf{Z} + \mathbf{M}\mathbf{Z} = 0. \tag{73}$$

Here $\mathbf{Z}$ is a column vector with components u, u', H_1 and K, and the matrix $\mathbf{M}$ contains the coupling among the perturbation functions and the background functions,

$$\bar{M} = \begin{bmatrix} u^-(I) & u^-(II) & u^+(I) & u^+(II) \\ u'^-(I) & u'^-(II) & u'^+(I) & u'^+(II) \\ H_1^-(I) & H_1^-(II) & H_1^+(I) & H_1^+(II) \\ K^-(I) & K^-(II) & K^+(I) & K^+(II) \end{bmatrix}_{r=r_c} . \tag{74}$$

As before, to obtain the quasinormal modes we need to solve the coupled set of Eq. (73) subject to the boundary conditions, and as before we do not have any incoming waves from infinity and all the modes are purely outgoing at infinity, $r^* \to \pm\infty$ (see Eqs. (61)–(64) and (67)–(70)), with components Z_i

$$Z_i \sim \begin{cases} e^{i\omega r^*}, & r \to +\infty \, , \\ e^{-i\omega r^*}, & r \to -\infty \, . \end{cases} \tag{75}$$

As before we divide space at some value r_c into two regions. In the region $r > r_c$ the asymptotic behaviour for $r \to +\infty$ is [23]

$$r > r_c \, , \quad Z_i^+(r) = e^{i\omega r^*} Z_i^P(r) \, , \tag{76}$$

and in the region $r < r_c$ the asymptotic behaviour for $r \to -\infty$ is

$$r < r_c \, , \quad Z_i^-(r) = e^{-i\omega r^*} Z_i^N(r) \, . \tag{77}$$

For some chosen value of ω we now generate independent solutions for the functions $Z_i^N(r)$ and $Z_i^P(r)$. Then we match these functions at $r = r_c$ and calculate the derivatives of the functions. The quasinormal modes are obtained when a linear combination of the two independent solutions in one region smoothly matches a linear combination of the solutions in the other region.

In order to combine the two distinct solutions, $Z_i^-(I)$ and $Z_i^-(II)$, on the left side in a linear manner, such that the linear combination smoothly matches a combination of the two distinct solutions, $Z_i^+(I)$ and $Z_i^+(II)$, on the right side, we require that the determinant of $\bar{M}$ is equal to zero at the specific point $r = r_c$,

$$\det \bar{M} = 0. \tag{78}$$

In the case when $C \neq 0$, the determinant of the matrix becomes zero at two distinct values of ω.[1] One of these values corresponds to branch 1, while the other belongs to branch 2. At each of these points where the determinant is zero, there exists a unique linear combination of perturbation functions that yields a smooth solution throughout the entire spacetime. For each ω value, the kernel of matrix $\bar{M}$ has a dimension of one.

In order to integrate numerically the equations subject to the corresponding boundary conditions, we again use the package Colsys [8].

[1] We know that the determinant has more than two zeroes, as there are excited modes, with larger imaginary part of ω, but here we just focus only on the fundamental modes.

4.2.3 Theoretical Set-Up for $l \geq 2$ and $C = 0$

In the scenario where there is no mass present, the first-order equations simplify to

$$K' = \frac{1}{2\omega\left(r^2 + r_0^2\right)}\Big(il\,(l+1)\,H_1 - 2\omega\,(rK - rH_0 - 2r_0 u)\Big), \tag{79}$$

$$H_0' = \frac{1}{2\omega\left(r^2 + r_0^2\right)}\Big[\left(-2i\omega^2\left(r^2 + r_0^2\right) + il\,(l+1)\right)H_1 - 2\omega\,(rK - rH_0 + 2r_0 u)\Big], \tag{80}$$

$$H_1' = -i\omega\,(H_0 + K)\,, \tag{81}$$

$$u' = -\frac{1}{4r_0}\Big[\left(2\omega^2\left(r^2 + r_0^2\right) + (l+2)(l-1)\right)K + (l+2)\,(l-1)\,H_0 - 2i\omega r H_1\Big] - \frac{2r}{r_0^2 + r^2}u\,. \tag{82}$$

For $C = 0$ the asymptotic expansions remain the same.

We calculate the derivative of (82) and obtain u'' in the form of a decoupled equation for the scalar perturbation,

$$u'' = -\frac{2r}{r_0^2 + r^2}u' - \frac{1}{(r_0^2 + r^2)^2}\Big(r_0^4\omega^2 + \left(2r_0^2\omega^2 - l^2 - l + 4\right)r_0^2 + \left(r^2\omega^2 - l^2 - l\right)r^2\Big)u\,. \tag{83}$$

Therefore we can solve this equation independent of the metric perturbations, and it has the non-trivial solution

$$u(r) = \frac{C_1\, r}{r_0^2 + r^2}\,\mathrm{HeunC}\Big(0, \frac{1}{2}, -2, -\frac{r_0^2\omega^2}{4}, \frac{1}{4}\left(r_0^2\omega^2 - l^2 - l + 5\right), -\frac{r^2}{r_0^2}\Big) + \frac{C_2}{r_0^2 + r^2}\,\mathrm{HeunC}\Big(0, -\frac{1}{2}, -2, -\frac{r_0^2\omega^2}{4}, \frac{1}{4}\left(r_0^2\omega^2 - l^2 - l + 5\right), -\frac{r^2}{r_0^2}\Big), \tag{84}$$

where HeunC is the Heun Confluent function, C_1 and C_2 are constants.

On the other hand, we obtain a second order equation for the spacetime perturbations for H_1,

$$H_1'' = \frac{8i\omega r_0 r u' + 16i\omega r_0 r^2 u/(r^2 + r_0^2) - 2r\omega^2\left(r^2 + r_0^2\right)H_1'}{\left(l^2 + l - 2 - \left(r^2 + r_0^2\right)\omega^2\right)\left(r^2 + r_0^2\right)} + \Big[\left(r^2 + r_0^2\right)^2\omega^4 + \left[6r^2 + 2r_0^2 - 2\left(r^2 + r_0^2\right)l\,(l+1)\right]\omega^2 + l^4 + 2l^3 - l^2 - 2l\Big] \times H_1 \times \Big[\left(l^2 + l - 2 - \left(r^2 + r_0^2\right)\omega^2\right)\left(r^2 + r_0^2\right)\Big]^{-1}. \tag{85}$$

We can represent the preceding perturbation equations using a single master equation, which aligns with the master equation governing the axial perturbations. It is important to highlight that if we describe the scalar perturbations in terms of the master variable $Z(r)$ as follows

$$u = \frac{1}{\sqrt{r^2 + r_0^2}} Z(r), \tag{86}$$

then Eq. (83) can be cast as a Schrödinger-like equation

$$\frac{d^2 Z(r)}{dr^2} + \left(\omega^2 - V(r)\right) Z(r) = 0, \tag{87}$$

with the effective potential

$$V_{C=0}^{polar}(r) = \frac{l(l+1)}{r^2 + r_0^2} - \frac{3r_0^2}{(r^2 + r_0^2)^2}. \tag{88}$$

Note that when $C = 0$, the tortoise coordinate (21) denoted as r^* coincides with r itself. Also, the potential is the same as the axial potential, Eq. (42), when $C = 0$.

Furthermore, it is feasible to reformulate the equation for the spacetime perturbations (85) as an identical master equation. Nonetheless, the process of transformation is more intricate. To accomplish this, we introduce the following definition

$$H_1(r) = A(r)\hat{Z}(r), \tag{89}$$

$$u(r) = B(r)\hat{Z}(r), \tag{90}$$

where $\hat{Z}(r)$ is the new master variable and $A(r)$ and $B(r)$ are defined as follows

$$\begin{aligned} A(r) = {} & \frac{(r^2 + r_0^2)^{\sqrt{3}-3/2}}{l^2 + l - 2 - \omega^2(r^2 + r_0^2)} \times \\ & \Bigg[D_1 r^3 \mathrm{HeunG}\Big(- r_0^2\omega^2\beta, -\frac{r_0^2\omega^2\beta}{2} + \frac{5\sqrt{3}}{2} + 3, \frac{5}{4} + \frac{\sqrt{149}}{4} + \sqrt{3}, \\ & \sqrt{3} + \frac{5}{4} - \frac{\sqrt{149}}{4}, \frac{5}{2}, 0, \beta\omega^2 r^2\Big) \\ & + D_2 \mathrm{HeunG}\Big(- r_0^2\omega^2\beta, -\frac{r_0^2\omega^2\beta}{2} - \frac{\sqrt{3}}{2} + \frac{3}{2}, -\frac{1}{4} + \frac{\sqrt{149}}{4} + \sqrt{3}, \\ & \sqrt{3} - \frac{1}{4} - \frac{\sqrt{149}}{4}, -\frac{1}{2}, 0, \beta\omega^2 r^2\Big) \Bigg], \end{aligned} \tag{91}$$

$$B(r) = -\frac{i\omega(r_0^2 + r^2)}{4r_0} A - \frac{i(r_0^2 + r^2)(l^2 + l - 2 - \omega^2(r_0^2 + r^2))}{4r_0 r\omega} A'. \tag{92}$$

In this case, we have $\beta^{-1} = l^2 + l - 1 - r_0^2\omega^2$, and the function HeunG corresponds to the Heun General function. Additionally, D_1 and D_2 represent two arbitrary constants. Subsequently, it can be demonstrated that Eq. (85) reduces to Eq. (41) when considering $\hat{Z}(r)$, resulting in the same potential as for the axial perturbations. As we will explicitly show in the subsequent analysis, this implies that the spectrum of polar quasinormal modes for $C = 0$ wormholes precisely matches the axial spectrum.

4.2.4 Numerical Method for $l \geq 2, C = 0$ and Isospectrality

The numerical method for this part is exactly like the one discussed in Sect. 4.2.2. But in the limit when C approaches zero, a remarkable phenomenon occurs. The determinant of the equation exhibits a unique double-zero at a specific value of ω. Consequently, at this particular eigenfrequency, it becomes possible to identify two distinct linear combinations of the perturbation functions that generate independent, smooth solutions to the perturbation equations. This is characterized by the fact that the kernel of matrix $\bar{M}$ has a dimension of two.

4.2.5 Spectrum

For values of l greater than or equal to 2, we encounter two distinct families of modes, as, for instance, discussed in Ref. [15]. When it comes to black holes, these modes can be classified as follows: gravitational-led modes, which are primarily influenced by gravitational perturbations and possess a dominant amplitude denoted as $A_g^{\pm}$, and scalar-led modes characterized by a dominant amplitude denoted as $A_s^{\pm}$. In the current context, it is challenging to make a clear distinction between these two families. Consequently, we refer to them as branch 1 and branch 2 for a given value of l, with our primary focus being on the fundamental branches. However, it is possible that such a classification could emerge if we were to consider larger wormhole masses, as in that limit, the wormhole modes are expected to approach those of a Schwarzschild black hole, as discussed in Ref. [16].

We first obtain the two fundamental branches of polar quadrupole ($l = 2$) modes. Figure 5 displays these two primary $l = 2$ branches. The left figure illustrates the relationship between the scaled frequency $\omega_R r_0$ and the scaled mass M/r_0, while the right figure shows the relation between the scaled decay rate $\omega_I r_0$ and the scaled mass. It is worth mentioning that the two branches intersect precisely at $M = 0$ due to the degeneracy of the eigenvalue ω in the massless scenario. However, as we move away from this crossing point, the frequencies and decay rates exhibit significant differences between the two branches.

Figure 6 displays the polar modes with $l = 3$, while Fig. 7 shows the polar modes with $l = 4$. Here again, we observe the intersection of the two branches in the massless case. However, as we move away from these crossing points, the frequencies ω_R of both branches tend to get closer for $l = 3$ and even closer for $l = 4$ than in the case of the quadrupole mode. Furthermore, the decay rates exhibit a smoother behavior for higher values of l as compared to the quadrupole mode.

To facilitate a comprehensive comparison among different values of l, we have gathered the branches for $l = 2, l = 3$, and $l = 4$ in Fig. 8. The left panel demonstrates

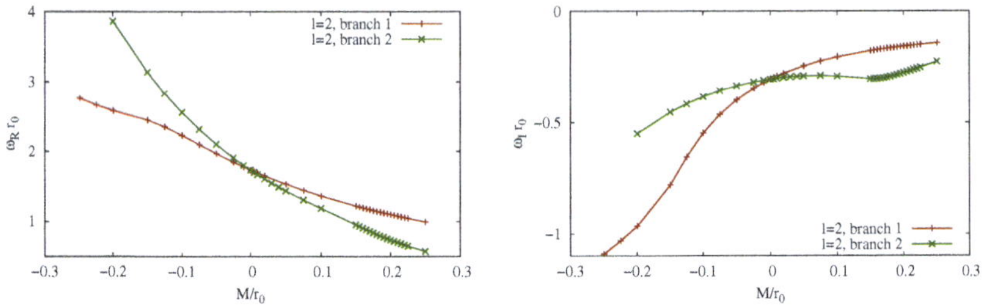

Fig. 5 Polar $l = 2$ quasinormal modes: dimensionless frequency $\omega_R r_0$ (left) and dimensionless decay rate $\omega_I r_0$ (right) versus dimensionless mass M/r_0 ($r_0 = 1$)

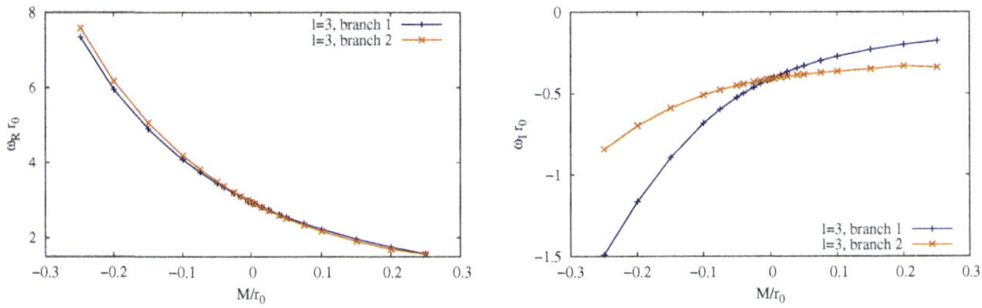

Fig. 6 Polar $l = 3$ quasinormal modes: dimensionless frequency $\omega_R r_0$ (left) and dimensionless decay rate $\omega_I r_0$ (right) versus dimensionless mass M/r_0 ($r_0 = 1$)

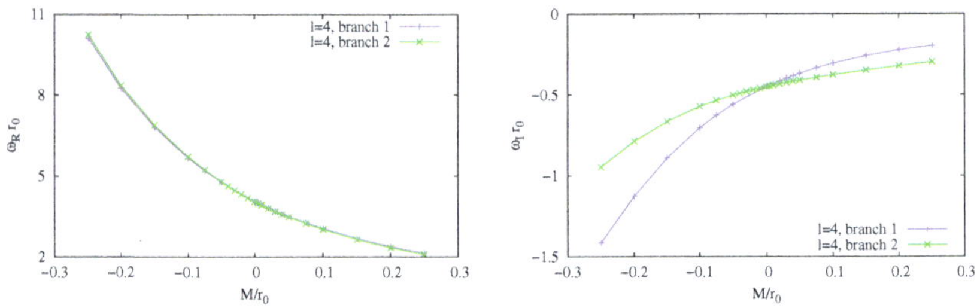

Fig. 7 Polar $l = 4$ quasinormal modes: dimensionless frequency $\omega_R r_0$ (left) and dimensionless decay rate $\omega_I r_0$ (right) versus dimensionless mass M/r_0 ($r_0 = 1$)

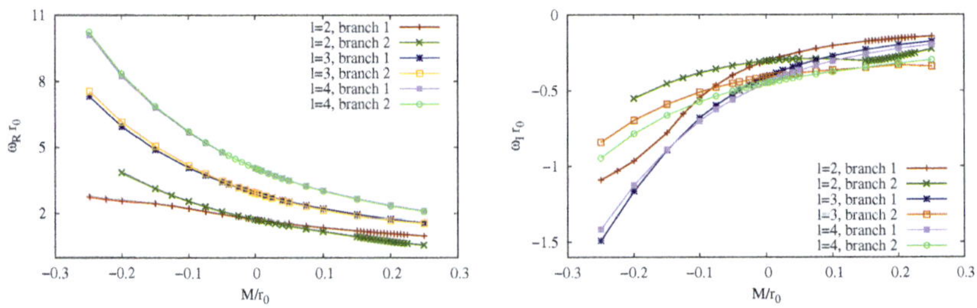

Fig. 8 Comparison of polar $l = 2, 3$ and 4 quasinormal modes: dimensionless frequency $\omega_R r_0$ (left) and dimensionless decay rate $\omega_I r_0$ (right) versus dimensionless mass M/r_0 ($r_0 = 1$)

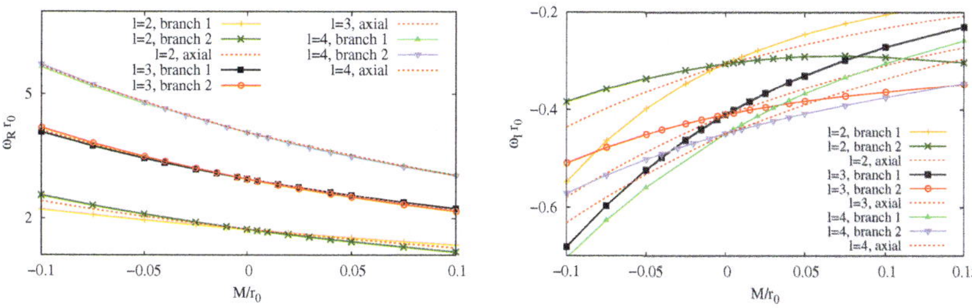

Fig. 9 Comparison of axial and polar $l = 2$, 3 and 4 quasinormal modes: dimensionless frequency $\omega_R r_0$ (left) and dimensionless decay rate $\omega_I r_0$ (right) versus dimensionless mass M/r_0 ($r_0 = 1$)

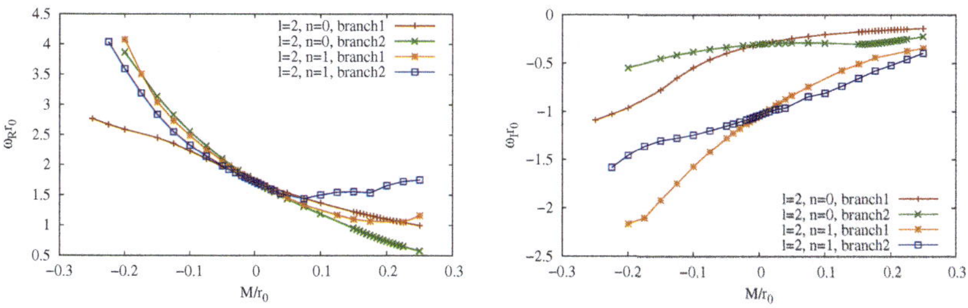

Fig. 10 Fundamental ($n = 0$) polar $l = 2$ quasinormal modes and their first overtone ($n = 1$): dimensionless frequency $\omega_R r_0$ (left) and dimensionless decay rate $\omega_I r_0$ (right) versus dimensionless mass M/r_0 ($r_0 = 1$)

a consistent upward trend in frequencies as l increases. In the right panel we observe an overall pattern of increasing decay rates with higher l. However, it is important to note that the branches for different l become intertwined. For significantly positive masses, the decay rate of branch 1 exhibits an increase with l and remains lower than the decay rate of branch 2, with the decay rate for $l = 4$ being smaller than that of $l = 3$. Conversely, for considerably negative masses, the decay rate of branch 2 rises with l and remains smaller than the decay rate of branch 1, with the decay rate for $l = 4$ being smaller than that of $l = 3$ once again. Additionally, it is noteworthy that the decay rates for branch 1 are very similar for $l = 3$ and $l = 4$.

When comparing the polar modes of the wormhole solutions with finite mass to the axial modes for $l \geq 2$, we find that the wormhole modes lack isospectrality. This means that the mode spectra are not identical. However, in the case of a massless wormhole, the modes exhibit degeneracy, resulting in all three fundamental branches for a given l sharing the same eigenvalue. This behavior is clearly seen in Fig. 9 for $l = 2$, $l = 3$, and $l = 4$. Isospectrality for $C = 0$ is also seen in Fig. 10 which exhibitis a comparison of the fundamental ($n = 0$) polar modes with the first overtone ($n = 1$) polar modes for $l = 2$. (The first overtone for the axial modes has not yet been obtained.)

4.2.6 $l = 1$

Now that we are done with the most general case $l \geq 2$, we are going to consider the cases $l = 1$ and $l = 0$, separately. Starting with $l = 1$ we observe that three of the Einstein's equations are identically zero,

$$\delta R_{t\varphi} = \delta R_{r\varphi} = \delta R_{\theta\varphi} = 0, \tag{93}$$

and the following three first order differential equations are obtained,

$$K' = \frac{1}{r_0^2 + r^2}\left(\frac{ie^f}{\omega} H_1 - \frac{C - 2r}{2} H_2 + (C - r)\, K + \sqrt{C^2 + 4r_0^2}\, u\right), \tag{94}$$

$$H_1' = -\frac{C}{r^2 + r_0^2} H_1 - i\omega e^{-f}\,(H_2 + K)\,, \tag{95}$$

$$H_0' = -\frac{C - r}{r^2 + r_0^2} H_0 - i\left(\omega e^{-f} - \frac{e^f}{\omega\,(r^2 + r_0^2)}\right) H_1 - \frac{C}{2\,(r^2 + r_0^2)} H_2 + \frac{C - r}{r^2 + r_0^2} K - \frac{\sqrt{C^2 + 4r_0^2}}{r^2 + r_0^2} u\,. \tag{96}$$

Furthermore, the *algebraic equation* reads as follows,

$$-\frac{C^2 - 3Cr - 2r_0^2}{2(r^2 + r_0^2)} H_0 + \frac{i}{2}\left(\frac{C}{\omega(r^2 + r_0^2)^2} + \frac{w(C - 2r)e^{-f}}{(r^2 + r_0^2)}\right) H_1 - \frac{c^2 + 4r_0^2}{4(r^2 + r_0^2)} H_2 + \left(\omega^2 e^{-f} + \frac{C(C - r)}{2(r^2 + r_0^2)^2}\right) K + \frac{\sqrt{C^2 + 4r_0^2}}{r^2 + r_0^2}\left(\frac{4r - C}{2(r^2 + r_0^2)} u + u'\right) = 0. \tag{97}$$

There is still a gauge freedom present in this set of equations, which allows us to further simplify the system. Let us explore two possibilities in this regard.

- Fixing the gauge by setting $H_1 = 0$:
  If we set $H_1 = 0$, we obtain from $\delta R_{t\theta}$

$$K = -H_2\,. \tag{98}$$

  The remaining perturbation equations lead to the following simplified system of first-order differential equations

$$H_2' = \frac{(4r - 3C)\, H_2 + 2\sqrt{C^2 + 4r_0^2}\, u}{2(r^2 + r_0^2)}\,, \tag{99}$$

$$H_0' = \frac{1}{r^2 + r_0^2}\left((r - C)\, H_0 + \frac{2r - 3C}{2} H_2 - \sqrt{C^2 + 4r_0^2}\, u\right), \tag{100}$$

and

$$
\begin{aligned}
u' &= \frac{C^2 - 3Cr - 2r_0^2}{2\left(r^2 + r_0^2\right)\sqrt{C^2 + 4r_0^2}} H_0 \\
&+ \frac{1}{\sqrt{C^2 + 4r_0^2}}\left(\omega^2 e^{-f} + \frac{r_0^2}{r^2 + r_0^2} + \frac{3C^2 - 2Cr}{4\left(r^2 + r_0^2\right)}\right) H_2 \\
&+ \frac{C - 4r}{2\left(r^2 + r_0^2\right)} u \,. \qquad (101)
\end{aligned}
$$

To calculate the quasinormal modes, we are thus left with three first order equations .

- Fixing the gauge by setting $H_0' + 2i\omega e^{-f} H_1 + H_2' - 2K' = 0$:
If we consider the change in the phantom field (34), this results in a second-order equation for the function u

$$
\begin{aligned}
u'' &= \frac{\sqrt{C^2 + 4r_0^2}}{4(r^2 + r_0^2)}\left(H_0' + 2i\omega e^{-f} H_1 + H_2' - 2K'\right) \\
&- \left(\omega^2 e^{-2f} + \frac{2}{r^2 + r_0^2}\right) u - \frac{2r}{r^2 + r_0^2} u' \,. \qquad (102)
\end{aligned}
$$

When we now choose the gauge

$$
H_0' + 2i\omega e^{-f} H_1 + H_2' - 2K' = 0 \,, \qquad (103)
$$

the equation simplifies to

$$
u'' = -\frac{2r}{r^2 + r_0^2} u' + \left[\frac{l(l+1)}{r^2 + r_0^2} - \omega^2 e^{-2f}\right] u. \qquad (104)
$$

Notice, that this equation is the same as Eq. (18) for the $l = 1$ scalar field perturbations in the background of the E-B wormhole, from which we obtained the scalar quasinormal modes.

We obtain the damped quasinormal modes for $l = 1$ by solving Eq. (104). As before, we can rewrite these equations as a second-order Schrödinger-like equation. Since this is the same equation as for the scalar part, the numerical method is the same as explained in the scalar perturbations section. Therefore the scalar branches obtained previously [16] correspond to the polar $l = 1$ quasinormal modes. It is evident that we only get one fundamental mode branch for $l = 1$. Moreover, as large masses are approached, the corresponding Schwarzschild scalar modes are reached.

4.2.7 $l = 0$

In this scenario, the function H_1 does not have any contribution in the Einstein equations. Therefore, to simplify the equations, we set H_1 to zero. Now, let us revisit the variation of the phantom field (34), resulting in an equation for u'',

$$u'' = \frac{\sqrt{C^2 + 4r_0^2}}{4(r^2 + r_0^2)} \left(H_0' + H_2' - 2K'\right) - \omega^2 e^{-2f}\, u - \frac{2r}{r^2 + r_0^2} u' \,. \tag{105}$$

We now choose the following gauge condition

$$H_0 = 2K - H_2 \,. \tag{106}$$

With this gauge the scalar field equation then decouples from the spacetime perturbations

$$u'' = -\omega^2 e^{-2f}\, u - \frac{2r}{r^2 + r_0^2} u' \,. \tag{107}$$

To obtain a second order equation we define

$$u(r) = A(r) Z(r^*) \tag{108}$$

with

$$A(r) = \frac{e^{\frac{C}{2r_0} \arctan(\frac{r}{r_0})}}{\sqrt{r^2 + r_0^2}} \tag{109}$$

and employ the tortoise coordinate r^*

$$\frac{dr^*}{dr} = e^{-f} \,. \tag{110}$$

We then rewrite Eq. (105) as a second order Schrödinger-like equation,

$$\frac{d^2 Z}{dr^{*2}} + \left(\omega^2 - V(r)\right) Z = 0, \tag{111}$$

with the effective potential

$$V(r) = -\frac{C^2 - 4Cr - 4r_0^2}{4(r^2 + r_0^2)^2} e^{2f} , \tag{112}$$

which is the same as the effective potential for the scalar perturbations, Eq. (23), when we set $l = 0$. The damped quasinormal modes are obtained by solving Eq. (107). Thus the non-trivial solutions of this equation yield the spectrum of scalar quasinormal modes with $l = 0$.

On the other hand, the perturbations of the spacetime can be expressed in terms of a second-order differential equation for the variable K''.

$$K'' = -\frac{(2Cr + 4r_0^2)K'}{(r^2 + r_0^2)(C - 2r)} + \frac{2CK}{(r^2 + r_0^2)(C - 2r)} - \omega^2 e^{-2f} K + \frac{4u\sqrt{C^2 + 4r_0^2}}{(r^2 + r_0^2)(C - 2r)}. \tag{113}$$

Again, we rewrite this equation as a second-order Schrödinger-like equation, and after obtaining the series expansion of the function at infinity we get one fundamental mode branch. Although the frequencies of the modes for $l = 1$ are higher than those for $l = 0$, their decay rates do not change much. As large masses are approached, the corresponding Schwarzschild scalar modes are reached.

It is well-known that E-B wormholes possess an unstable radial ($l = 0$) mode [16,26,33,34,63]. For solutions with $u = 0$ the unstable mode is found. By defining the function Z,

$$K = \frac{(C - 2r)}{\sqrt{r^2 + r_0^2}} e^{\frac{C}{2r_0}\arctan(\frac{r}{r_0})} Z, \tag{114}$$

and the tortoise coordinate r^* (Eq. (21)) we rewrite the equation into a Schrödinger-like equation,

$$\frac{d^2 Z}{dr^*} + \left(\omega^2 - V_r(r)\right) Z = 0 \tag{115}$$

with the effective potential $V_r(r)$

$$V_r = -\left(C^4 - 8rC^3 + 12(r^2 - r_0^2)C^2 - 16r(r^2 - r_0^2)C - 16r_0^2(3r^2 + 2r_0^2)\right)\times \frac{e^{2f}}{4(C - 2r)^2(r^2 + r_0^2)}. \tag{116}$$

Given that the effective potential exhibits a singularity at $C = 2r$, we examine its behavior in the vicinity of this singularity and find

$$V_r(r \to \frac{C}{2}) = \frac{2e^{\frac{2C}{r_0}\left(\tan^{-1}(\frac{C}{2r_0}) - \frac{\pi}{2}\right)}}{\left(r - \frac{C}{2}\right)^2} + \frac{8Ce^{\frac{2C}{r_0}\left(\tan^{-1}(\frac{C}{2r_0}) - \frac{\pi}{2}\right)}}{\left(C + 4r_0^2\right)\left(r - \frac{C}{2}\right)} + \cdots. \tag{117}$$

For $r \to +\infty$ the effective potential has the expansion

$$V_r(r \to +\infty) = \frac{C}{r^3} - \frac{-\frac{7C^2}{4} + 3r_0^2}{r^4} + \cdots, \tag{118}$$

and for $r \to -\infty$

$$V_r(r \to -\infty) = \frac{e^{-\frac{2\pi C}{r_0}} C}{r^3} - \frac{e^{-\frac{2\pi C}{r_0}} C^2 + \frac{e^{-\frac{2\pi C}{r_0}}(12C^2 - 48r_0^2)}{16}}{r^4} + \cdots . \quad (119)$$

For $l = 0$ Eq. (115) has been used to determine this unstable radial mode of the E-B wormholes [16]. The strength of this instability decreases as the wormhole mass increases, resulting in a more stable wormhole. Nevertheless, the wormhole still retains a radial instability as long as its mass is finite.

5 Conclusions

In conclusion, this chapter has focused on the study of the linear perturbations of the Ellis-Bronnikov wormholes. The introduction provided an overview of the research topic, highlighting the significance of investigating perturbations of wormholes, while Chap. 2 has delved into the details of the Ellis-Bronnikov wormhole metric, providing a comprehensive understanding of the spacetime structure.

Subsequently the perturbations of the Ellis-Bronnikov wormhole have been explored, emphasizing the significance of understanding the behavior of the various perturbation modes. Chapter 3 has specifically focused on the scalar perturbations, presenting the theoretical set-up and the numerical methods employed for the analysis. The obtained spectrum of scalar quasinormal modes has revealed valuable insights into the behavior of the wormhole under scalar perturbations. We have shown that in this case there is one fundamental mode branch.

In Chap. 4 we have addressed the full set of perturbations of the Ellis-Bronnikov wormhole, starting with axial perturbations. First the theoretical framework and the numerical techniques utilized for obtaining axial modes have been described, which correspond to purely metric modes. The resulting spectrum of axial quasinormal modes has provided crucial information about the behavior of the wormhole under axial perturbations.

The discussion has then turned to polar perturbations, where metric and scalar perturbations are in general coupled. Here the general cases $l \geq 2$ with $C \neq 0$ (massive wormholes) and $l \geq 2$ with $C = 0$ (massless wormholes) have been addressed first, and the theoretical set-ups and the numerical methodologies have been discussed in detail, as well as the obtained spectrum of polar quasinormal modes. In particular, we have discussed the distinct characteristics of the different branches and highlighted the isospectrality obtained in the massless case, $C = 0$. The observed isospectrality in fact corresponds to a threefold degeneracy, since not only the two polar modes but also the axial mode are the same for $C = 0$. The chapter has ended with an exploration of the $l = 1$ and $l = 0$ perturbations, where we have seen that in these cases the polar perturbation can be considered the same as the scalar perturbations.

Acknowledgements I would like to gratefully acknowledge support by DAAD and thank Jose Luis Blázquez-Salcedo, Luis Manuel González-Romero, Fech Scen Khoo, Jutta Kunz and Francisco Navarro-Lérida for collaboration and discussions.

References

1. B.P. Abbott et al. [LIGO Scientific and Virgo], Phys. Rev. Lett. **116**, 061102 (2016)
2. B.P. Abbott et al. [LIGO Scientific and Virgo], Phys. Rev. Lett. **119**, 161101 (2017)
3. B.P. Abbott et al., Astrophys. J. Lett. **848**, L12 (2017)
4. F. Abe, Astrophys. J. **725**, 787 (2010)
5. M. Alcubierre, F.S.N. Lobo, Fundam. Theor. Phys. **189** (2017)
6. S. Aneesh, S. Bose, S. Kar, Phys. Rev. D **97**, 124004 (2018)
7. G. Antoniou, A. Bakopoulos, P. Kanti, B. Kleihaus, J. Kunz, Phys. Rev. D **101**, 024033 (2020)
8. U. Ascher, J. Christiansen, R.D. Russell, Math. Comp. **33**, 659 (1979)
9. B. Azad, F. Loran, A. Mostafazadeh, Eur. Phys. J. C **80**, 1097 (2020)
10. B. Azad, J.L. Blázquez-Salcedo, X.Y. Chew, J. Kunz, D.h. Yeom, Phys. Rev. D **107**, 084024 (2023)
11. C. Bambi, Phys. Rev. D **87**, 084039 (2013)
12. C. Bambi, Phys. Rev. D **87**, 107501 (2013)
13. B.J. Barros, F.S.N. Lobo, Phys. Rev. D **98**, 044012 (2018)
14. E. Berti, V. Cardoso, A.O. Starinets, Class. Quant. Grav. **26**, 163001 (2009)
15. J.L. Blázquez-Salcedo, C.F.B. Macedo, V. Cardoso, V. Ferrari, L. Gualtieri, F.S. Khoo, J. Kunz, P. Pani, Phys. Rev. D **94**, 104024 (2016)
16. J.L. Blázquez-Salcedo, X.Y. Chew, J. Kunz, Phys. Rev. D **98**, 044035 (2018)
17. J.L. Blázquez-Salcedo, C. Knoll, Eur. Phys. J. C **80**, 174 (2020)
18. J.L. Blázquez-Salcedo, C. Knoll, E. Radu, Phys. Rev. Lett. **126**, 101102 (2021)
19. J.L. Blázquez-Salcedo, C. Knoll, E. Radu, Eur. Phys. J. C **82**, 533 (2022)
20. M. Bouhmadi-López, C.Y. Chen, X.Y. Chew, Y.C. Ong, D.h. Yeom, JCAP **10** (2021), 059
21. K.A. Bronnikov, Acta Phys. Polon. B **4**, 251 (1973)
22. S. Chandrasekhar, S.L. Detweiler, Proc. Roy. Soc. Lond. A **344**, 441 (1975)
23. S. Chandrasekhar, *The Mathematical Theory of Black Holes* (Clarendon Press, Oxford, 1998)
24. M.S. Churilova, R.A. Konoplya, A. Zhidenko, Phys. Lett. B **802**, 135207 (2020)
25. J.G. Cramer, R.L. Forward, M.S. Morris, M. Visser, G. Benford, G.A. Landis, Phys. Rev. D **51**, 3117 (1995)
26. F. Cremona, F. Pirotta, L. Pizzocchero, Gen. Rel. Grav. **51**, 19 (2019)
27. E. Deligianni, J. Kunz, P. Nedkova, S. Yazadjiev, R. Zheleva, Phys. Rev. D **104**, 024048 (2021)
28. E. Deligianni, B. Kleihaus, J. Kunz, P. Nedkova, S. Yazadjiev, Phys. Rev. D **104**, 064043 (2021)
29. A. Einstein, N. Rosen, Phys. Rev. **48**, 73 (1935)
30. G.F.R. Ellis, J. Math. Phys. **14**, 104–118 (1973)
31. H.G. Ellis, Gen. Rel. Grav. **10**, 105 (1979)
32. V. Ferrari, L. Gualtieri, P. Pani, *General Relativity and its Applications: Black Holes, Compact Stars and Gravitational Waves*, 1st edn. (CRC Press, 2020)
33. J.A. Gonzalez, F.S. Guzman, O. Sarbach, Class. Quant. Grav. **26**, 015010 (2009)
34. J.A. Gonzalez, F.S. Guzman, O. Sarbach, Class. Quant. Grav. **26**, 015011 (2009)
35. P.A. González, E. Papantonopoulos, Á. Rincón, Y. Vásquez, Phys. Rev. D **106**, 024050 (2022)
36. M. Guerrero, G.J. Olmo, D. Rubiera-Garcia, D. Gómez Sáez-Chillón, Phys. Rev. D **105**, 084057 (2022)
37. G. Gyulchev, P. Nedkova, V. Tinchev, S. Yazadjiev, Eur. Phys. J. C **78**, 544 (2018)
38. T. Harko, Z. Kovacs, F.S.N. Lobo, Phys. Rev. D **78**, 084005 (2008)
39. T. Harko, Z. Kovacs, F.S.N. Lobo, Phys. Rev. D **79**, 064001 (2009)
40. K. Jusufi, Gen. Rel. Grav. **53**, 87 (2021)
41. P. Kanti, B. Kleihaus, J. Kunz, Phys. Rev. Lett. **107**, 271101 (2011)

42. P. Kanti, B. Kleihaus, J. Kunz, Phys. Rev. D **85**, 044007 (2012)
43. S.W. Kim, Prog. Theor. Phys. Suppl. **172**, 21 (2008)
44. K.D. Kokkotas, B.G. Schmidt, Living Rev. Rel. **2**, 2 (1999)
45. R.A. Konoplya, C. Molina, Phys. Rev. D **71**, 124009 (2005)
46. R.A. Konoplya, A. Zhidenko, Phys. Rev. D **81**, 124036 (2010)
47. R.A. Konoplya, A. Zhidenko, Rev. Mod. Phys. **83**, 793 (2011)
48. R.A. Konoplya, A. Zhidenko, JCAP **1612**, 043 (2016)
49. R.A. Konoplya, Phys. Lett. B **784**, 43 (2018)
50. R.A. Konoplya, A.F. Zinhailo, Z. Stuchlík, Phys. Rev. D **99**, 124042 (2019)
51. R.A. Konoplya, A. Zhidenko, Phys. Rev. Lett. **128**, 091104 (2022)
52. P.K.F. Kuhfittig, Eur. Phys. J. C **74**, 2818 (2014)
53. F. Lamy, E. Gourgoulhon, T. Paumard, F.H. Vincent, Class. Quant. Grav. **35**, 115009 (2018)
54. M.S. Morris, K.S. Thorne, Am. J. Phys. **56**, 395–412 (1988)
55. K. Nakajima, H. Asada, Phys. Rev. D **85**, 107501 (2012)
56. K.K. Nandi, Y.Z. Zhang, A.V. Zakharov, Phys. Rev. D **74**, 024020 (2006)
57. P.G. Nedkova, V.K. Tinchev, S.S. Yazadjiev, Phys. Rev. D **88**, 124019 (2013)
58. T. Ohgami, N. Sakai, Phys. Rev. D **91**, 124020 (2015)
59. V. Perlick, Phys. Rev. D **69**, 064017 (2004)
60. T. Regge, J.A. Wheeler, Phys. Rev. **108**, 1063 (1957)
61. M. Safonova, D.F. Torres, G.E. Romero, Phys. Rev. D **65**, 023001 (2002)
62. R. Shaikh, Phys. Rev. D **98**, 024044 (2018)
63. H.A. Shinkai, S.A. Hayward, Phys. Rev. D **66**, 044005 (2002)
64. R. Takahashi, H. Asada, Astrophys. J. **768**, L16 (2013)
65. Y. Toki, T. Kitamura, H. Asada, F. Abe, Astrophys. J. **740**, 121 (2011)
66. N. Tsukamoto, T. Harada, K. Yajima, Phys. Rev. D **86**, 104062 (2012)
67. N. Tsukamoto, T. Harada, Phys. Rev. D **95**, 024030 (2017)
68. M. Visser, *Lorentzian Wormholes: From Einstein to Hawking, Woodbury* (AIP, USA, 1995)
69. S.H. Völkel, K.D. Kokkotas, Class. Quant. Grav. **35**, 105018 (2018)
70. M. Zhou, A. Cardenas-Avendano, C. Bambi, B. Kleihaus, J. Kunz, Phys. Rev. D **94**, 024036 (2016)

Stationary Models of Relativistic Viscous Torus

Sayantani Lahiri

Abstract

This chapter will be devoted to describing the findings of the study of stationary configurations of a geometrically thick viscous torus in Schwarzschild spacetime. A viscous torus is modelled by a relativistic non-ideal fluid. Using the hyperbolic theory of relativistic hydrodynamics of non-ideal fluids developed using the gradient expansion scheme which is discussed at length in this chapter, it is revealed that morphologies of a viscous torus—both hydrodynamical as well as the magnetised, are significantly modified by shear viscosity and the curvature of the Schwarzschild black hole.

1 Introduction

An accretion disk is a cloud of dust or of another substrate orbiting around a central compact object. As an example, such a disk-like luminous structure may form from a close binary system consisting of a massive primary compact object, and a less massive star such that the less massive companion exceeds its Roche lobe initiating a gaseous outflow towards the compact object. In this sense, accretion disks are "feeding" the central object. Typically, these astrophysical disks are modelled by a relativistic hydrodynamical fluid. During the process of accretion, inward mass transfer takes place towards the central object, for example, the black hole, resulting into an outward transfer of the orbital angular momentum in the radial direction, as a consequence, the disk substrate loses energy, part of which contributes to the luminous appearance of the disk and radiation. The accretion mechanism is also accompanied by dissipative processes, for example, viscous effects and heat flow. This process is one of the most efficient processes of energy release in the universe

S. Lahiri (✉)
Center of Applied Space Technology and Microgravity (ZARM), Universität Bremen, Am Fallturm 2, 28359 Bremen, Germany
e-mail: sayantani.lahiri@zarm.uni-bremen.de

B. Hartmann and J. Kunz (eds.), *Gravity, Cosmology, and Astrophysics*, Lecture Notes in Physics 1022, https://doi.org/10.1007/978-3-031-42096-2_2

and such processes can be observed in various systems as protoplanetary disks, X-ray binaries, gamma-ray bursts, active galactic nuclei, and quasars [13].

Often an ideal fluid is considered to describe the physics of an accretion disk, however, it is a crude assumption that can easily be violated in actuality. Moreover, in realistic accretion flows, dissipative processes cannot be entirely neglected. Most importantly, during the outward angular momentum transport in a binary system, the velocity gradients giving rise to underlying differential rotation of the adjacent fluid layers within an accretion disk can generate shear viscous effects. In this direction, a common viewpoint of the origin of shear viscosity is turbulence. Such turbulence in presence of magnetic fields gives rise to the magneto rotational instability which aids in radially outward angular momentum transport. The standard prescription for computing the shear viscosity generated due to turbulence is famously known as α-viscosity prescription developed by Shakura and Sunyeav [33]. This formula is prescribed for disks obeying Keplar's rotation law which are usually geometrically thin accretion disks. However, the exact origin of shear viscosity is not completely understood in case of geometrically thick disks or torus, also known as Polish doughnuts. Hence, it calls for our attention to investigate the origin and consequences of viscous effects within an accretion disk in the strong gravity regime.

In this chapter, we will restrict ourselves to discussing shear viscosity effects on stationary solutions of torus. In the following a step-by-step discussion is presented that illustrates the influences of shear viscosity on non-magnetised and magnetised torus in the Schwarzschild spacetime. However, for a self-consistency, a brief description of a hydrodynamical torus composed of an ideal fluid will be presented the next section.

Notations and conventions: The signature of Riemannian geometry is taken as $-, +, +, +$ and the geometrized units ($G = c = 1$) are followed throughout the chapter.

2 Nutshell Description of Thick Equilibrium Torus

2.1 Tori Supported by Ideal Fluid

Out of all accretion disk models, the geometrically thick disks, also known as 'Polish doughnuts' are the simplest, analytical, stationary model of accretion disk rotating in circular motion around a central compact object, for example, a black hole. The matter model of such disks are conventionally considered to be the ideal fluid characterised by energy density ϵ and fluid pressure p, and is additionally assumed to be stationary and axially symmetric. These simplest models are pressure-supported, where gravity and the ideal disk fluid remain in hydrostatic equilibrium and thus are non-accreting.

Undoubtedly, constructing the stationary solutions of geometrically thick tori in the framework of ideal fluid description is an over-simplified approach. In realistic astrophysical scenarios, dissipative effects like viscosity, heat flux, vorticity are no longer completely negligible and therefore it is worthwhile to explore the effects of the dissipative effects, in particular, the viscosity. While investigations with bulk, shear viscosity with resulting heat flux in their full glory is a monumental task to begin with, specially, if one takes into account of the hyperbolic nature of the corresponding hydrodynamical equations of motion, nevertheless, it still makes sense to include viscous effects in studying stationary solutions of Polish doughnuts, as a starting point. On the other hand, a rigorous treatment would involve discarding the circular motion of the fluid around a central compact object, and picking up a non-zero radial velocity. For example, a very well-known model of viscosity is the phenomenological α viscosity prescription proposed by Shakura & Sunyeav in thin-disk models where it was argued that magnetic fields are most likely responsible for generating shear viscosity which enable the angular momentum transport thereby facilitating the accretion process within a disk. This concept gained further momentum with the realisation of magnetorotational instability [6].

The simplest model of an accretion disk, developed by Paczyński, Abramowicz and their collaborators in late 1970 and early 1980, consists of the ideal fluid in equilibrium undergoes uniform circular motion four-velocity u^μ around a rotating black hole. In a spherical polar coordinate system (t, r, θ, ϕ), the metric of the stationary, axially symmetric Kerr black hole (with the mass parameter M and spin parameter $a = J/M$) is given by,

$$ds^2 = g_{tt}dt^2 + 2g_{t\phi}dt\, d\phi + g_{rr}dr^2 + g_{\theta\theta}d\theta^2 + g_{\phi\phi}d\phi^2 \tag{1}$$

and the associated Killing vectors are $\eta^\mu = (1, 0, 0, 0)$ and $\xi^\mu = (0, 0, 0, 1)$. Since the matter model also shares the same symmetries, the energy momentum tensor of the ideal fluid and the four-velocity are given by,

$$T^{\mu\nu} = (\epsilon + p)u^\mu u^\nu + pg^{\mu\nu}, \tag{2}$$

and with

$$u^r = u^\theta = 0 \qquad \text{(purely circular motion)} \tag{3}$$

$$u^\mu = \frac{dx^\mu}{d\tau} = u^t(\eta^\mu + \Omega\xi^\mu), \qquad \Omega(r, \theta) = \frac{u^\phi(r, \theta)}{u^t(r, \theta)}, \tag{4}$$

where τ is the proper time of the observer coming with the fluid. Here ϵ and p are functions of radial and polar coordinates only. Using (1) and the normalisation condition $u^\alpha u_\alpha = -1$, u^t is determined in the following way,

$$\frac{1}{(u^t)^2} = -(g_{tt} + \Omega^2 g_{\phi\phi} + 2\Omega g_{t\phi}) \tag{5}$$

Since the energy momentum tensor is conserved, following the condition $\nabla_\nu T^{\mu\nu} = 0$, the momentum conservation law can be recast in terms of u_t, Ω and specific angular momentum l in the following way,

$$\partial_\mu \ln(-u_t) - \frac{\Omega\partial_\mu l}{1 - \Omega l} = -\frac{\partial_\mu p}{\epsilon + p} \tag{6}$$

where the specific angular momentum is defined as $l = -\frac{u_t}{u_\phi}$. Commonly known as the relativistic Euler equation, (6) depicts the condition of hydrostatic equilibrium between the pressure-gradient and gravity. For a barotropic fluid whose equation of state (EOS) is characterised by $p = p(\epsilon)$, the integrability condition of (6) exists, which is given by,

$$\Omega = \Omega(l). \tag{7}$$

apart from the other trivial case of rigid rotation. The condition (7) is famously known as relativistic von-Zeipel theorem [1,7,8]. It states the surfaces of constant velocity Ω coincide with surfaces of constant specific angular momentum l (known as von-Zeipel cylinders), provided the ideal fluid undergoing circular motion obeys a barotropic EOS in a stationary and axisymmetric spacetime. Alternatively stated, iso-density surfaces coincide with iso-pressure surfaces, provided constant l surfaces coincide with constant Ω surfaces. The underlying reason for the existence of the integrability condition can be appreciated from the fact that the term on the right hand side of (6) i,e. $\frac{\partial_\mu p}{\epsilon + p}$ is an exact differential, as a result of which, the integrability condition is obtained in a straightforward way by imposing the compatibility condition $\partial_\nu \partial_\mu p = \partial_\mu \partial_\nu p$. Due to the existence of integrability condition given by (7), the integral form of the Euler equation can be easily expressed as follows,

$$W - W_{\text{in}} = \ln(-u_t) - \ln(-u_t)_{\text{in}} - \int_{l_{\text{in}}}^{l} \frac{\Omega}{1 - \Omega l} \partial_\mu l \tag{8}$$

with the total differential defined in the following way,

$$W - W_{\text{in}} = -\int_0^p \frac{dp}{(\epsilon + p)} \tag{9}$$

where 'in' refers to the inner edge of the disk at the equatorial plane where the pressure and the corresponding energy density vanish. It is to be noted that $W(r)$, in the Newtonian limit, represents a total potential of a test particle (centrifugal plus gravitational).

It is necessary to determine a closed-form relation between l and Ω (which individually are functions of spacetime metric of the background geometry) for determining the shape of a torus. In principle, such a relation is accurately prescribed only if the dissipative effects are taken into consideration. However, due to an ambiguity in proposing an accurate closed form relation between l and Ω with inputs from dissipative processes, one may consider a simplest assumption of constant specific angular momentum as $l = l_0$, where l_0 is a constant. Such a consideration not only simplifies the Euler equation considerably but also simplifies the computation of equipotential surfaces. Moreover, the angular velocity now becomes a plain relation of metric coefficients. In this simplified approach, the equipotential surfaces are easily computed using the relation, $W(r, \theta) = \ln(-u_t)$, if one also sets $W_{\text{in}} = \ln(-u_t)_{\text{in}}$ for a given stationary and axially symmetric black hole spacetime. The equilibrium torus obtained under this assumption are also known as marginally stable disks as they are marginally stable under axisymmetric perturbations. In general, W_{in} can either be closed ($W < 0$) or open ($W > 0$) at any point on the (r, θ) plane whereas $W = 0$ corresponds to an equipotential surface that is closed at infinity. It is worthwhile to note that closed equipotential surfaces at the equatorial plane are characterised by two radial positions, namely the cusp r_{cusp} and the centre r_{cen}, whereas the inner edge r_{in} is the free parameter chosen between the cusp and the centre of the torus. At the cusp and the at the centre, the gravitational force of the central compact object is exactly balanced by the pressure gradient forces of the fluid element resulting into vanishing acceleration. The situation when $a_\mu = 0$ implies $\partial_\mu p = 0 = \partial_\mu W$, as a result the fluid does not experience any net force at these locations which further means the fluid elements undergo geodesic motion. The specific angular momentum is then given by the Keplarian angular momentum l_k. For the case of constant specific angular momentum distributions, the radial positions of the centre and the cusp at the equatorial plane can be determined from the following relation,

$$l_k(r) - l_0 = 0 \tag{10}$$

In particular, r_{cen} corresponds to the position of pressure maximum and r_{cusp} corresponds to location of self-crossing equipotentials. In the context of a binary system, the cusp is viewed as the Roche lobe or Lagrangian point L_1 which corresponds to the last equipotential surface from which mass of the fluid remains in equilibrium and does not overflow towards the central compact object. If corresponding to the marginally bound orbit r_{mb} and the marginally stable orbit r_{ms}, the respective values of specific angular momentum are l_{mb} and l_{ms}, then torus configurations with closed equipotential surfaces, marked with distinct radial positions $r_{\text{cusp}}, r_{\text{cen}}, r_{\text{in}}, r_{\text{out}}$, are obtained provided $l_{\text{ms}} < l < l_{\text{mb}}$. Note here that r_{out} corresponds to the outer surface of the torus and can be determined from the condition $W_{\text{in}}(r) = W_{\text{out}}(r)$.

Since the specification of inner edge of the torus is somewhat arbitrary, it makes sense to incorporate the idea of mass transfer of the fluid in assigning the radial location of r_{in} and thus define the potential barrier ΔW, also known as the potential gap between the cusp and the inner edge. The case $\Delta W < 0$ corresponds to the torus within its Roche lobe, for which no mass transfer is possible. On the other hand, $\Delta W > 0$ corresponds to the case when fluid from the torus overflows the Roche lobe towards the central compact object in absence of shear viscosity. Finally, $\Delta W = 0$ corresponds to the situation when the fluid exactly fills the Roche lobe.

Although the assumption of constant specific angular momentum simplifies the calculations and promotes a detailed understanding of the disk morphology of geometrically thick torus but it is an idealised and oversimplified consideration.

In this direction, a much realistic assumption is considered in Font, where the distribution of angular momentum of the torus is taken to be the power-law at the equatorial plane which is given by,

$$l_{\text{eq}}(r) = K r^\alpha \tag{11}$$

The constraining condition $\alpha \geq 0$ is supported by Rayleigh stability condition and also leads to stable configuration of the torus with constant angular momentum distribution. Nevertheless, the value of α is further constrained in the range $0 \leq \alpha < \frac{1}{2}$ by demanding the existence of closed configuration of a torus described by its cusp and centre. The prograde torus corresponds to $K > 0$ whereas the retrograde torus corresponds to $K < 0$. In this case, the potential is found to be,

$$W(r,\theta) = W_{\rm eq}(r_0) + \ln\left(\frac{-u_t(r,\theta)}{-u_t(r_0,\frac{\pi}{2})}\right) \tag{12}$$

where r_0 is the radius at which the von Zeipel's cylinder passing at (r,θ) intersects the equatorial plane. Similar to constant angular momentum tori, W can be positive, negative or zero. Further discussions on torus with non-constant angular momentum can be found in [9,24].

It should be emphasised here that these simple analytical solutions of thick tori hold their own importance as these solutions serve as starting point for more complicated numerical studies for describing time-dependent accretion flows.

2.2 Magnetised Tori

In [19], Komissarov extended the equilibrium stationary solution of tori by adding azimuthal magnetic field in presence of constant specific angular momentum distributions. It was possible to determine the integrability conditions because the magnetic field enters the solution as a pressure-like term. In presence of a toroidal magnetic field b^α with $b^r = b^\theta = 0$, the momentum conservation equation becomes,

$$\partial_\mu \ln(-u_t) - \frac{\Omega \partial_\mu l}{1-\Omega l} + \frac{\partial_\mu p}{w} + \frac{\partial_\mu(\mathcal{L}b^2)}{2\mathcal{L}w} = 0 \tag{13}$$

where the enthalpy is $w = \epsilon + p$ and $\mathcal{L} = g_{t\phi}^2 - g_{tt}g_{\phi\phi}$. The integral form of the momentum conservation equation can be written as,

$$\ln(-u_t) + \int_0^p \frac{dp}{w} - \int_0^l \frac{\Omega dl}{1-\Omega l} + \int_0^{\tilde{p}_m} \frac{d\tilde{p}_m}{\tilde{w}} = \text{constant} \tag{14}$$

such that $\tilde{p}_m = \mathcal{L}p_m$ and $\tilde{w} = \mathcal{L}w$. In terms of the total potential the above equation reduces to,

$$W - W_{\rm in} + \int_0^p \frac{dp}{w} + \int_0^{\tilde{p}_m} \frac{d\tilde{p}}{\tilde{w}} = 0 \tag{15}$$

In [14] the Komissarov's solution is extended for non-constant specific angular momentum in Kerr geometry.

3 Relativistic Non-ideal Fluids

The angular momentum transport triggered by shear viscosity may typically arise from differential rotation of adjacent fluid layers within a disk. In that case, the torus is modelled by relativistic non-ideal fluid with non-zero shear viscosity. In this chapter, we discuss the findings related to shear viscosity effects on stationary solutions of a torus by using hyperbolic equation of motion. On a different note, it is well-known that conservation laws of relativistic non-ideal hydrodynamics, initially developed by Landau-Lifschitz and Eckart, are not causal and therefore do not reproduce hyperbolic equations of motion. Furthermore, the equilibrium states are dynamically unstable under linear perturbations. A relativistic theory plagued with the drawbacks associated to a causality and unstable equilibrium states is therefore clearly unacceptable. This limitations of Landau-Eckart's first order theory was successfully evaded by including second-order gradients, a formalism first developed much later in 1970s, by Müller in the non-relativistic setup and later extended by Israel and Stewart for relativistic non-ideal fluids.

Next, a brief account of the Muller- Isreal-Stewart formalism is described in self-consistent fashion by reporting the advantages and disadvantages.

3.1 Causal Theory of Relativistic Hydrodynamics of Non-ideal Fluids

The well-established theory of relativistic hydrodynamics of ideal fluids is discussed extensively in the existing literature (cite: Landau, Rezzolla), therefore, in the present chapter, we include only those aspects which are relevant for a self-consistent description of hydrodynamics of relativistic non-ideal fluids.

The ideal fluid hydrodynamics is formulated under thermodynamic equilibrium configuration of the fluid. Assuming a simple picture, when an ideal fluid is composed of a single species of conserved charge characterized by its number density n, equilibrium total energy density e and equilibrium fluid pressure p, the constitutive relations in terms of four-velocity $u^\mu(x)$ of the fluid are given by,

$$N^\mu_{\text{ideal}} = nu^\mu, \qquad T^{\mu\nu}_{(0)} = e\, u^\mu u^\nu + p\, \Delta^{\mu\nu} \tag{16}$$

where the four-velocity is $u^\mu = (\gamma, \gamma \vec{v})$ and γ is the Lorentz factor. The variables p, e and n are slowly varying fields. In the curved spacetime, $\Delta^{\mu\nu} = g^{\mu\nu} + u^\mu u^\nu$ is the projection tensor satisfying the condition $u_\mu \Delta^{\mu\nu} = 0$ and the normalization condition is given by $u^\rho u_\rho = -1$. If the covariant derivative is decomposed as $\nabla^\mu = -u^\mu D + \mathcal{D}^\mu_\perp$ where the longitudinal part is $D = u^\alpha \nabla_\alpha$ and the transverse parts is $\mathcal{D}^\mu_\perp = \Delta^{\mu\nu}\nabla_\nu$, then using (16), the conservation of the particle current and that of the energy-momentum tensor give rise respectively to the continuity equation, energy conservation equation and the Euler equation as follows,

$$Dn + n\nabla.u = 0, \tag{17}$$

$$De + (e+p)\nabla.u = 0, \tag{18}$$

$$a^\mu = Du^\mu = -\frac{1}{(e+p)}\mathcal{D}^\mu_\perp p \tag{19}$$

It is worthwhile to note that the constitutive relations of the ideal fluid do not contain derivatives/gradients of hydrodynamical variables, therefore, the ideal fluid hydrodynamics is also known as zeroth order hydrodynamics. However, in presence of dissipative fluxes, the idea of global thermodynamic equilibrium breaks down and the hydrodynamic variables are no longer constant anymore. The hydrodynamical variables namely $u^\mu(x)$, temperature $T(x)$ and the chemical potential $\mu(x)$ are not uniquely defined in out-of-the equilibrium situations. This leads to a picture, where one can have several local values of the hydrodynamical variables at every point in spacetime, each differing by gradients but leads to their respective equilibrium values in absence of any dissipation, thus gives to ambiguities in the definition of these variables. To get rid of any ambiguities one brings forth the idea of *choice* of frames. In general two choices namely the Eckart frame (no charge diffusion) and the Landau frame (no heat flow) are the broadly defined frames that exist in the literature. We will adopt to the Eckart frame where we have $u_\mu N^\mu = n$, with n being identified with the conserved charge in the system. The discussion in presence of conserved charges on relativistic viscous hydrodynamics can be found in [29].

A non-ideal fluid characterised by a single species of charge n in the Eckart frame is described by two following conserved currents,

$$N^\mu = nu^\mu, \qquad T^{\mu\nu} = eu^\mu u^\nu + (p+\Pi)\Delta^{\mu\nu} + q^\mu u^\nu + q^\nu u^\mu + \pi^{\mu\nu}$$
$$= T^{\mu\nu}_{(0)} + T^{\mu\nu}_{(\text{non-ideal})}, \tag{20}$$

where

$$T^{\mu\nu}_{(\text{non-ideal})} = \Pi\Delta^{\mu\nu} + q^\mu u^\nu + q^\nu u^\mu + \pi^{\mu\nu}. \tag{21}$$

The dissipative quantities are given by the heat flow vector q^α, bulk viscosity scalar Π (trace part) and the shear viscosity tensor $\pi^{\mu\nu}$(symmetric transverse traceless part). The Eckart frame, where no charge flow takes place, the continuity equation remains unchanged as that of an ideal fluid and in presence of a single conserved charge and the four-velocity of the fluid is expressed as, $u^\mu = \dfrac{N^\mu}{\sqrt{-N^\alpha N_\alpha}}$. The corresponding conservation laws are then given by,

$$\Delta^\lambda_\nu \nabla_\mu T^{\mu\nu} = 0 \quad \text{Navier-Stokes equation (momentum conservation)}, \tag{22}$$

$$u_\nu \nabla_\mu T^{\mu\nu} = 0 \quad \text{energy conservation equation}, \tag{23}$$

$$\nabla_\mu N^\mu = 0 \quad \text{continuity equation}. \tag{24}$$

The relativistic Navier-Stokes equation in terms of dissipative quantities is given by,

$$(\epsilon + p + \Pi)a^{\mu} + \mathcal{D}^{\mu}_{\perp}(p + \Pi) + \pi^{\mu\alpha}a_{\alpha} + \Delta^{\mu}_{\ \beta}\mathcal{D}^{\perp}_{\alpha}\pi^{\beta\alpha} + \Delta^{\mu}_{\ \alpha}Dq^{\alpha} + \left(\sigma^{\mu}_{\alpha} + \Omega^{\mu}_{\alpha} + \frac{4}{3}\Delta^{\mu}_{\alpha}(\nabla.u)\right)q^{\alpha} = 0 \tag{25}$$

where the vorticity tensor is defined as $\Omega_{\mu\nu} = \frac{1}{2}(\nabla_{\mu}u_{\nu} - \nabla_{\nu}u_{\mu})$. The general forms of the dissipative quantities are then determined respecting the second law of thermodynamics using the entropy current defined as,

$$S^{\mu} = nu^{\mu} + \frac{q^{\mu}}{T} \tag{26}$$

$$\Pi = -\zeta(\nabla.u), \tag{27}$$

$$\pi^{\mu\nu} = -2\eta\sigma^{\mu\nu} = -2\eta\left[\Delta^{\mu\alpha}\Delta^{\nu\beta}\left(\frac{\nabla_{\alpha}u_{\beta} + \nabla_{\beta}u_{\alpha}}{2}\right) - \frac{1}{3}\Delta^{\mu\nu}\Delta^{\alpha\beta}\nabla_{\alpha}u_{\beta}\right], \tag{28}$$

$$q^{\mu} = -\kappa T\,(\mathcal{D}^{\mu}_{\perp}\ln T + Du^{\mu}) \tag{29}$$

It is noted that the dissipative fluxes are first order gradients of hydro-variables (i,e. u^{α}, T and μ), as a result, the relativistic hydrodynamics developed by Landau-Liftshitz and Eckart are known as first-order theory of hydrodynamics or *Classical Irreversible Thermodynamics*. The bulk viscosity coefficient ζ, shear viscosity coefficient η and the coefficient of thermal conductivity satisfy the conditions $\zeta \geq 0$, $\eta \geq 0$, $\kappa \geq 0$ following the second law of thermodynamics and are also collectively known as the first order transport coefficients. In this case, as the thermodynamic fluxes are expressed in terms of first order gradients of hydrodynamical variables therefore that the energy momentum tensor contains terms only up to first order in gradients, resulting into the formalism, known called the first order hydrodynamics.

3.1.1 Müller-Isreal-Stewart Approach

The relativistic Navier-Stokes equation and the heat conduction equation both violate causality due to the parabolic nature of the equations of motion allowing superluminal velocities of propagating signals under linear perturbations [25,31]. In addition, the associated equilibrium states are also dynamically unstable [15,16].

In order to fix the non-causal behaviour of dissipative fluids, Müller [27], Israel and Stewart [17,18] (who generalised for relativistic version of the work by [27]) put forward a phenomenological approach. They extended the Grad's 14-moment approximation in curved background and successfully cured the causality violating nature of the conservation laws of relativistic non-ideal fluids by considering second order gradients of hydrodynamic variables. Then the general structures of fluxes, i.e. Π, q^{β} and $\pi^{\mu\nu}$ contain additional coefficients which arise due to second order gradients terms. The corresponding associated are broadly known as second order transport coefficients which serve as input parameters of the theory. Essentially, the coefficients are determined by using an underlying microscopic theory describing the system. The first and second order transport coefficients have been extensively computed in the context of conformal/non-conformal theories and kinetic theory [11,20,26,28,30,32].

All in all, the MIS approach is based on the assumption that the entropy 4-current of the non-ideal fluid will be modified by incorporating quadratic terms of dissipative quantities, see also [29]. From the positivity of the new entropy 4-current, one obtains dynamical equations of dissipative flux quantities and new corrections terms to constitutive relations and hence to the energy momentum tensor. The Müller, Israel-Stewart (MIS) formalism produces an hyperbolic theory, causal formulation of relativistic dissipative hydrodynamics and is perfectly stable under linear perturbations [15].

3.1.2 Gradient Expansion Scheme in Eckart Frame

Though the MIS formalism theory came up with a hyperbolic theory of non-ideal hydrodynamics devoid of instabilities, it did not capture the full picture, in the sense it missed out possible spacetime curvature terms at the second order, all of which are likely to contribute in a curved background, if all possible combinations of second order gradients are taken into consideration. In this direction, Baier, Romatschke et al. [4,32] reformulated MIS theory by incorporating all possible second order terms including the curvature terms, specifically in the Landau frame for both conformal and non-conformal theories with the primary motivation to implement their approach in the field of relativistic heavy-ion collisions. However, in the astrophysical context, Eckart frame of reference is more favoured over the Landau frame. For example, an accretion disk involves radiation transfer processes which may be attributed to shear viscosity and heat flow. Once again, the effects of non-negligible heat flow can be addressed only in the context of Eckart frame of reference. Here we describe the formalism in which general forms of bulk viscosity, shear viscosity and heat flow vector are obtained up to second order in gradients by adopting a similar approach taken in [4,32]. Consequently, the obtained general relativistic hydrodynamical equations in the Eckart frame are causal in in the Eckart frame.

In the realm of a phenomenological approach, the construction involves systematic addition of gradients of second order of hydrodynamical variables where the corrections to the energy momentum tensor of the ideal fluid, are associated to dissipative fluxes namely the viscosity and heat flow. By systematically writing down all possible second order terms (and a most general entropy 4- current) one constructs the most general second order viscosity tensor and other dissipative flux quantities in both the Landau [4] and Eckart frame [21] descriptions. This procedure is known as the *gradient expansion method* and each dissipative flux quantity like the viscosity tensor contain all possible second order terms which are not present in the MIS formalism. The important consequence is that curvature terms appear in the equations of motion. Additionally, first and second order transport coefficients are present together with relaxation time coefficients. These transport coefficients can be determined within the framework of a particular microscopic theory under consideration, for example, using kinetic theory and strongly coupled theories.

This approach has important advantages. First, the resulting conservation laws are hyperbolic. Second, the existence of curvature terms (one of many second order terms) in each of dissipative flux quantities allows to study curvature effects which could be essential for assessing the role of spacetime curvature, particularly in the early universe or in accretion disks. Based on this method, the most general form of fluid dynamical equations can be derived in [5,21].

Procedure: To begin with, the conservation laws of an ideal fluid given by (18) and (19) can be re-expressed using thermodynamical identities $e + p = Ts + \mu n$, $de = Tds + \mu dn$ and Gibbs-Duhem relation $dp = sdT + nd\mu$ as follows,

$$D \ln s = -\nabla.u, \tag{30}$$

$$Du^{\mu} = -\mathcal{D}_{\perp}^{\mu} \ln T - \frac{n}{(s + \alpha n)} \mathcal{D}_{\perp}^{\mu} \left(\frac{\mu}{T}\right) \tag{31}$$

where s, T, μ are respectively the entropy density, temperature, the chemical potential of the fluid and $\alpha = \frac{\mu}{T}$. Since in the non-relativistic limit, $D \simeq \partial_t$ and $\mathcal{D}_{\perp}^{\mu} \simeq \partial_i$ (up to higher order gradients), (30) and (31) indicate time derivatives can be recast into space derivatives thus implying all first order gradients are not independent. To construct a general form of energy momentum tensor, second order gradients of hydrodynamical variables i,e. u^{μ}, $T(x)$ and $\mu(x)$ are then used for constructing scalars (for bulk viscosity), vectors orthogonal to u^{μ} (for heat flow vector) and symmetric traceless tensors orthogonal to u^{μ} (for shear tensor) in the Eckart frame However, in absence of charge diffusion, the particle current is unchanged in the Eckart frame.

In a non-ideal fluid with a single conserved charge, the first order linearly independent components giving rise to second order corrections to $T^{\mu\nu}$ are: a scalar $\mathcal{D}_{\perp}^{\alpha} u_{\alpha}$, two vectors $\mathcal{D}_{\perp}^{\nu} \mu$, $\mathcal{D}_{\perp}^{\nu} \ln T$ and a tensor $\mathcal{D}_{\alpha}^{\perp} u_{\beta}$. Then all dissipative quantities up to second order of gradients are constructed out of linearly independent terms consisting of comoving spatial first order gradients namely $\mathcal{D}_{\perp}^{\alpha} u_{\alpha}$, $\mathcal{D}_{\perp}^{\nu} \mu$ and $\mathcal{D}_{\perp}^{\nu} \ln T$ which appear in (30) and (31) of the ideal fluid. Let us mention two important points here that are considered for obtaining the second order theory using the gradient expansion. First, the rank-two tensor $\mathcal{D}_{\alpha}^{\perp} u_{\beta}$ can be easily expressed in terms of trace part plus symmetric and antisymmetric trace-less parts orthogonal to u^{μ} as shown below,

$$\mathcal{D}_{\alpha}^{\perp} u_{\beta} = \frac{\nabla.u}{3} \Delta_{\alpha\beta} + \frac{1}{2} \sigma_{\alpha\beta} + \Omega_{\alpha\beta} \tag{32}$$

Second, the curvature tensor term arises as a consequence of non-commutative property of covariant derivatives in the curved background. In this case, the Riemann tensor is obtained when two covariant derivatives acting on the four-velocity vector do not commute as a result of which one obtains,

$$R^{\lambda}{}_{\mu\alpha\beta} u_{\lambda} = \nabla_{\alpha} \nabla_{\beta} u_{\mu} - \nabla_{\beta} \nabla_{\alpha} u_{\mu} \tag{33}$$

where the Riemann tensor is: $R^{i}_{\ jnl} = \frac{\partial}{\partial x^n}\Gamma^i_{lj} - \frac{\partial}{\partial x^l}\Gamma^i_{nj} + \Gamma^s_{lj}\Gamma^i_{ns} - \Gamma^s_{nj}\Gamma^i_{ls}$. Considering all possible second order terms, then general forms of the dissipative fluxes Π, q^μ and $\pi^{\mu\nu}$ at the second order can be presumed to be as follows,

$$\Pi = -\zeta(\nabla.u) + \Sigma_{i=1}^{i=10}\alpha_i \mathcal{M}_i \tag{34}$$

$$q^\mu = -\frac{\kappa n T^2}{(e+p)}\mathcal{D}^\mu_\perp\left(\frac{\mu}{T}\right) + \Sigma_{i=1}^{i=11}\beta_i \mathcal{N}^\mu_i \tag{35}$$

$$\pi^{\mu\nu} = -2\eta\sigma^{\mu\nu} + \Sigma_{i=1}^{i=10}\lambda_i \mathcal{O}^{\mu\nu}_i \tag{36}$$

where α_i, β_i and λ_i are coefficients appearing at the second order. Described below is the method for obtaining Eqs. (34)–(36) using second order gradients. The possible second gradient order terms are given by,

$$\begin{array}{llll} (\nabla.u)^2, & (\mathcal{D}^\mu_\perp \ln T)(\mathcal{D}^\nu_\perp \ln T), & (\mathcal{D}^\mu_\perp u^\alpha)(\mathcal{D}^\nu_\perp \ln T), & (\mathcal{D}^\mu_\perp u^\alpha)(\mathcal{D}^\nu_\perp u^\beta), \\ R^\lambda_{\ \mu\alpha\beta}, & \mathcal{D}^\mu_\perp \mathcal{D}^\nu_\perp u^\alpha, & \mathcal{D}^\mu_\perp \mathcal{D}^\nu_\perp \ln T, & \mathcal{D}^\alpha_\perp \mu \mathcal{D}^\beta_\perp \mu, \\ \mathcal{D}^\alpha_\perp \mathcal{D}^\beta_\perp \mu, & & & \end{array} \tag{37}$$

which are utilised to construct possible scalars, vectors and tensors for constructing the dissipative flux quantities at the second order . Note that the traceless symmetric part of a tensor is $A^{<\mu\nu>} = \frac{1}{2}\Delta^{\mu\alpha}\Delta^{\nu\beta}(A_{\alpha\beta} + A_{\beta\alpha}) - \frac{1}{3}\Delta^{\mu\nu}\Delta^{\alpha\beta}A_{\alpha\beta}$. Furthermore, each scalar is denoted by $\mathcal{M}_i$, each vector by $\mathcal{N}_i$ and each tensor by $\mathcal{O}^{\mu\nu}_i$, where i labels a particular term. The possible terms corresponding to the scalars $\mathcal{M}_I$, the vectors $\mathcal{N}_i$, and the tensors $\mathcal{O}^{\mu\nu}_i$ are

$$\begin{array}{lll} \mathcal{M}_1 = \mathcal{D}^\perp_\alpha \mathcal{D}^\alpha_\perp \ln T, & \mathcal{M}_2 = (\mathcal{D}^\alpha_\perp \ln T)\mathcal{D}^\perp_\alpha \ln T, & \mathcal{M}_3 = (\nabla.u)^2, \\ \mathcal{M}_4 = R, & \mathcal{M}_5 = u^\alpha u^\beta R_{\alpha\beta}, & \mathcal{M}_6 = \sigma^{\alpha\beta}\sigma_{\alpha\beta}, \\ \mathcal{M}_7 = \Omega^{\alpha\beta}\Omega_{\alpha\beta}, & \mathcal{M}_8 = (\mathcal{D}^\alpha_\perp \mu)\mathcal{D}^\perp_\alpha \mu, & \mathcal{M}_9 = \mathcal{D}^\perp_\alpha \mathcal{D}^\alpha_\perp \mu, \\ \mathcal{M}_{10} = (\mathcal{D}^\alpha_\perp \mu)\mathcal{D}^\perp_\alpha \ln T, & & \end{array} \tag{38}$$

$$\begin{array}{lll} \mathcal{N}^\nu_1 = (\mathcal{D}^\perp_\alpha \ln T)\frac{(\nabla.u)}{3}\Delta^{\alpha\nu}, & \mathcal{N}^\nu_2 = (\mathcal{D}^\perp_\alpha \ln T)\sigma^{\alpha\nu}, & \mathcal{N}^\nu_3 = (\mathcal{D}^\perp_\alpha \ln T)\Omega^{\alpha\nu}, \\ \mathcal{N}^\nu_4 = (\mathcal{D}^\perp_\alpha \mu)\frac{(\nabla.u)}{3}\Delta^{\alpha\nu}, & \mathcal{N}^\nu_5 = (\mathcal{D}^\nu_\perp \mu)(\nabla.u), & \mathcal{N}^\nu_6 = (\mathcal{D}^\perp_\alpha \mu)\sigma^{\alpha\nu}, \\ \mathcal{N}^\nu_7 = \Delta^{\alpha\nu}u^\gamma R_{\alpha\gamma}, & \mathcal{N}^\nu_8 = \mathcal{D}^\perp_\alpha \sigma^{\alpha\nu}, & \mathcal{N}^\nu_9 = \mathcal{D}^\perp_\alpha \Omega^{\alpha\nu}, \\ \mathcal{N}^\nu_{10} = (\mathcal{D}^\perp_\alpha \mu)\Omega^{\alpha\nu}, & \mathcal{N}^\nu_{11} = (\mathcal{D}^\perp_\alpha \mu)R^{\alpha\nu}, & \end{array} \tag{39}$$

and

$$\begin{aligned}
\mathcal{O}_1^{\mu\nu} &= (\mathcal{D}_\perp^{<\mu} \ln T)(\mathcal{D}_\perp^{\nu>} \ln T), & \mathcal{O}_2^{\mu\nu} &= \mathcal{D}_\perp^{<\mu}\mathcal{D}_\perp^{\nu>} \ln T, \\
\mathcal{O}_3^{\mu\nu} &= (\mathcal{D}_\perp^{<\mu} \mu)(\mathcal{D}_\perp^{\nu>} \mu), & \mathcal{O}_4^{\mu\nu} &= \mathcal{D}_\perp^{<\mu}\mathcal{D}_\perp^{\nu>} \mu, \\
\mathcal{O}_5^{\mu\nu} &= \mathcal{D}_\perp^{<\mu} \ln T \mathcal{D}_\perp^{\nu>} \mu, & \mathcal{O}_6^{\mu\nu} &= \sigma^{\gamma<\mu}\Omega^{\nu>}{}_\gamma, \\
\mathcal{O}_7^{\mu\nu} &= \sigma^{\gamma<\mu}\sigma^{\nu>}{}_\gamma, & \mathcal{O}_8^{\mu\nu} &= \Omega^{\gamma<\mu}\Omega^{\nu>}{}_\gamma, \\
\mathcal{O}_9^{\mu\nu} &= R^{<\mu\nu>}, & \mathcal{O}_{10}^{\mu\nu} &= u_\alpha u_\beta R^{\alpha<\mu\nu>\beta}.
\end{aligned} \tag{40}$$

Notably, (34)–(36) are of algebraic form responsible for violating causality and lead to unstable equations. In order to produce a causal as well as a stable theory in the similar spirit of MIS formalism dynamical equations for dissipative flux quantities are essential. A viable way to facilitate this is, by introducing relaxation-time coefficients for Π, q^μ and $\pi^{\mu\nu}$. Based on all considerations and requirements, the general form of the shear viscosity tensor up to second order in gradients is presented as follows,

$$\begin{aligned}
\pi^{\mu\nu} &= -2\eta\sigma^{\mu\nu} - \tau_\pi^< D(-2\eta)\sigma^{\mu\nu>} - 2\eta\tau_2\left[\frac{D\eta}{\eta} - \frac{(\nabla.u)}{4} + \frac{D\ln T}{4}\right]\sigma^{\mu\nu} \\
&+ \xi_2\, \mathcal{D}_\perp^{<\mu}\mathcal{D}_\perp^{\nu>} \ln T + \xi_3\, (\mathcal{D}_\perp^{<\mu} \mu)(\mathcal{D}_\perp^{\nu>} \mu) + \xi_4\, \mathcal{D}_\perp^{<\mu}\mathcal{D}_\perp^{\nu>} \mu \\
&+ \xi_5\, \mathcal{D}_\perp^{<\mu} \ln T \mathcal{D}_\perp^{\nu>} \mu + \xi_6\, \sigma^{\gamma<\mu}\Omega^{\nu>}{}_\gamma + \xi_7\, \sigma^{\gamma<\mu}\sigma^{\nu>}{}_\gamma \\
&+ \xi_8\, \Omega^{\gamma<\mu}\Omega^{\nu>}{}_\gamma + \kappa_1\, R^{<\mu\nu>} + \kappa_2\, u_\alpha u_\beta R^{\alpha<\mu\nu>\beta}.
\end{aligned} \tag{41}$$

In fact, (41) constructed using gradient expansion scheme reduces to the relaxation-type theory first introduced in MIS formalism to combat the acausality aspect of the first-order theories. It is therefore necessary to express (41) in a form that will give rise to dynamical equations of $\pi^{\mu\nu}$ while keeping the essence of MIS theory at the same time. A similar procedure as employed in [10,32] will be considered here. For constructing a dynamical equation for $\pi^{\mu\nu}$ with a goal to correctly reproduce a relaxation-type equation, a new basis is chosen (now with coefficients c_i) and the combined term $^< D\sigma^{\mu\nu>} + T\nabla_\lambda \left(\frac{u^\lambda}{4T}\right)\sigma^{\mu\nu}$[1] is expressed in the new basis with $\mathcal{O}_i^{\mu\nu}$ as follows,

$$^< D\sigma^{\mu\nu>} + T\nabla_\lambda \left(\frac{u^\lambda}{4T}\right)\sigma^{\mu\nu} = c_1\mathcal{O}_1^{\mu\nu} + c_2\mathcal{O}_2^{\mu\nu} + \ldots\ldots.. c_9\mathcal{O}_9^{\mu\nu} \tag{42}$$

where c_i's are coefficients of the newly chosen basis. Eliminating $\mathcal{O}_1^{\mu\nu}$ from (42)

$$\mathcal{O}_1^{\mu\nu} = \frac{1}{c_1}\left[^< D\sigma^{\mu\nu>} + T\nabla_\lambda \left(\frac{u^\lambda}{4T}\right)\sigma^{\mu\nu} - c_2\mathcal{O}_2^{\mu\nu} - \ldots\ldots.. - c_9\mathcal{O}_9^{\mu\nu}\right] \tag{43}$$

[1] The particular combination consisting of second order terms is also present in MIS theory for $\pi^{\mu\nu}$.

and thereby substituting (43) in (36), one obtains,

$$\pi^{\mu\nu} = -2\eta\sigma^{\mu\nu} + \frac{\lambda_1}{c_1}{}^{<}D\sigma^{\mu\nu>} + \frac{\lambda_1}{c_1}T\nabla_\lambda\left(\frac{u^\lambda}{4T}\right)\sigma^{\mu\nu} + \Sigma_{i=2}^{i=9}\left(\lambda_i - \frac{c_i}{c_1}\lambda_1\right)\mathcal{O}_i^{\mu\nu} \tag{44}$$

To actually reproduce the relaxation-type equation, finally the first-order solution[2] i,e. $\sigma^{\mu\nu} = -\frac{1}{2\eta}\pi^{\mu\nu}$ is substituted in all second-order terms of Eq. (44) to obtain,

$$\begin{aligned}\tau_\pi {}^{<}D\pi^{\mu\nu>} + \pi^{\mu\nu} &= -2\eta\sigma^{\mu\nu} + \tau_\pi\left[\frac{D\eta}{\eta} - \frac{(\nabla.u)}{4} + \frac{D\ln T}{4}\right]\pi^{\mu\nu}\\ &+ \xi_2\,\mathcal{D}_\perp^{<\mu}\mathcal{D}_\perp^{\nu>}\ln T + \xi_3\,(\mathcal{D}_\perp^{<\mu}\mu)(\mathcal{D}_\perp^{\nu>}\mu) + \xi_4\,\mathcal{D}_\perp^{<\mu}\mathcal{D}_\perp^{\nu>}\mu\\ &+ \xi_5\,\mathcal{D}_\perp^{<\mu}\ln T\mathcal{D}_\perp^{\nu>}\mu - \frac{\xi_6}{2\eta}\pi^{\gamma<\mu}\Omega^{\nu>}{}_\gamma + \frac{\xi_7}{4\eta^2}\pi^{\gamma<\mu}\pi^{\nu>}{}_\gamma\\ &+ \xi_8\,\Omega^{\gamma<\mu}\Omega^{\nu>}{}_\gamma + \kappa_1\,R^{<\mu\nu>} + \kappa_2\,u_\alpha u_\beta R^{\alpha<\mu\nu>\beta}\end{aligned} \tag{45}$$

where $\tau_\pi = \dfrac{\lambda_1}{2c_1\eta}$ is the transport coefficient, also known as relaxation-time coefficient for $\pi^{\mu\nu}$. It preserves causality of the Navier-Stokes equation in absence of bulk viscosity, heat flux and charge diffusion. By following a similar procedure the structure of bulk viscosity up to second order with relaxation time coefficient τ_Π becomes,

$$\begin{aligned}\Pi &= -\zeta(\nabla.u) - \tau_\Pi D(-\zeta\nabla.u)\\ &+ \zeta_2\,(\mathcal{D}_\perp^\alpha\ln T)\mathcal{D}_\alpha^\perp\ln T + \zeta_3\,(\nabla.u)^2 + \zeta_4\,R + \zeta_5\,u^\alpha u^\beta R_{\alpha\beta}\\ &+ \zeta_6\sigma^{\alpha\beta}\sigma_{\alpha\beta} + \zeta_7\,\Omega^{\alpha\beta}\Omega_{\alpha\beta} + \zeta_8\,(\mathcal{D}_\perp^\alpha\mu)\mathcal{D}_\alpha^\perp\mu + \zeta_9\,\mathcal{D}_\alpha^\perp\mathcal{D}_\perp^\alpha\mu\\ &+ \zeta_{10}\,(\mathcal{D}_\perp^\alpha\mu)\mathcal{D}_\alpha^\perp\ln T\end{aligned} \tag{46}$$

and in the same way heat-flow vector may be written as,

$$\begin{aligned}q^\mu &= -\kappa\left(\mathcal{D}_\perp^\mu T + T\,Du^\mu\right) + \tau_q D\left[\mathcal{D}_\perp^\mu T + T\,Du^\mu\right]\\ &+ \chi_2\,(\mathcal{D}_\alpha^\perp\ln T)\sigma^{\alpha\nu} + \chi_3\,(\mathcal{D}_\alpha^\perp\ln T)\Omega^{\alpha\nu} + \chi_4\,(\mathcal{D}_\alpha^\perp\mu)\frac{(\nabla.u)}{3}\Delta^{\alpha\nu}\\ &+ \chi_5\,(\mathcal{D}_\perp^\nu\mu)(\nabla.u) + \chi_6\,(\mathcal{D}_\alpha^\perp\mu)\sigma^{\alpha\nu} + \chi_7\,\Delta^{\alpha\nu}u^\gamma R_{\alpha\gamma} + \chi_8\,\mathcal{D}_\alpha^\perp\sigma^{\alpha\nu}\\ &+ \chi_9\,\mathcal{D}_\alpha^\perp\Omega^{\alpha\nu} + \chi_{10}\,(\mathcal{D}_\alpha^\perp\mu)\Omega^{\alpha\nu} + \chi_{11}\,(\mathcal{D}_\alpha^\perp\mu)R^{\alpha\nu}\end{aligned} \tag{47}$$

where τ_q is the relaxation-time coefficient corresponding to the heat flux. It is worthwhile to note that each of these coefficients measure the time, the hydrodynamical

[2] As $D\sigma^{\mu\nu}$ and $D\pi^{\mu\nu}$ are both second order in gradients, it is possible to substitute the first order results of dissipative fluxes in second order theories. As a result, one can switch between $D\pi^{\mu\nu}$ and $D(-2\eta\sigma^{\mu\nu})$ thereby connecting to the MIS theory without a loss of accuracy at the second order, see for eg. [32].

fluid takes to return to the equilibrium state in absence of dissipative quantities. Here Eqs. (41), (46) and (47) are respectively the general forms of the shear viscosity tensor, bulk viscosity and the heat flow vector in the Eckart frame up to second order in gradients. In total, there are nine second order transport coefficients namely corresponding to the shear viscosity. Together with τ_Π, there are ten second order transport coefficients corresponding to the bulk viscosity and the heat flow count to total eleven second order transport coefficients.

4 Stationary Relativistic Viscous Torus

It is well-known that real-time accretion process is accompanied by an outward orbital angular momentum transport as well as an inward mass-transfer of the accreting matter towards the central compact object together with radiation transfer processes. Other than magnetorotational instability, one of the governing factors of the angular momentum transport due to the differential rotation of the adjacent fluid layers may be thought to be contributed by the shear viscosity.

Let us therefore investigate the effects of shear viscosity on stationary solutions of a geometrically thick torus in the context of hyperbolic theory of relativistic hydrodynamics. For simplicity, we will only discuss with the Schwarzschild black hole. The effects of viscosity on the disk morphology is then studied using conservation laws constructed using the gradient expansion scheme. Apart from hyperbolic nature of the conservational laws, an interesting feature of (45) is, the curvature of the central compact object also takes part and contributes to the shear viscosity tensor. As a result, the morphology of the torus will be modified in different way for each and individual geometry of the central compact object, for example, a black hole. In addition, the information of the transport coefficients will also be crucial for determining the stationary solutions of the viscous torus.

4.1 Purely Hydrodynamical Viscous Torus

In order to investigate effects of the shear viscosity and the spacetime curvature of the Schwarzschild black hole on the morphology of a geometrically thick torus, several simplifying assumptions are adopted and are listed as follows,

(a) Only test-fluid assumptions is considered.
(b) The shear viscosity is assumed to be small in comparison to equilibrium pressure and energy and acts in the form of perturbation.
(c) With the consideration that shear viscosity only brings about perturbative corrections to the fluid pressure and does not give rise to radial velocity and velocity perturbations, the fluid within the torus describes circular orbits around the Schwarzschild black hole. Therefore the four-velocity of the constituents of the disk fluid has the following form,

$$u^{\mu} = (u^t, 0, 0, u^{\phi}) \tag{48}$$

where u^t and u^{ϕ} are functions of r and θ only with the normalization condition $u^{\alpha}u_{\alpha} = -1$. Due to circular motion of the fluid, we have,

$$\Omega_{\mu\nu} = 0, \quad \sigma_{\mu\nu} \neq 0 \tag{49}$$

(d) The specific orbital angular momentum distribution of the torus is constant.
(e) In a more generalized framework, the transport coefficients can in principle be functions of pressure or of space and time co-ordinates. However, these coefficients are taken as constant parameters of the theory.
(f) In this simple model, the heat flow is also assumed to be small compared to leading order pressure perturbations and hence the effects are not taken into consideration.
(g) All non-linear terms in (45) are also neglected.

Thus in the simplified scenario, the shear viscosity tensor consists of the curvature tensor term and the causality preserving term i,e. $2\tau_2^{<} D\eta\sigma^{\alpha\beta >}$ and therefore takes the following form,

$$\pi^{\mu\nu} = \left[-2\eta\sigma^{\mu\nu} - \eta\tau_2^{<} D(-2\sigma^{\mu\nu})^{>} + \kappa_2 u_{\alpha} u_{\beta} R^{\alpha<\mu\nu>\beta} \right] \tag{50}$$

where $R^{<\mu\nu>} = 0$ in the Schwarzschild background which is then substituted in the general relativistic causal form of momentum conservation equation given by,

$$(e+p)a^{\mu} + \Delta^{\mu\rho}\nabla_{\rho} p + \pi^{\mu\alpha} a_{\alpha} + \Delta^{\mu}_{\gamma}\Delta_{\kappa\tau}\nabla^{\tau}\pi^{\gamma\kappa} = 0 \tag{51}$$

where the four acceleration is denoted by $a^{\mu} = u^{\rho}\nabla_{\rho}u^{\mu}$ and the angular brackets have the following definitions,

$$\begin{aligned}
\sigma^{\mu\nu} &= \Delta^{\mu\alpha}\Delta^{\nu\beta}\left(\frac{\nabla_{\alpha}u_{\beta} + \nabla_{\beta}u_{\alpha}}{2}\right) - \frac{1}{3}\Delta^{\mu\nu}\Delta^{\alpha\beta}\nabla_{\alpha}u_{\beta} \\
{}^{<}D(\eta\sigma^{\mu\nu})^{>} &= \Delta^{\mu\alpha}\Delta^{\nu\beta}\left(\frac{D(\eta\sigma_{\alpha\beta}) + D(\eta\sigma_{\beta\alpha})}{2}\right) - \frac{1}{3}\Delta^{\mu\nu}\Delta^{\alpha\beta}D(\eta\sigma_{\alpha\beta}) \\
R^{\alpha<\mu\nu>\beta} &= \Delta^{\mu\rho}\Delta^{\nu\sigma}\left(\frac{R^{\alpha}{}_{\rho\sigma\gamma}g^{\beta\gamma} + R^{\alpha}{}_{\sigma\rho\gamma}g^{\beta\gamma}}{2}\right) - \frac{1}{3}\Delta^{\mu\nu}\Delta^{\rho\sigma}R^{\alpha}{}_{\rho\sigma\gamma}g^{\beta\gamma}
\end{aligned} \tag{52}$$

It is to be noted that the symmetry of the spacetime also implies $D\eta = u^{\alpha}\nabla_{\alpha}\eta = 0$. Restricting up to the first order in perturbation and taking into account of the symmetries of the background black hole geometry, the fluid pressure and the energy density in presence of viscosity are expanded in the following way,

$$\begin{aligned}
e(r,\theta) &= e_{(0)}(r,\theta) + \lambda e_{(1)}(r,\theta) \\
p(r,\theta) &= p_{(0)}(r,\theta) + \lambda p_{(1)}(r,\theta)
\end{aligned} \tag{53}$$

where $p_{(0)}$ and $e_{(0)}$ denote the pressure and the energy density of the ideal fluid whereas $p_{(1)}$ and $e_{(1)}$ denote the respective first-order perturbations that purely arise due to the shear viscosity. Also, $e_{(1)} << e_{(0)}$ and $p_{(1)} << p_{(0)}$ are always satisfied in this picture. Note that λ is the perturbation parameter. The regime of vanishing coefficients where $\eta = \kappa_2 = \tau_2 = 0$ corresponds to the ideal fluid case where (51) reduces to the Euler equation. As a result, both $p_{(1)} = 0$ and $e_{(1)} = 0$ in absence of viscous effects.

In general, the angular velocity Ω and the specific angular momentum l are functions of r and θ co-ordinates and in the Schwarzschild spacetime, these two quantities are related in the following way,

$$\Omega = -\frac{g_{tt}}{g_{\phi\phi}} l, \tag{54}$$

For the entire analysis, l will be taken as constant. Furthermore, the internal energy density is assumed to be very small, therefore, the total energy approximately equals to the rest-mass density i,e. $e \approx \rho$. To finally determine the stationary solutions of the viscous torus, consider the Schwarzschild spacetime which is described by the following metric,

$$ds^2 = -\left(1 - \frac{2M}{r}\right) dt^2 + \left(1 - \frac{2M}{r}\right)^{-1} dr^2 + r^2 d\theta^2 + r^2 \sin^2\theta d\phi^2 \tag{55}$$

Then both the temporal and the azimuthal components of (51) give rise to a single relation given by,

$$\frac{2\eta l_0 r \left[Mr(r - 3M)\sin^4\theta - l_0^2(1 - 2M/r)^2(r - 3M\sin^2\theta)\right]}{\sin\theta\sqrt{r - 2M}\left(r^3\sin^2\theta + l_0^2(2M - r)\right)^{5/2}} = 0 \tag{56}$$

where $l(r, \theta) = l_0$ is the constant specific angular momentum distribution and at the equatorial plane (56) reduces to,

$$(r - 3M)\left[Mr - l_0^2\left(1 - \frac{2M}{r}\right)^2\right] = 0 \tag{57}$$

consisting of the root $l_0 = \dfrac{r\sqrt{Mr}}{(r - 2M)} \equiv l_k$ which is the Keplarian angular momentum. The transport coefficients η and κ_2 are small and act as perturbations with the perturbation parameter λ as follows,

$$\eta = \lambda m_1, \qquad \kappa_2 = \lambda m_2 \tag{58}$$

where m_1, m_2 are constant input parameters and $\lambda = 1$ is set without a loss of generality. Assuming a barotropic equation of state relating p, e and $p_{(1)}$, $e_{(1)}$, the only undetermined variable to be solved from the radial and polar components of

(51) then becomes $p_{(1)}$. To find the corresponding relations between pressure and energy densities, let us expand the barotropic equation of state in the first order of λ as follows,

$$p_{(0)} + p_{(1)} = K(e_{(0)} + e_{(1)})^{\gamma} \tag{59}$$

where K is a constant and γ is the constant polytropic exponent. The respective equation of state at the zeroth order and the first order are as follows,

$$p_{(0)} = Ke^{\gamma}_{(0)}, \qquad p_{(1)} = \gamma K e^{\gamma-1}_{(0)} e_{(1)} \tag{60}$$

Setting the mass of the black hole $M = 1$ and $K = 1$, the radial component and the polar components of (51) are as follows,

$$\begin{aligned}
&\frac{(\tau_2 m_1) l_0^2 (r-3)}{2r^2\left(r^3\sin^2\theta + l_0^2(2-r)\right)^3}\Big[r^3\cos 4\theta(10r-21) \\
&\qquad + \cos 2\theta\left\{4r^3(2r^2-14r+21) - 8l_0^2(r-3)(r-2)\right\} \\
&\qquad - r^3(2r-7)(4r-9) - 8l_0^2(r-2)(r^2-3r+3)\Big] \\
&+ m_2\Big[3r^6 + r^6\cos 4\theta + 2r^3\cos 2\theta\left\{l_0^2(r-2)(5r-14) - 2r^3\right\} \\
&\qquad - 2r^3 l_0^2(r-2)(5r-14) - 4l_0^4(r-2)^3\Big] \times \left[4r^5\left(r^3\sin^2\theta + l_0^2(2-r)\right)^2\right]^{-1} \\
&- \frac{\left\{r^3\sin^2\theta - l_0^2(2-r)^2\right\}\left(\gamma K + e_{(0)}^{1-\gamma}\right)}{\gamma K r^2\left(r^3\sin^2\theta + l_0^2(2-r)\right)} p_{(1)} + \frac{(r-2)}{r}\frac{\partial p_{(1)}}{\partial r} = 0
\end{aligned} \tag{61}$$

$$\begin{aligned}
&\tau_2 m_1 \frac{4l_0^2\cot\theta\left\{r^3(4r-9)\sin^2\theta + (r-2)\left(2l_0^2(4r-9) - rl_0^2(r-2)\csc^2\theta - 2r^4\right)\right\}}{r^2\left(r^3\sin^2\theta + l_0^2(2-r)\right)^3} \\
&+ m_2\frac{2l_0^2(r-2)\cot\theta\left(2r^3\sin^2\theta + l_0^2(r-2)\right)}{r^3\left(r^3\sin^2\theta + l_0^2(2-r)\right)^2} - \frac{(r-2)l_0^2\cot\theta\left(\gamma K + e_{(0)}^{1-\gamma}\right)}{\gamma K\left(r^3\sin^2\theta + l_0^2(2-r)\right)} p_{(1)} \\
&+ \frac{\partial p_{(1)}}{\partial\theta} = 0
\end{aligned} \tag{62}$$

The energy density and the fluid pressure in a ideal torus can be expressed in terms of the total potential $W(r,\theta)$ as follows (Font & Daigne, 2002),

$$e_{(0)} = \left[\frac{(\gamma-1)}{\gamma}\frac{(e^{W_{in}-W(r,\theta)}-1)}{K}\right]^{\frac{1}{\gamma-1}}, \tag{63}$$

$$p_{(0)} = K\left[\frac{(\gamma-1)}{\gamma}\frac{(e^{W_{in}-W(r,\theta)}-1)}{K}\right]^{\frac{\gamma}{\gamma-1}} \tag{64}$$

with the boundary condition defined to be $W_{in} \to 0$ as $r \to \infty$. Then from (62), $p_{(1)}$ is eliminated and substituted back in (61) to obtain a first order partial differential equation for $p_{(1)}$ involving the coefficients τ_2, m_1 and m_2,

$$\begin{aligned}
&2(\tau_2 m_1)\cot\theta\Big[r^6(4r-9)\cos 6\theta + 2r^6(4r(3r-11)+45)\\
&+32l_0^4(r-2)^2(r-3)(2r-3) - \cos 2\theta\{r^6(4r(8r-31)+135)\\
&-16r^3l_0^2(r-2)(7r^2-28r+27)+32l_0^4(r-2)^2(3r^2-11r+9)\}\\
&+2r^3\cos 4\theta\left\{r^3(4r^2-20r+27)-2l_0^2(r-2)\left(6r^2-26r+27\right)\right\}\\
&-4r^3l_0^2(r-2)\left(22r^2-86r+81\right)\Big]+\left(r^3\cos 2\theta - r^3 + 2(r-2)l_0^2\right)^3\\
&\times\left[2r(r-2)^2\cot\theta\frac{\partial p_{(1)}}{\partial r} - \frac{r^3\cos 2\theta - r^3 + 2l_0^2(r-2)^2}{l_0^2}\frac{\partial p_{(1)}}{\partial\theta}\right]\\
&+3m_2(r-2)\cot\theta\Big[r^6\cos 6\theta - 6r^3\cos 4\theta\left(r^3 - l_0^2(r-2)^2\right)\\
&+\cos 2\theta\left\{15r^6 - 24r^3l_0^2(r-2)^2 + 16l_0^4(r-3)(r-2)^2\right\}\\
&+2\left\{9r^3l_0^2(r-2)^2 - 8l_0^4(r-3)(r-2)^2 - 8\,r\,l_0^6(1-2/r)^4 - 5r^6\right\}\Big]\\
&=0
\end{aligned} \tag{65}$$

Since τ_2 is the relaxation time coefficient in the same spirit of MIS formalism which preserves causality of the relativistic Navier-Stokes equation. It does not possess perturbative character while η, κ_2 are treated as perturbations. With constant coefficients η, κ_2 and τ_2 and with l =constant, (65) is solved numerically to determine $p_{(1)}$ with $\gamma = 5/3$, $K = 1$ and $\tau_2 = 0.2$. Moreover, the solutions are obtained when the potential gap satisfies $\Delta W_{\text{in}} > 0$.

The effects of shear viscosity and the curvature are determined by comparing the constant pressure contours of the ideal fluid and that of the constant pressure surfaces in presence of viscosity. As shown in Fig. 2, the comparison is performed between $p_{(0)}$ of the ideal fluid and $p_{(0)} + p_{(1)}$ of the viscous fluid for $l_0 = 3.8$. To furnish a comparison, same set of fluid profiles are considered. The Fig. 1 shows that the locations of the cusps due to viscous effects. In fact, there were no cusps for the ideal fluid case, whereas with subsequent increase in m_1 and m_2 results into the formation of cusps. The shift in the locations of cusps is therefore directly related to cumulative effects of shear viscosity and the spacetime curvature of the Schwarschild black hole, all of which appear through $p_{(1)}$.

We now turn our attention to stationary configurations of magnetized tori. Let us now understand how shear viscosity and curvature affects their morphological features.

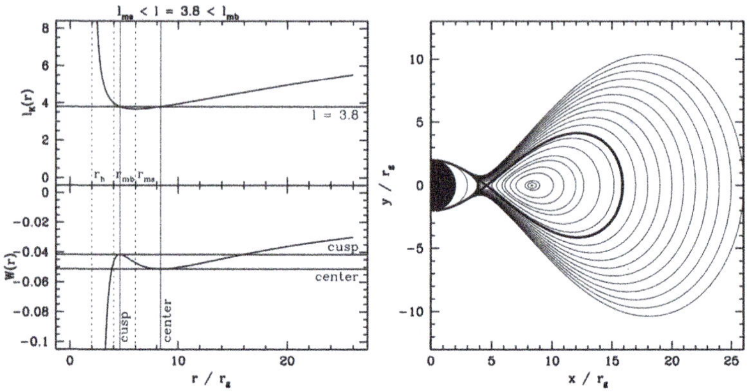

Fig. 1 The effective potential $W(r)$ and the Keplarian angular momentum distribution $l_k(r)$ are shown on the left panel at the equatorial plane. On the right panel, closed equipressure surfaces are shown at equilibrium for a constant angular momentum distribution described by its cusp and centre where the effective acceleration on the fluid vanishes. The cusp is marked with self-crossing constant pressure surfaces where the centre of the torus is characterised by maximum pressure. Taken from [2]

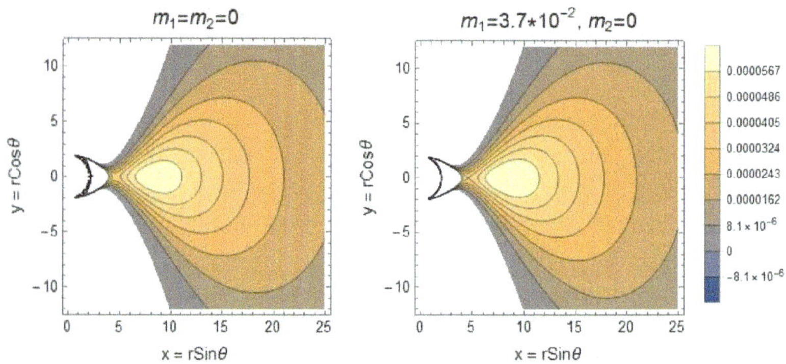

Fig. 2 Comparison of constant pressure surfaces of ideal fluid and viscous fluid for $l_0 = 3.8$. Taken from [22]

4.2 Magnetised Viscous Torus

It is a general consensus that both shear viscosity and magnetic fields play a key roles for angular momentum transport in accretion disks. As a result, it make sense to understand the impacts of both the magnetic field and shear viscosity on stationary models of a torus. For this, let us now describe stationary configurations of magnetized viscous tori by assuming a toroidal distribution of the magnetic field. In this simple set-up, all the assumptions considered for a purely hydrodynamical viscous torus are also valid. In essence, the purely hydrodynamical solutions presented in [23] are extended by incorporating toroidal magnetic fields by further assuming the torus is endowed with constant angular momentum distribution. The new solutions are built using the second-order gradient expansion scheme in the Eckart frame description [21].

The energy-momentum tensor of the viscous fluid in the presence of a magnetic field is given by,

$$T^{\mu\nu} = (w + b^2)u^\mu u^\nu + \left(p + \frac{1}{2}b^2\right) g^{\mu\nu} - b^\mu b^\nu + \pi^{\mu\nu} . \tag{66}$$

where the enthalpy density is given by $w = e + p$, p is the fluid pressure, e is the total energy, and $\pi^{\mu\nu}$ denotes the shear viscosity tensor. The dual of the Faraday tensor relative to an observer with four-velocity u^μ is [3] given by,

$$^*F^{\mu\nu} = b^\mu u^\nu - b^\nu u^\mu , \tag{67}$$

where b^μ is the magnetic field in that frame, that obeys the relation $b^2 = b^\alpha b_\alpha$ and gives rise to the conservation law $\nabla_\nu {}^*F^{\mu\nu} = 0$. In the fluid frame, $b^\mu = (0, \mathbf{B})$ where $\mathbf{B}$ denotes the three-vector of the magnetic field and satisfies the orthogonality condition $u^\alpha b_\alpha = 0$. In presence of purely toroidal magnetic field distributions, it follows that

$$b^r = b^\theta = 0, \qquad b^\mu = (b^t, 0, 0, b^\phi) . \tag{68}$$

From the condition $u^\alpha b_\alpha = 0$ one obtains,

$$b^t = l b^\phi, \qquad b_t = -\Omega b_\phi , \tag{69}$$

and

$$b^2 = (1 - \Omega l) b^\phi b_\phi = 2 p_m , \tag{70}$$

where the magnetic pressure is defined to be $p_m \equiv b^2/2$. Similar to the purely hydrodynamical case discussed in the previous section, the general form of the shear viscosity tensor is [21],

$$\pi^{\mu\nu} = -2\eta\sigma^{\mu\nu} - \tau_2^< D(-2\eta\sigma^{\mu\nu})^> + \kappa_2 u_\alpha u_\beta R^{\alpha<\mu\nu>\beta} , \tag{71}$$

where $D \equiv u^\alpha \nabla_\alpha$, η is the constant shear viscosity coefficient, τ_2, κ_1 and κ_2 are the second-order transport coefficients. The angular brackets indicate traceless symmetric combinations (see (52)). Substituting (66) in $\Delta_{\mu\nu}\nabla_\lambda T^{\lambda\nu} = 0$, one can easily obtain the following conservation equation

$$(e + p)a_\mu + \Delta^\rho_\mu \nabla_\rho p + \frac{\partial_\mu(\mathcal{L} b^2)}{2\mathcal{L}} + g_{\mu\rho}\pi^{\rho\nu} a_\nu + \Delta_{\mu\gamma}\Delta_{\kappa\tau}\nabla^\tau \pi^{\gamma\kappa} = 0 , \tag{72}$$

which is the general form of the momentum conservation equation in the presence of the toroidal magnetic field. The four-acceleration is given by $a^\mu = u^\rho \nabla_\rho u^\mu = D u^\mu$ and $\mathcal{L} \equiv -g_{tt} g_{\phi\phi}$.

Since the magnetised torus is characterised by constant specific angular momentum distributions, let us take again $l(r) \equiv l_0$. Similar to the non-magnetised torus, the internal energy density is assumed to be very small, so that the total energy is approximately equal to the rest-mass density. As the shear viscosity is small and acts as perturbation, as before, the transport coefficients η and κ_2 are considered as perturbations in the disk fluid. For simplicity, once again they are assumed to be constants and in accordance to (58) are expressed in a similar way,

$$\eta = \lambda m_1, \qquad \kappa_2 = \lambda m_2 . \tag{73}$$

The shear viscosity in the torus fluid generates linear perturbations in the energy density, pressure, and magnetic field. Each of these quantities expressed up to linear order become,

$$e(r,\theta) = e_{(0)}(r,\theta) + \lambda e_{(1)}(r,\theta), \tag{74}$$
$$p(r,\theta) = p_{(0)}(r,\theta) + \lambda p_{(1)}(r,\theta), \tag{75}$$
$$b^t(r,\theta) = b^t_{(0)}(r,\theta) + \lambda b^t_{(1)}(r,\theta), \tag{76}$$
$$b^\phi(r,\theta) = b^\phi_{(0)}(r,\theta) + \lambda b^\phi_{(1)}(r,\theta), \tag{77}$$

where, as usual, index (0) denotes background quantities (ideal fluid) and index (1) quantities at linear perturbation order (in presence of viscosity). The magnetic pressure can be read off using (70) and (77),

$$p_m^{(0)} = \frac{1}{2}(1 - \Omega l)\, b^\phi_{(0)} b^{(0)}_{(\phi)}, \tag{78}$$
$$p_m^{(1)} = \frac{1}{2}\left[b^\phi_{(0)}\left(l b_t^{(1)} + b_\phi^{(1)}\right) + b_\phi^{(0)}\left(b^\phi_{(1)} - \Omega b^t_{(1)}\right)\right]. \tag{79}$$

Let us now define the magnetization parameter as $\beta_m \equiv p/p_m$. Then the zeroth-order and first-order changes in this parameter are,

$$\beta_m^{(0)} = \frac{p_{(0)}}{p_m^{(0)}}, \qquad \beta_m^{(1)} = \frac{p_{(1)}}{p_m^{(0)}} - \beta_m^{(0)} \frac{p_m^{(1)}}{p_m^{(0)}}. \tag{80}$$

Using the relations for the equation of state involving linear corrections of pressure and energy density, one obtains $p_{(0)}$, $p_{(1)}$ whereas $p_m^{(0)}$ and $p_m^{(1)}$ are related by,

$$p_{(1)} = p_{(0)} \frac{\gamma e_{(1)}}{e_{(0)}}, \tag{81}$$
$$p_m^{(1)} = p_m^{(0)} \frac{\gamma e_{(1)}}{e_{(0)}}. \tag{82}$$

So from the momentum conservation equation (72), there are four unknown quantities to be determined, namely, $p_{(1)}, e_{(1)}, b_t^{(1)}$ and $b_\phi^{(1)}$. Note that the variables $p_{(1)}$ and

$e_{(1)}$ are not independent under the assumption of a barotropic equation of state. The equations of state corresponding to the fluid pressure p and the magnetic pressure p_m, with the same polytropic index γ, are given by [14,19],

$$p = Ke^{\gamma}, \tag{83}$$

and

$$p_m = K_m \mathcal{L}^{\gamma-1} e^{\gamma}. \tag{84}$$

After expanding the equations of state up to linear order and substituting in the definition of magnetization parameter, the linear correction of which reduces to,

$$\beta_m^{(1)} = 0 \tag{85}$$

From the orthogonality relation $u^{\alpha} b_{\alpha} = 0$ one obtains,

$$b^t_{(0)} = l b^{\phi}_{(0)}, \qquad b^t_{(1)} = l b^{\phi}_{(1)}, \tag{86}$$

that imply $b^{\phi}_{(1)}$ and $b^t_{(1)}$ are not independent variables. Using the relations, $b^2 = (1 - l\Omega) b^{\phi} b_{\phi}$ and $b^2 = 2p_m$, one can easily write down the zeroth-order and first-order terms of the magnetic field as,

$$b^{\phi}_{(0)} = \sqrt{\frac{2\beta_m}{p_{(0)}(1 - l\Omega) g_{\phi\phi}}}, \qquad b^{\phi}_{(1)} = \frac{p_{(1)}}{\beta_m} \sqrt{\frac{p_{(0)}}{2\beta_m (1 - \Omega l) g_{\phi\phi}}}, \tag{87}$$

Thus the variables $p_m^{(1)}$, $b^t_{(1)}$ and $b^{\phi}_{(1)}$ are all related to $p_{(1)}$ which is then determined by solving (72)

$$(e_{(1)} + p_{(1)}) a_{\mu} + \Delta^{\rho}_{\mu} \nabla_{\rho} p_{(1)} + \frac{\partial_{\mu} \left[\frac{\mathcal{L}}{\beta_m} p_{(1)} \right]}{\mathcal{L}} + g_{\mu\rho} \pi^{\rho\nu} a_{\nu}$$
$$+ \Delta_{\mu\gamma} \Delta_{\kappa\tau} \nabla^{\tau} \pi^{\gamma\kappa} = 0 \tag{88}$$

The effects of the shear viscosity on the magnetic pressure $p_m^{(1)}$ can be easily understood with the determination of $p_{(1)}$.

Similar to the pure hydrodynamical case, $M = 1$ is set for the rest of the analysis. The temporal and the azimuthal parts give rise to exactly the same condition as is obtained for the solution of the torus without shear viscosity. For $\eta \neq 0$, one obtains the relation $r^3 - l_0^2 (r-2)^2 = 0$ in the equatorial plane where l_0 is the constant angular momentum of the torus in presence of the magnetic field. Correspondingly, the radial and angular components are

$$
\begin{aligned}
&\frac{(\tau_2 m_1) l_0^2 (r-3)}{2r^2 \sin^6\theta \left(r^3 + l_0^2(2-r)\csc^2\theta\right)^3}\Big[r^3 \cos 4\theta (10r - 21) \\
&\quad + \cos 2\theta \left\{4r^3(2r^2 - 14r + 21) - 8l_0^2(r-3)(r-2)\right\} \\
&\quad - r^3(2r-7)(4r-9) - 8l_0^2(r-2)(r^2 - 3r + 3)\Big] \\
&+ m_2 \times \Big[3r^6 + r^6 \cos 4\theta + 2r^3 \cos 2\theta \left\{l_0^2(r-2)(5r-14) - 2r^3\right\} \\
&\qquad - 2r^3 l_0^2 (r-2)(5r-14) - 4l_0^4(r-2)^3\Big] \\
&\quad \times \Big[4r^5 \sin^4\theta \left(r^3 + l_0^2(2-r)\csc^2\theta\right)^2\Big]^{-1} \\
&+ \left(1 + \frac{1}{\beta_m}\right) \frac{(r-2)}{r} \frac{\partial p_{(1)}}{\partial r} \\
&+ \left[\frac{2(r-1)}{r^2 \beta_m} + \frac{\left\{r^3 - l_0^2(2-r)^2 \csc^2\theta\right\}\left(\gamma K + e_{(0)}^{1-\gamma}\right)}{\gamma K r^2 \left(r^3 + l_0^2(2-r)\csc^2\theta\right)}\right] p_{(1)} = 0\,, \qquad (89)
\end{aligned}
$$

$$
\begin{aligned}
&\tau_2 m_1 \frac{4l_0^2 \cot\theta \left\{r^3(4r-9)\sin^2\theta + (r-2)\left(2l_0^2(4r-9) - r l_0^2 (r-2)\csc^2\theta - 2r^4\right)\right\}}{r^2 \sin^4\theta \left(r^3 + l_0^2(2-r)\csc^2\theta\right)^3} \\
&+ m_2 \frac{2l_0^2(r-2)\cot\theta \left(2r^3 \sin^2\theta + l_0^2(r-2)\right)}{r^3 \sin^4\theta \left(r^3 + l_0^2(2-r)\csc^2\theta\right)^2} + \left(1 + \frac{1}{\beta_m}\right)\frac{\partial p_{(1)}}{\partial \theta} \\
&+ \left[\frac{2\cot\theta}{\beta_m} - \frac{(r-2) l_0^2 \cot\theta \left(\gamma K + e_{(0)}^{1-\gamma}\right)}{\gamma K \sin^2\theta \left(r^3 + l_0^2(2-r)\csc^2\theta\right)}\right] p_{(1)} = 0\,. \qquad (90)
\end{aligned}
$$

In the limit $\beta_m \to \infty$, (89), (90) reduce to the corresponding equations obtained in the previous section for a purely hydrodynamical viscous thick disk without the magnetic field. Finally, using (90) and (89), the differential equation for $p_{(1)}$ is found as,

$$
\begin{aligned}
&2l_0^2 \cot\theta \left(\tilde{A} + \frac{\tilde{B} k_1}{C}\right)(\tau_2 m_1) + \frac{1}{2}\cot\theta \left(\tilde{f}_1 - \frac{\tilde{f}_2 k_1}{C}\right) m_2 \\
&+ \sin^6\theta \left[r^3 + l_0^2(2-r)\csc^2\theta\right]^3 \frac{(1+\beta_m)}{\beta_m}\left[r(r-2)\cot\theta \frac{\partial p_{(1)}}{\partial r} - \frac{4k_1}{C}\frac{\partial p_{(1)}}{\partial \theta}\right] = 0\,, \qquad (91)
\end{aligned}
$$

with the following definitions,

$$
\begin{aligned}
\tilde{A} = -2(r-3)\Big[&r^3(10r-21)\cos 4\theta \\
&+ \cos 2\theta \left\{4r^3(2r^2 - 14r + 21) - 8l_0^2(r-3)(r-2)\right\} \\
&- r^3(2r-7)(4r-9) - 8l_0^2(r-2)(r^2 - 3r + 3)\Big], \qquad (92)
\end{aligned}
$$

$$\tilde{B} = 4\left[2rl_0^2(r-2)^2 - \sin^2\theta\left\{r^3(2r-3) + 4l_0^2(4r-9)(r-2) - r^3\cos 2\theta(4r-9)\right\}\right], \tag{93}$$

$$k_1 = -2\beta_m \sin^2\theta\left(r^3 + l_0^2(2-r)\csc^2\theta\right)e_{(0)} + K\gamma e_{(0)}^{\gamma}\left\{8l_0^2(1+\beta_m)\right.$$
$$\left. +2r^2l_0^2(\beta_m+2) - 2r^4 - r^3(\beta_m-2) - 4l_0^2r(3+\beta_m) + r^3\cos 2\theta(2r+\beta_m-2)\right\}, \tag{94}$$

$$C = (r-2)l_0^2\beta_m e_{(0)} + K\gamma e_{(0)}^{\gamma}\left[l_0^3(r-2)(2+\beta_m) - 2r^3\sin^2\theta\right], \tag{95}$$

$$\tilde{f}_1 = \left[10r^6 - 6r^3l_0^2(5r-12)(r-2) + 32l_0^4(r-3)(r-2)^2 + 16rl_0^6\left(1-\frac{2}{r}\right)^4\right.$$
$$\left. + \cos 2\theta\left\{8(5r-12)r^3l_0^2 - 15r^6 - 32l_0^4(r-3)(r-2)^2\right\}\right], \tag{96}$$

$$\tilde{f}_2 = 4r^3\left(1-\frac{2}{r}\right)\left[3r^6 - 2r^3l_0^2(r-2) + 4l_0^4(r-2)^2\right.$$
$$\left. +2r^3\cos 2\theta(2r^3-(r-2)l_0^2) + r^6\cos 4\theta\right]. \tag{97}$$

Once the magnitudes of the parameters m_1, τ_2, l_0 and β_m are chosen, (91) can be readily solved using the appropriate boundary conditions.

It is already noted in (85) that magnetization parameter can be completely expressed in terms of the zeroth-order quantities i,e. $\beta_m(r,\theta) = p_{(0)}/p_m^{(0)}$. Using (83) and (84),

$$\beta_m(r,\theta) = \frac{K}{K_m\mathcal{L}^{\gamma-1}(r,\theta)}. \tag{98}$$

In addition, the magnetization parameter at the center of the torus expressed as $\beta_{m,c} \equiv \beta_m(r_c, \pi/2)$ is,

$$\beta_{m,c} = \frac{K}{K_m\mathcal{L}^{\gamma-1}(r_c,\pi/2)}. \tag{99}$$

The magnetization parameter for the Schwarzschild metric in terms of $\beta_{m,c}$ is,

$$\beta_m(r,\theta) = \beta_{m,c}\left(\frac{r_c(r_c-2)}{r(r-2)\sin^2\theta}\right)^{\gamma-1}, \tag{100}$$

where r_c, $\beta_{m,c}$ and K are constant parameters. In the Schwarzschild geometry, it is already known that the total potential $W(r,\theta)$ for constant angular momentum distributions is,

$$W(r,\theta) = \frac{1}{2}\ln\frac{r^2(r-2)\sin^2\theta}{r^3\sin^2\theta - l_0^2(r-2)}. \tag{101}$$

At the equatorial plane, taking $\partial_r W = 0$ leads to,

$$r^3 - l_0^2(r-2)^2 = 0. \tag{102}$$

The largest root of the above equation corresponds to the disk center, r_c which is determined by finding the extrema of the effective (gravitational plus centrifugal) potential W, as the center of the torus is located at a minimum of the potential (see, e.g. [12] for details). In the absence of any dissipative terms, the relativistic momentum conservation equation, with the choices of equation of state, become [14]

$$W - W_s + \frac{\gamma}{\gamma - 1}\left(\frac{p_{(0)}}{e_{(0)}} + \frac{p_m^{(0)}}{e_{(0)}}\right) = 0, \tag{103}$$

which is rewritten as

$$W - W_s + \frac{\gamma K e_{(0)}^{\gamma-1}}{\gamma - 1}\left(1 + \frac{1}{\beta_m(r,\theta)}\right), \tag{104}$$

where W_s is the potential at the surface of the disk where $p_{(0)} = p_m^{(0)} = e_{(0)} = 0$. Finally, from the above expression, the zeroth-order energy density can be obtained and it reads as

$$e_{(0)} = \left(\frac{1}{K}\right)^{\frac{1}{\gamma-1}}\left(\frac{\gamma(1+\beta_m(r,\theta))}{(1-\gamma)\beta_m(r,\theta)(W - W_s)}\right)^{\frac{1}{1-\gamma}}, \tag{105}$$

and the zeroth-order pressure is,

$$p_{(0)} = K^{\frac{1}{\gamma-1}}\left(\frac{\gamma\left(1 + \beta_{m,c}\left(\frac{r_c(r_c-2)}{r(r-2)\sin^2\theta}\right)^{\gamma-1}\right)}{(1-\gamma)\beta_{m,c}\left(\frac{r_c(r_c-2)}{r(r-2)\sin^2\theta}\right)^{\gamma-1}(W - W_s)}\right)^{\frac{\gamma}{1-\gamma}}, \tag{106}$$

which corresponds to the fluid pressure of the magnetized fluid without viscosity.

Now (91) is solved with the domain of definition set by the conditions $W(r,\theta) \leq W_s$, $r_{in} \leq r \leq r_{out}$ where r_{in} and r_{out} are the inner and the outer boundary of the torus at the equatorial plane. Let us focus on the torus with slightly overflowing Roche lobe (i.e. $W_s \gtrsim W(r_{cusp}, \pi/2)$ where r_{cusp} corresponds to the location of the self-crossing point of the critical equipotential surface). Under this situation, it is crucial to note that the disks do not possess an inner edge (i.e. the outermost equipotential surface is attached to the event horizon of the black hole). Here, the value of r_{in} is chosen such that $r_{in} \lesssim r_{cusp}$ thereby allowing to study the cusp region, and exclude the region closest to the black hole, as it is irrelevant for our study (the reason will become clear later in this section). Additionally, the funnel region is excluded along the symmetry axis ($\theta = 0$) by further restricting the domain of investigation containing equipotential surfaces that cross the equatorial plane at least once.

It can be observed that (91) can be rewritten in a more compact form as shown,

$$(r,\theta) \cdot \nabla_{(r,\theta)} p_{(1)} - \tilde{c}(r,\theta) = 0, \tag{107}$$

with the given definitions,

$$\begin{aligned}
\alpha_r(r,\theta) &= \sin^6\theta\left[r^3 + l_0^2(2-r)\csc^2\theta\right]^3 \frac{(1+\beta_m)}{\beta_m}\left(r(r-2)\cot\theta\right),\\
\alpha_\theta(r,\theta) &= r\sin^6\theta\left[r^3 + l_0^2(2-r)\csc^2\theta\right]^3 \frac{(1+\beta_m)}{\beta_m}\left(-\frac{4k_1}{C}\right),\\
\tilde{c}(r,\theta) &= -2l_0^2\cot\theta\left(\tilde{A} + \frac{\tilde{B}k_1}{C}\right)(\tau_2 m_1) - \frac{m_2}{2}\cot\theta\left(\tilde{f}_1 - \frac{\tilde{f}_2 k_1}{C}\right).
\end{aligned} \tag{108}$$

A careful examination of the coefficients in (108) reveals that, at the equatorial plane, (107) reduces to,

$$\frac{\partial p_{(1)}}{\partial\theta} = 0. \tag{109}$$

The above equation shows that (i) surfaces of constant $p_{(1)}$ are orthogonal to the equatorial plane, (ii) it is not possible to extract information of the distribution of $p_{(1)}$ at the equatorial plane directly from (107). Therefore, in order to determine the values of $p_{(1)}$ at the equatorial plane we must look for the solution $p_{(1)}(r,\theta)$ that belongs to the domain of the θ coordinate. In order to maximize the accuracy of the solution, (107) will be solved in the Cartesian coordinates. Rewriting this equation as,

$$'(x,y)\cdot\nabla_{(x,y)}p_{(1)} - c'(x,y) = 0, \tag{110}$$

in which the change of coordinates are defined by $x = r\sin\theta$, $y = r\cos\theta$, and the new expressions for the coefficients reduce to,

$$\alpha'_x(x(r,\theta),y(r,\theta)) = \alpha_r(r,\theta)\sin\theta + \alpha_\theta(r,\theta)\cos\theta\,, \tag{111}$$

$$\alpha'_y(x(r,\theta),y(r,\theta)) = \alpha_r(r,\theta)\cos\theta - \alpha_\theta(r,\theta)\sin\theta\,, \tag{112}$$

$$c'(x(r,\theta),y(r,\theta)) = \tilde{c}(r,\theta)\,, \tag{113}$$

Here it is noted that α'_x and α'_y denote the x and y components of the vector of coefficients $'(x,y)$. Considering the fact that $\alpha'_y(x,y) \neq 0$ in domain of investigation, all the coefficients can be redefined as,

$$a(x,y) = \alpha'_x(x,y)/\alpha'_y(x,y)\,, \tag{114}$$

$$b(x,y) = 1\,, \tag{115}$$

$$c(x,y) = c'(x,y)/\alpha'_y(x,y)\,. \tag{116}$$

Therefore, the final form of the partial differential equation reads as,

$$a(x,y)\frac{\partial p_{(1)}}{\partial x} + \frac{\partial p_{(1)}}{\partial y} - c(x,y) = 0\,. \tag{117}$$

The method of characteristics will be employed to solve (117), in which the partial differential equation (PDE) is reduced to a set of ordinary differential equations (ODEs), one for each initial value defined at the boundary of the domain. The final form of the characteristic equations is given by,

$$\frac{dx}{dt} = a(x, y)\,, \tag{118}$$

$$\frac{dy}{dt} = 1\,, \tag{119}$$

$$\frac{dp_{(1)}}{dt} = c(x, y)\,. \tag{120}$$

Procedure: We start from a point (x_0, y_0) in the boundary of the domain (i.e. $\{(x_0, y_0)\ /\ W(x_0, y_0) = W_s\}$), then, the system of ODEs are integrated as follows. Here, note that the solution of Eq. (119) is trivially $y(t) = t + y_0$ which can be used to rewrite Eq. (118) as

$$\frac{dx}{dy} = a(x, y)\,. \tag{121}$$

The above equation can be integrated numerically starting from the selected point (x_0, y_0). The solution of the equation $(x(y))$ gives a characteristic curve of the problem—a curve along which the solution of the PDE coincides with the solution of the ODE. Finally, (120) can be rewritten in the same way as the previous one,

$$\frac{dp_{(1)}}{dy} = c(x(y), y)\,. \tag{122}$$

Then $p_{(1)}$ is obtained by integrating,

$$p_{(1)}(y) = \int_{y_0}^{y} c(x(y), y)dy + p_{(1)_0}\,, \tag{123}$$

where $p_{(1)}(x_0, y_0) = p_{(1)_0}$. It can be immediately observed that $p_{(1)}(x, y)$ can be recovered by using both (123) and the expression for the characteristic curve $x(y)$. Repetition of this three-step procedure over a sufficiently large and well-chosen sample of initial points produces a mapping of the domain and hence, the solution of the PDE for the whole domain. For more details about the numerical implementation, the readers are referred to [23].

The stationary solutions of magnetized viscous tori are constructed for a set of values of the parameters τ_2, m_1, m_2 and the magnetization parameter at the center of the disk, $\beta_{m,c}$. Additionally, the polytropic exponent is chosen as $\gamma = 5/3$. The magnitude of the zeroth-order correction to the energy density at the center is taken as $e_{(0),c} = e_{(0)}(r_c, \pi/2) = 1$. For the sake of convenience, we have defined a new parameter $s_1 = \tau_2\, m_1$ and have set $\tau_2 = 1$ without loss of generality. The solutions

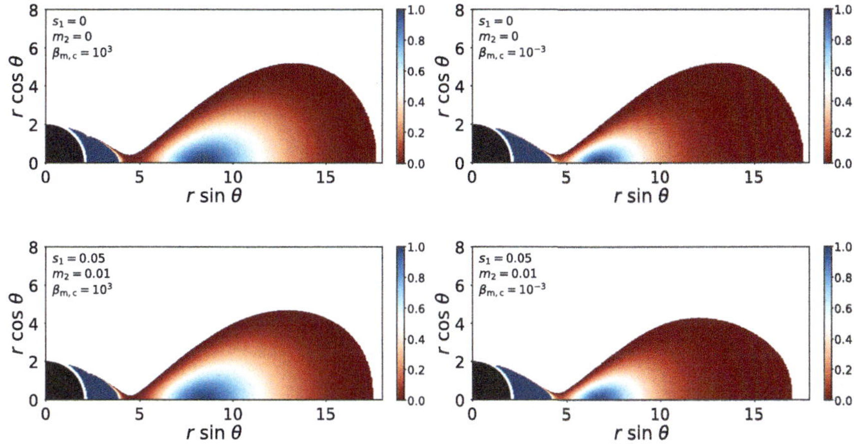

Fig. 3 The colour gradient in all four panels refers to the distributions of the total fluid pressure where we have considered $W_s = -0.039$. The top panels show ideal fluid solutions and the bottom panels shows viscous solutions for $\tau_2 = 1, m_1 = 0.05$ and $m_2 = 0.01$. The left panels correspond to non-magnetized torus for which $\beta_{m,c} = 10^3$ and the right panels correspond to highly-magnetized torus ($\beta_{m,c} = 10^{-3}$). The noticeable difference in the morphology of the torus is associated to a small decrease in size at the very low pressure region for the viscous cases. Taken from [23]

are constructed for two values of the magnetization parameter at the center of the disk, namely $\beta_{m,c} = 10^3$ (low magnetization, almost a purely hydrodynamical model) and $\beta_{m,c} = 10^{-3}$ (high magnetization) which are sufficient to bring out the effects of a toroidal magnetic field on the viscous torus.

Then, for a given choice of parameters, the corrections to the pressure $p_{(1)}$ and to the energy density $e_{(1)}$ are determined by solving (91) numerically, using the method of characteristics as described earlier. The results reveal that the effects of the shear viscosity are particularly noticeable only close to the cusp of the torus leaving the large-scale morphology of the torus remains essentially unaltered which can be immediately concluded from distribution of the pressure in Fig. 3 obtained in the entire domain for a set of illustrative stationary models. Note that the physical solution is attached to the black hole, even though in Fig. 3 there is a gap between the torus and the event horizon of the Schwarzschild black hole for the reason that (117) is singular at the event horizon, as a result, the solution cannot be extended to it.

Figures 4 and 5 present the radial plots showing the zeroth-order and first-order corrections of the pressure and of the energy density at the equatorial plane, corresponding to the low and high value of the magnetization parameter, respectively. It is to be noted that, in contrary to purely hydrodynamical tori, the location of the center of a magnetized tori r_c does not exactly coincide with the location of the maximum of the pressure. In fact it is slightly shifted towards the black hole [14]. It can be observed for the highly magnetized case in Fig. 5. Corresponding to both low and high values of $\beta_{m,c}$, the corrections $p_{(1)}$ and $e_{(1)}$ near the cusp due to viscosity remain small in comparison to their respective equilibrium values $p_{(0)}$ and

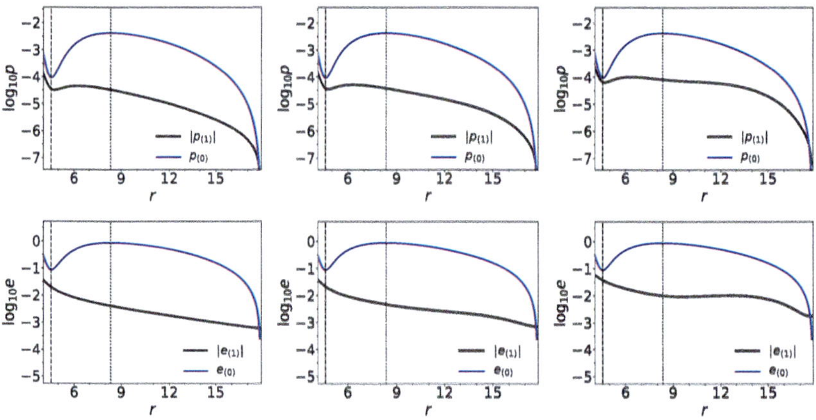

Fig. 4 Radial plots of $\log p_{(0)}$ and $\log |p_{(1)}|$ (top row) and $\log e_{(0)}$ and $\log |e_{(1)}|$ (bottom row) at the equatorial plane for $W_s = -0.039$, $s_1 = 0.05$, $\beta_{m,c} = 10^3$ and $m_2 = (0, 0.001, 0.01)$. Each column corresponds to an increasing value of the parameter m_2. The vertical dashed line denotes the location of the self-crossing pressure isocontour r_{cusp} and the vertical dotted line represents the location of the maximum of the pressure r_{cen}. Note that r_{cen} coincides to the center of the disk r_c for non-magnetized disks. Taken from [23]

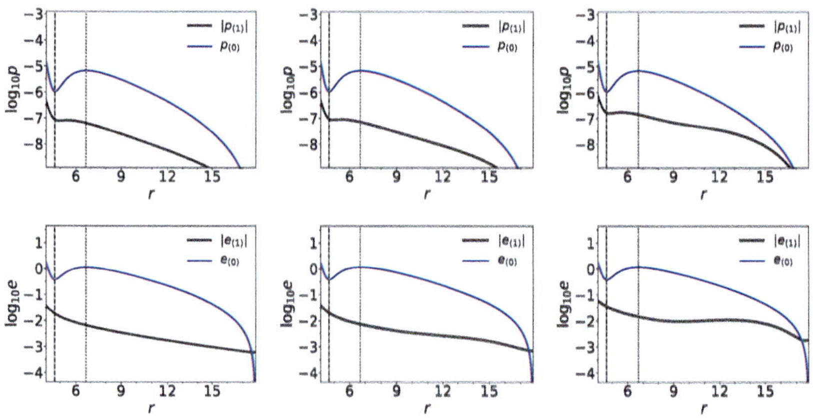

Fig. 5 Same as Fig. 4 but for the parameter $\beta_{m,c} = 10^{-3}$. Taken from [23]

$e_{(0)}$. As one approaches the outer edge of the torus, the difference between $p_{(0)}$ and $p_{(1)}$ diminishes. This tendency is mostly noticeable for low magnetized torus, as shown in Fig. 4. In addition, by increasing the value of m_2, while keeping m_1 fixed, the difference between $p_{(0)}$ and $p_{(1)}$ also decreases near the cusp, until when $p_{(1)}/p_{(0)} \sim \mathcal{O}(1)$ and $e_{(1)}/e_{(0)} \sim \mathcal{O}(1)$ are obtained. Under these situations, neither m_1 nor m_2 can further be increased. This suggests that we are no longer in the regime of the validity of near-equilibrium hydrodynamics where gradients are small. Using this argument, one can set an upper limit on the contributions of curvature effects and shear viscosity on stationary solutions of the magnetized viscous torus before non-equilibrium effects are not ignorable.

Due to shear viscosity and curvature effects, the change in pressure $\Delta p_{\rm cusp}$ at the newly formed cusp $\Delta r_{\rm cusp}$ in comparison to the inviscid case is computed in the following way,

$$\Delta r_{\rm cusp} = \frac{r_{\rm cusp,new} - r_{\rm cusp}}{r_{\rm cusp}}, \tag{124}$$

$$\Delta p_{\rm cusp} = \frac{p_{\rm t,cusp} - p_{(0),\rm cusp}}{p_{(0),\rm cusp}}, \tag{125}$$

where the total pressure is $p_{\rm t} = p_{(0)} + p_{(1)}$ and $r_{\rm cusp,new}$ is the new location of the cusp at the equatorial plane. The new cusp positions are computed by fitting the values of $p_{\rm t}$ using a third-order order spline interpolation. The values of $r_{\rm cusp,new}$ and $p_t(r_{\rm cusp,new})$ have also been obtained at the same time using this technique.

A more concrete estimation of the allowed values of the parameters s_1 and m_2 with $\beta_{m,c}$ is obtained from the 2D plot of $|\Delta p_{\rm cusp}|$ shown in Fig. 6.

The black contour in some of the plots in Fig. 6 indicates a cut-off value of s_1 and m_2 for which $\log_{10}|\Delta p_{\rm cusp}| = 0$. Corresponding to low magnetized viscous torus (top panels with $\beta_{m,c} = 10^3$), the permissible values of s_1 and m_2 are large for $\Delta W_{\rm s} > 0$ and that the allowed parameter space of (s_1, m_2) appreciably decreases as the potential gap $\Delta W_{\rm s} \to 0$ indicating that with $\Delta W_{\rm s} \approx 0$ does not allow for large shear viscosity and curvature effects in comparison to $\Delta W_{\rm s} > 0$. On the other hand, for highly magnetized tori ($\beta_{m,c} = 10^{-3}$, bottom panels), stationary viscous models exist over the entire choice of the parameter space and in the considered regions

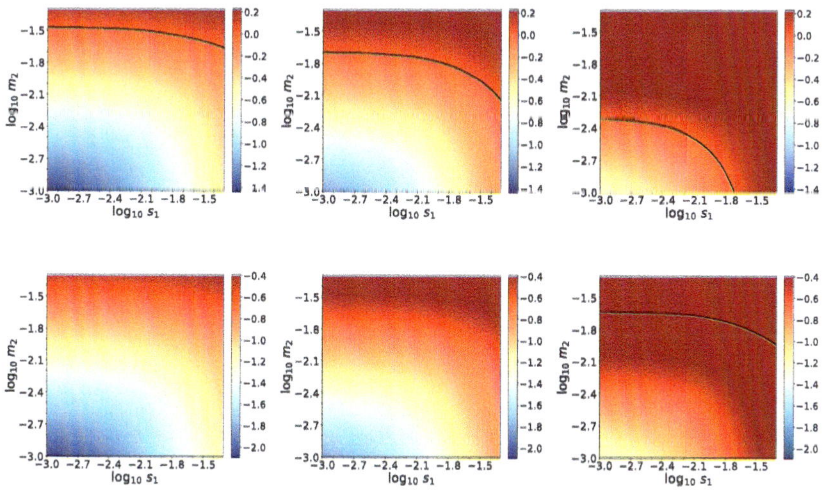

Fig. 6 2D plots for $\log_{10}|\Delta p_{\rm cusp}|$. The first row corresponds to the stationary models with $\beta_{\rm m,c} = 10^3$. The second row corresponds to the models with $\beta_{\rm m,c} = 10^{-3}$. From left to right, the columns correspond to three different values of $W_{\rm s}$, namely -0.039, -0.040 and -0.041. The black contour which appear in some of the plots corresponds to $\log_{10}|\Delta p_{\rm cusp}| = 0$. Taken from [23]

of the potential gap i.e. $\Delta W_s > 0$ and $\Delta W_s \approx 0$. Therefore, within the perturbative approach, the stationary models are restricted up to maximum values of $m_1 = s_1 = 0.05$ and $m_2 = 0.05$. For more such values of s_1 and m_2, the readers are referred to [23].

Isocontours of the total pressure p_t of the stationary viscous tori are shown in Figs. 7 and 8 for both low and high values of the central magnetization parameter, respectively.

These figures display the regions close to the cusp as it is in those regions where the effects of the shear viscosity are most manifest. The self-intersecting contours of p_t possessing a cusp are shown by the blue dashed curves in the figures for the values of s_1 and m_2 indicated in the captions. The red isocontours correspond to surfaces of constant pressure of magnetized ideal fluid tori which would self-intersect, had there been no shear viscosity. For a given value of s_1 and W_s it is observed that when m_2 increases, the location of the newly formed cusp moves towards the Schwarzschild black hole. Also, the thickness of the cusp region in the disk also diminishes which

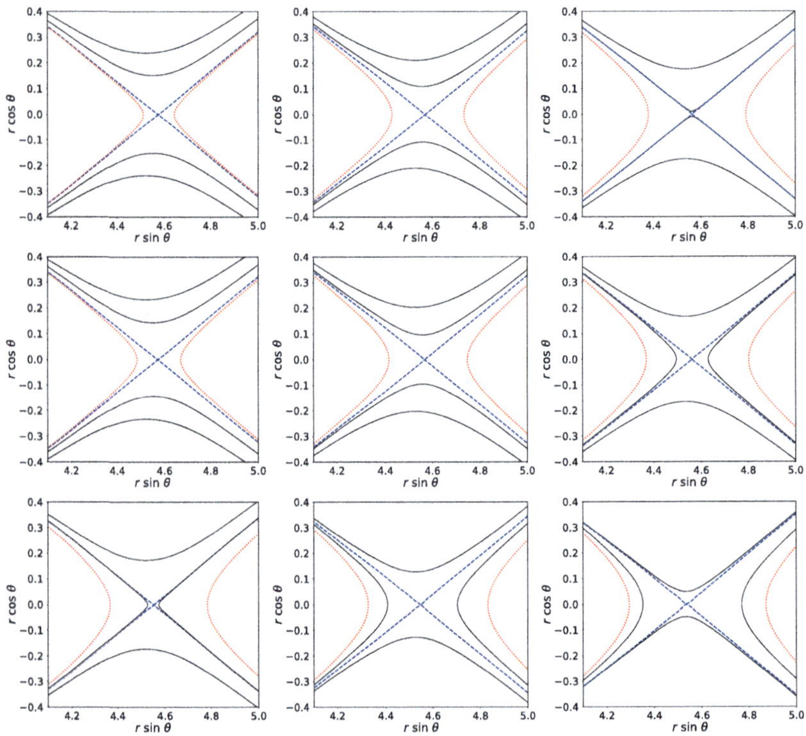

Fig. 7 Isocontours of $p_{(t)} = p_{(0)} + p_{(1)}$ for $W_s = -0.039$ and $\beta_{m,c} = 10^3$ in the cusp region. From top to bottom the rows correspond to $m_2 = (0, 0.005, 0.01)$. From left to right the columns correspond to $s_1 = (0.005, 0.01, 0.05)$. Red isocontours denote cusp-generating constant pressure surfaces without viscosity and blue isocontours represent newly-formed self-intersecting constant pressure surfaces in presence of viscosity and curvature effects. The two black isocontours depict the values $p_t = 2p_{0,\mathrm{cusp}}/3$ and $p_t = p_{0,\mathrm{cusp}}/3$. Taken from [23]

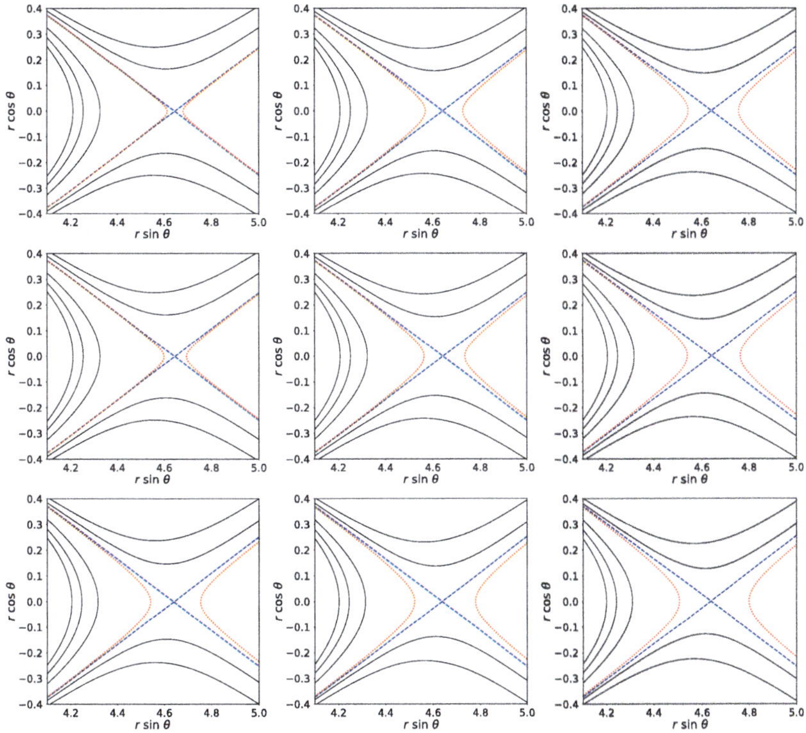

Fig. 8 Same as Fig. 7 but for the parameter $\beta_{\mathrm{m,c}} = 10^{-3}$. The three black isocontours in the left part of all plots correspond values of the total pressure equal to $p_{\mathrm{t}_i} = (p_{(0),\max} i - p_{(0),\mathrm{cusp}})/4$ for $i = 1, 2, 3$, where $p_{(0),\max}$ is the value of $p_{(0)}$ at the maximum of the pressure. Taken from [23]

can be observed by looking at the change of location of the black isocontours located above and below the cusp region in Figs. 7 and 8 respectively. Particularly, Fig. 7 shows that, in the bottom row and in the right column, the isocontour corresponding to $p_{\mathrm{t}} = 2p_{0,\mathrm{cusp}}/3$, changes its position (from above and below the cusp, to the left and right of the cusp) which implies that for these cases, $p_{\mathrm{t,cusp}} < 2p_{0,\mathrm{cusp}}/3$. Additionally, the isocontour corresponding to $p_{\mathrm{t}} = p_{0,\mathrm{cusp}}/3$ moves significantly closer to the self-crossing surface.

Hence, one can infer that in the framework based on causal relativistic hydrodynamics, the role of shear viscosity and the curvature of the Schwarzschild black hole spacetime are apparent through a noticeable rearrangement of the constant pressure surfaces of magnetized viscous disks when compared to the purely inviscid case [23]. In addition, the Figs. 7 and 8 shows that as the magnetic field strength increases, the shift of the location of the cusp towards the black hole also increases. This might mitigate the development of the so-called runaway instability that affects inviscid constant angular momentum tori [12].

5 Concluding Remarks

This chapter presents the findings associated to stationary solutions of viscous torus. The results demonstrates the new modifications in the morphology of a geometrically thick torus, both purely hydrodynamical and magnetized in the presence of shear viscosity as compared to the inviscid case in the context of a simple setup where stationary viscous disks with constant angular momentum distributions are built around a Schwarzschild black hole. The shear viscosity is assumed to only induce perturbative effects on the fluid so that the fluid in the torus can still move in circular orbits. The analysis of isopressure and isodensity surfaces of this constrained system provide evidences showing that the shear viscous and curvature effects in the stationary disk models are only tractable using the causal approach.

Acknowledgements A large part of this chapter is based on the collaborative works based on [21–23] with Claus Lämmerzahl, José A. Font, Sergio Gimeno-Soler and Alejandro Mus Mejías. The author gratefully thanks all her collaborators for agreeing on the collaborations on the topics of viscosity and accretion disks. The author is supported by the Deutsche Forschungsgemeinschaft (DFG) with grant/4040115. During the completion of projects, the author was supported by DFG with grant/40401089 and by the ERC Synergy Grant "BlackHoleCam: Imaging the Event Horizon of Black Holes" (Grant No. 610058).

References

1. M.A. Abramowicz, ACTA Astron. **21** (1971)
2. M.A. Abramowicz, P.C. Fragile, Living Rev. Relativ. **16**, 1 (2013)
3. A.M. Anile, *Relativistic Fluids and Magneto-Fluids: With Applications in Astrophysics and Plasma Physics* (Cambridge University Press, 1990)
4. R. Baier, P. Romatschke, D.T. Son, A.O. Starinets, M.A. Stephanov, JHEP **04**, 100 (2008)
5. R. Baier, S. Lahiri, P. Romatschke, arXiv:1907.02974 [gr-qc]
6. S.A. Balbus, J.F. Hawley, Rev. Mod. Phys. **70**, 1–53 (1998)
7. J.M. Bardeen, Astrophys. J. **162**, 71 (1970)
8. R. Boyer, Math. Proc. Camb. Philos. Soc. **61**(2), 527–530 (1965)
9. F. Daigne, J.A. Font, Mon. Not. R. Astron. Soc. **349**, 841 (2004)
10. G.S. Denicol, J. Noronha, H. Niemi, D.H. Rischke, J. Phys. G **38**, 124177 (2011)
11. J. Erdmenger, M. Haack, M. Kaminski, A. Yarom, JHEP **01**, 055 (2009)
12. J.A. Font, F. Daigne, Mon. Not. R. Astron. Soc. **334**, 383 (2002)
13. J. Frank, A. King, D. Raine, *Accretion Power in Astrophysics* (Cambridge University Press, 2002)
14. S. Gimeno-Soler, J.A. Font, Astron. Astrophys. **607**, A68 (2017)
15. W.A. Hiscock, L. Lindblom, Ann. Phys. **151**, 466 (1983)
16. W.A. Hiscock, L. Lindblom, Phys. Rev. D **31**, 725 (1985)
17. W. Israel, J.M. Stewart, Ann. Phys. **118**, 341 (1979)
18. W. Israel, J.M. Stewart, Proc. R. Soc. Lond. A - Math. Phys. Eng. Sci. **365**, 43 (1979)
19. S.S. Komissarov, Mon. Not. R. Astron. Soc. **368**, 993 (2006)
20. P. Kovtun, D.T. Son, A.O. Starinets, Phys. Rev. Lett. **94**, 111601 (2005)
21. S. Lahiri, Class. Quantum Gravity **37**(7), 075010 (2020)
22. S. Lahiri, C. Lämmerzahl, arXiv:1909.10381 [gr-qc]
23. S. Lahiri, S. Gimeno-Soler, J.A. Font, A.M. Mejías, Phys. Rev. D **103**(4), 044034 (2021)

24. Q. Lei, M.A. Abramowicz, P.C. Fragile, J. Horak, M. Machida, O. Straub, Astron. Astrophys. **498**, 471 (2009)
25. R. Maartens, arXiv:astro-ph/9609119 [astro-ph]
26. G.D. Moore, K.A. Sohrabi, Phys. Rev. Lett. **106**, 122302 (2011)
27. I. Muller, Z. Phys. **198**, 329 (1967)
28. A. Muronga, Phys. Rev. Lett. **88**, 062302 (2002)
29. A. Muronga, Phys. Rev. C **69**, 034903 (2004)
30. A. Nakamura, S. Sakai, Phys. Rev. Lett. **94**, 072305 (2005)
31. P. Romatschke, Int. J. Mod. Phys. E **19**, 1 (2010)
32. P. Romatschke, Class. Quantum Gravity **27**, 025006 (2010)
33. N. Shakura, R. Sunyaev, A & A **24**, 337 (1973)

Shadows and Accretion Disk Images of Compact Objects

Petya Nedkova

Abstract

The black hole shadow was predicted theoretically already in the 1960s as a strong gravitational lensing phenomenon. Recently, its observation became feasible and opened a major experimental channel for probing the gravitational interaction in the strong field regime. Although considered historically as a property of the black hole spacetimes, the shadow is not exclusively a black hole effect and does not require the presence of an event horizon. Other compact objects can also cast a shadow if their gravitational field is sufficiently strong bringing up the issue of how we can differentiate between self-gravitating systems based on their shadow images. In this chapter we discuss the analytical or semi-analytical methods for obtaining the observable images produced by compact objects in some basic physical settings such as a uniform spherical distribution of distant light sources or the presence of a thin accretion disk. We review the calculation of the shadow boundary for the Kerr black hole and focus on recent research on the images created by wormholes and naked singularities. These compact objects can look qualitatively very similar to black holes in some cases, but they can also possess clear-cut observational signatures.

1 Introduction

General relativity has been tested in the weak field regime with a remarkable precision [93,94], but its validity in strong gravitational fields is still hypothetical. A central question in this respect is whether the Kerr solution indeed represents the unique stationary and axisymmetric black hole in vacuum. Testing this statement known as the Kerr hypothesis is equivalent to confirming whether general relativity describes

P. Nedkova (✉)
Faculty of Physics, Sofia University, 5 James Bourchier Blvd, 1164 Sofia, Bulgaria
e-mail: pnedkova@phys.uni-sofia.bg

B. Hartmann and J. Kunz (eds.), *Gravity, Cosmology, and Astrophysics*, Lecture Notes in Physics 1022, https://doi.org/10.1007/978-3-031-42096-2_3

the black holes in our Universe or another more refined gravitational theory takes its place in the extreme regime.

In the last decade we witnessed a major experimental breakthrough in this direction with the detection of gravitational waves which opened a new window for exploring gravity [1,2]. On the other hand, the Event Horizon Telescope (EHT) collaboration provided the first direct images of the supermassive compact objects in the nearby galactic targets M87 and Sgr A with a horizon-scale resolution [7,9,10]. These developments launched a new era in fundamental physics allowing to combine information from different observational channels in order to test its predictions. Thus, the investigation of compact objects turned for the first time from a predominantly theoretical field into an active area in observational astrophysics.

Black hole imaging experiments such as the Event Horizon Telescope have their roots in the foundations of general relativity governing the propagation of particles and light in curved spacetime. These phenomena are described by the theory of gravitational lensing which explains the observational effects caused by the interplay of light with extreme curvature. In particular, the strong gravitational field in the vicinity of black holes gives rise to a specific phenomenon known as the black hole shadow. This effect arises when the black hole is surrounded by a distribution of light sources and we observe a dark spot on the luminous sky. It can be easily predicted by considering the definition of black hole. Assuming a bundle of photon trajectories with a broad range of initial condition, there will be always such ones that will enter the black hole region and consequently never come back to us. In this way they leave dark directions in the observer's sky, which form the black hole shadow.

The shadow provides a means to map the black hole vicinity into an optical image which encodes essential information about the properties of the underlying spacetime. The main observational characteristic is the shadow boundary and its size and shape determine the black hole spin and other relevant charges [6,44,52,58,86]. Ideally, if the boundary curve is measured with a very high precision, it enables us to specify the black hole giving rise to it. This correspondence can be further elaborated and extended into a procedure for testing the Kerr hypothesis and imposing constraints on the modified theories of gravity [17,54,60,75].

The study of the black hole shadow has a long history dating back to the classical works of general relativity. The phenomenon was described theoretically already in the 1960s in the early works [82,100], which obtained the viewing angle of the shadow for the Schwarzschild black hole. In the next decade Bardeen investigated the shadow of the Kerr black hole and taking advantage of the separability of the null geodesic equations developed a general formalism for obtaining the shadow boundary [16]. Although of fundamental importance, these developments considered the black hole shadow as a purely theoretical phenomenon which is unlikely to be experimentally detected. The idea that observing the shadow of the black hole at the center of our galaxy may be feasible was first suggested in [38], where the necessary experimental conditions were also discussed. This seminal paper put the foundations for the development of the global interferometer Event Horizon Telescope which recently produced the first black hole images.

The analytical construction of the black hole shadow boundary has some fundamental consequences. Revealing the explicit mechanism for the shadow formation it became evident that the phenomenon is not limited exclusively to black holes but a much broader class of compact objects will lead to a similar image [48,64,77,80]. The reason is that the mathematical structure which determines the shadow is not the event horizon but another fundamental surface called a photon sphere. The photon sphere represents a separatrix between two families of infalling null geodesics. The first class consists of trajectories which are reflected by the gravitational field and manage to reach a distant observer. The second class of trajectories get trapped in the gravitational potential and fail to scatter away to infinity, thus forming the shadow. For black holes these are the geodesics which enter the event horizon. In other spacetimes such behavior can develop for different physical reasons. For example, wormhole geometries suggest that part of the geodesics will pass through the wormhole throat and continue to propagate in another universe. In naked singularity spacetimes shadows are formed due to the geodesics which end at the singularity.

Most of the classical results in the compact objects astrophysics were developed considering the Kerr black hole. However, ideas from fundamental physics suggest that general relativity may not be the final theory of gravity. It will probably need to be modified in order to explain mysterious phenomena such as the dark energy and dark matter and describe the gravitational interaction at the Planck scale. In order to address these issues various alternative theories of gravity were proposed motivated as a low-energy limit of unification theories or effective theories of quantum gravity. Modifying the properties of the gravitational interaction they allow for a much greater variety of compact objects than general relativity including black holes with different kinds of hair and wormholes which do not violate the energy conditions. Other solitonic self-gravitating configurations are also studied such as regular black holes, boson stars and gravastars. These exotic compact objects from the perspective of general relativity are no longer purely theoretical ideas. Their experimental detection has become a solid part of the agenda of the major astrophysical missions both in the gravitational wave and the electromagnetic spectrum inspiring a range of works investigating and predicting their observational signatures.

The aim of this review is to describe the theoretical foundations of the black hole shadow and to demonstrate how the effect can be extended to other compact objects such as wormholes and naked singularities. We will consider spacetimes with high degree of symmetries which allow for the integrability of the geodesic equations and analytical construction of the shadow boundary. The classical calculation of the black hole shadow assumes a uniform distribution of the light sources around the compact object. This set-up is a simplified toy-model but it allows to extract the most essential information about the phenomenon without additional complications from more restricting initial conditions. In realistic astrophysical scenarios the compact objects are surrounded by accretion disks which represent the main source of the electromagnetic emission. Therefore, as a next step we will demonstrate how the observational picture is modified if we adopt a more realistic light sources distribution. We will consider the basic model of a geometrically thin and optically thick disk around a spherically symmetric compact object and construct the observable image for the Schwarzschild black hole and certain types of naked singularities. Although simplified, such models provide valuable intuition how the observational signatures arise and how the properties of the compact object influence the images which can be useful for interpreting the results of more complicated simulations.

2 Black Hole Shadow in Static Spherically Symmetric Spacetime

In order to describe the effect of black hole shadow we will discuss initially static spherically symmetric spacetime in vacuum. This simple geometrical setting allows to illustrate clearly the theoretical ideas which we will generalize subsequently in more complicated scenarios. On the relevant scales for gravitational lensing light propagation is described by the geometric optics approximation. Light follows null geodesics and the geodesic equations can be derived from the least action principle by introducing the Lagrangian

$$\mathcal{L} = \frac{1}{2} g_{\mu\nu} \dot{x}^\mu \dot{x}^\nu, \tag{1}$$

where $g_{\mu\nu}$ is the metric, x^μ are the spacetime coordinates, and the dot denotes differentiation with respect to the affine parameter along the geodesics.[1] In this way we obtain a system of four ordinary differential equations and we have in addition the constraint

$$g_{\mu\nu} \dot{x}^\mu \dot{x}^\nu = 0, \tag{2}$$

which ensures that the type of geodesic is preserved for any value of the affine parameter. The geodesic equations can be simplified if the spacetime possesses higher degree of symmetries. Stationary and axisymmetric spacetimes are particularly important since they describe the quasi-equilibrium configurations of the astrophysical objects such as black holes surrounded by an accretion disk or galaxies with a supermassive central compact object. These symmetries are manifested by the presence of a Killing vector associated with time translations $\partial/\partial t$ and a Killing vector associated with rotations with respect to the symmetry axis $\partial/\partial\varphi$ which induce conservation laws on the geodesics. In particular they lead to the conservation of the photon's specific energy E and specific angular momentum L.

Spherical symmetry increases the number of constants of motion since it generates two additional Killing vectors which lead to conservation of the plane of motion. Thus, geodesics propagate in a single plane $\theta = \text{const.}$ which we assume for convenience to correspond to the equatorial plane. In this way the geodesic equations become completely integrable and reduce to a one-dimensional problem. The photon trajectories can be obtained by either integrating the radial geodesic equation, or more conveniently considering the constraint given by Eq. (2).

[1] For convenience we introduce a spherical coordinate system $x^\mu = \{t, r, \theta, \varphi\}$.

Let us introduce a static spherically symmetric metric in the general form

$$ds^2 = -A(r)dt^2 + A^{-1}(r)dr^2 + r^2 D(r)(d\theta^2 + \sin^2\theta d\varphi^2), \tag{3}$$

where the metric functions $A(r)$ and $D(r)$ depend only on the radial coordinate. Then, the constants of motion and the constraint equation take the form

$$\begin{aligned} &E = A(r)\dot{t}, \qquad L = r^2 D(r)\dot{\varphi}, \\ &\dot{r}^2 + V_{\text{eff}} = E^2, \qquad V_{\text{eff}} = L^2 \frac{A(r)}{r^2 D(r)}. \end{aligned} \tag{4}$$

We see that the geodesic equations reduce effectively to a familiar problem representing the motion in the field of a spherically symmetric potential V_{eff}. Although this analogy is only formal, we can analyse the qualitative behaviour of the photon trajectories by taking advantage of our intuition from classical mechanics. In general we can rescale the affine parameter by the photon's specific energy E. Thus, we see than there is a single dynamically important parameter defined by the radio of the specific energy and angular momentum $b = L/E$. It is called an impact parameter and the type of the photon trajectories is determined by its value and the particular form of the effective potential.

Let's examine this problem for the Schwarzschild black hole. The effective potential is given explicitly by

$$V_{\text{eff}} = \frac{b^2}{r^2}\left(1 - \frac{R_s}{r}\right), \tag{5}$$

where $R_s = 2M$ is the Schwarzschild radius and M is the mass of the black hole. The effective potential tends to zero at the spacetime infinity $r \to \infty$ and behaves as $V_{\text{eff}} \to -\infty$ at the curvature singularity $r = 0$ (see Fig. 1). It possesses a single maximum in between which determines an unstable circular photon orbit by the conditions

$$V_{\text{eff}} = 1, \qquad V'_{\text{eff}} = 0, \qquad V''_{\text{eff}} < 0, \tag{6}$$

where the prime denotes derivative with respect to the radial coordinate. Solving these equations we obtain that the circular orbit is located at $r = 3M$ and corresponds to the value of the impact parameter $b_{\text{crit}} = 3\sqrt{3}M$. Let us consider photon trajectories approaching the black hole from infinity. They will separate in two qualitatively different classes. Photon trajectories with impact parameters larger than that of the circular orbit $b > b_{\text{crit}}$ will be reflected by the effective potential and scatter away to infinity, while those with impact parameters $b < b_{\text{crit}}$ will manage to cross the potential barrier and end up plunging into the black hole. Thus, the unstable circular orbit serves as a critical curve which separates gravitational scattering from gravitational capture. Its impact parameter represents the limiting value for the impact parameters of infalling trajectories which are capable of reaching back to a distant observer. Hence, it determines the boundary of the black hole shadow.

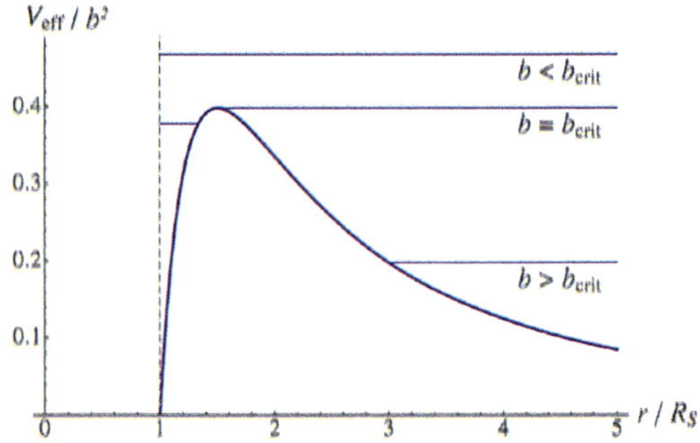

Fig. 1 Effective potential for the null geodesics for the Schwarzschild black hole. Infalling photon trajectories from large distances scatter away from the potential if their impact parameter satisfies $b > b_{\text{crit}}$, and plunge into the black hole if $b < b_{\text{crit}}$. Reprinted from [23]. with the permission of AIP Publishing

Let's imagine the collection of all the unstable circular orbits in all the possible planes of motion for the null geodesics. They will build up a sphere with a radius $r = 3M$, which is called a photon sphere. The shadow of the Schwarzschild black hole represents a lensed image of the photon sphere projected on the observer's sky. By the described argument we see that any static spherically symmetric compact object which allows for both families of scattering and plunging photon trajectories will possess a photon sphere and therefore cast a shadow. Thus, the shadow is not exclusively a black hole phenomenon but an observational characteristic of compact objects possessing a photon sphere. These are also called ultra-compact objects and include soliton-like self-gravitating configurations of various physical nature like wormholes, naked singularities, boson stars, gravastars as particular examples.

The photon sphere is a fundamental surface which is important not only for its association with observational signatures. Compact objects spacetimes can be classified using this structure similar to the black hole classification based on the properties of the event horizon. In this way it was demonstrated that the Schwarzschild spacetime is the unique static and asymptotically flat solution to the Einstein equations in vacuum possessing a photon sphere [21]. This result was extended to electro-vacuum spacetimes [22,97] and to the Einstein-scalar field theory where it was proven that any static and asymptotically flat solution which possesses a photon sphere is spherically symmetric and isometric to the Janis–Newman–Winicour weakly naked singularity [96]. Further generalizations were developed considering the Einstein–Maxwell-dilaton theory [98] and multiple scalar fields [99]. These theorems provide a much broader classification than the black hole uniqueness theorems assuming an analytical event horizon since they include also horizonless compact objects.

Let us get more intuition about the properties of the photon sphere by considering some explicit solutions of the geodesic equations for the Schwarzschild black hole. When our aim is to obtain particular solutions for the trajectories it is convenient to express the constraint given by Eq. (2) as a differential equation for the variation of the radial coordinate with respect to the azimuthal angle. Thus, using the constants of motion we obtain

$$\left(\frac{dr}{d\varphi}\right)^2 = r^2 D(r)\, A(r) \left(\frac{r^2 D(r)}{b^2 A(r)} - 1\right) = f(r), \tag{7}$$

for a general spherically symmetric spacetime. For the Schwarzschild black hole it is convenient to make the substitution $u = R_s/r$. Plugging in the particular form of the metric functions, we see that the right-hand side of the trajectory equation reduces to a polynomial in the new variable

$$\left(\frac{du}{d\varphi}\right)^2 = u^3 - u^2 + \frac{R_s^2}{b^2} = f(u). \tag{8}$$

This form of the equation allows for straightforward qualitative analysis of the photon motion without introducing an effective potential. The possible types of photon trajectories are determined by the root structure of the function $f(u)$ at the right-hand side of the equation since its roots correspond to the possible turning points of the trajectories. We can demonstrate that the function always possesses a negative root which is irrelevant for the geodesic motion. In addition it possesses two distinct positive real roots for large values of the impact parameter b, which approach each other when b declines, merge into a double real root for a certain critical value b_{crit}, and turn into a pair of complex conjugate roots for lower impact parameters $b < b_{\text{crit}}$. The first type of root structure containing two distinct real roots implies that infalling null geodesics from infinity will always possess a turning point at the smaller root of $f(u)$ and therefore will scatter away from the black hole. On the contrary, complex roots correspond to the absence of turning points so geodesics with such impact parameters will always plunge into the black hole.

Let's examine the marginal case between these two types of behaviour. Having a double real root implies that the equations

$$f(u) = 0, \qquad f^{'}(u) = 0, \tag{9}$$

are satisfied. Consequently, this case represents a circular orbit. We can further demonstrate that $f^{''}(u) > 0$ is satisfied at this point, i.e. the circular orbit is unstable, and solving Eq. (9) we obtain its position $r = 3M$ and the corresponding impact parameter $b_{\text{crit}} = 3\sqrt{3}M$. In this way we have obtained the photon sphere from a different perspective.

The photon sphere consists of unstable circular orbits so photons propagating on such orbits with either scatter away or fall into the black hole by the slightest perturbation. Thus, the photon sphere should be rather interpreted as a limiting surface for two families of infalling and outgoing geodesics with the impact parameter $b_{\text{ph}} = 3\sqrt{3}M$. We can derive the explicit form of these trajectories by solving Eq. (8) for the value of the impact parameter $b = b_{\text{ph}}$, however requiring that $\frac{du}{d\varphi} \neq 0$. Assuming that we have infalling geodesics with initial conditions $r > r_{\text{ph}}$ we obtain [24] (see also [23])

$$u = -\frac{1}{3} + \tanh^2 \frac{1}{2}(\varphi - \varphi_0), \tag{10}$$

where φ_0 is an integration constant, while the solution

$$u = -\frac{1}{3} + \coth^2 \frac{1}{2}(\varphi - \varphi_0), \tag{11}$$

corresponds to outgoing trajectories which originate at radial distances between the horizon and the photon sphere ($R_s < r < r_{\rm ph}$). Both solutions represent geodesics which approach the photon sphere either from its exterior or its interior and spiral infinitely around its surface. We see that when $r \to 3M$, i.e. $u \to 2/3$, the azimuthal angle tends to infinity, i.e. the trajectory performs an infinite number of turns around the photon sphere.

In lensing problems we frequently consider the deflection angle defined as the variation of the azimuthal angle along the photon trajectory. For static spherically symmetric spacetimes it is given by

$$\varphi = \int_{r_{source}}^{r_{obs}} \frac{dr}{r\sqrt{D(r)\,A(r)}\sqrt{\left(\frac{r^2 D(r)}{b^2 A(r)} - 1\right)}}, \tag{12}$$

where we integrate between the radial positions of the light source $r_{\rm source}$ and the observer $r_{\rm obs}$. For the Schwarzschild black hole the deflection angle can be expressed explicitly in terms of elliptic functions for a general value of the impact parameter (see e.g. [24]). However, the behaviour in the vicinity of the photon sphere can be represented by the approximate solution [43,59]

$$\begin{aligned} &\varphi = \log \frac{C_\pm}{|b - b_{ph}|}, \quad b \to b^{\pm}_{ph}, \\ &C_\pm = const. \end{aligned} \tag{13}$$

which is valid for impact parameters approaching the photon sphere with lower values $b \to b^{-}_{\rm ph}$ or higher ones $b \to b^{+}_{\rm ph}$. Both solutions for the deflection angle are illustrated in Fig. 2. As we already know, the deflection angle diverges logarithmically at the photon sphere. In addition, we can see that a hierarchy of infalling trajectories is formed in the neighbourhood of the photon sphere which circle around it an arbitrary large but finite number of turns n as the winding number n grows when the trajectory's impact parameter approaches $b_{\rm ph}$. These trajectories ultimately scatter away to infinity and carry important information for the observations. Even for comparatively low values of the winding number $n \sim 1.25$ they probe sufficiently well the photon sphere and provide a reasonable estimate for its image [43].

Section summary: We introduced the notion of photon sphere in the simplest geometrical set-up of static spherically symmetric spacetimes using the Schwarzschild black hole as a particular example. We illustrated its role for the light propagation from different perspectives discussing the effective potential for the null geodesics, particular explicit solutions for the photon trajectories and the properties of the deflection angle. These concepts can be generalized and provide intuition in technically more involved geometries such as stationary axisymmetric spacetimes and light sources distributions such as accretion disks. We will consider such generalizations in the next sections.

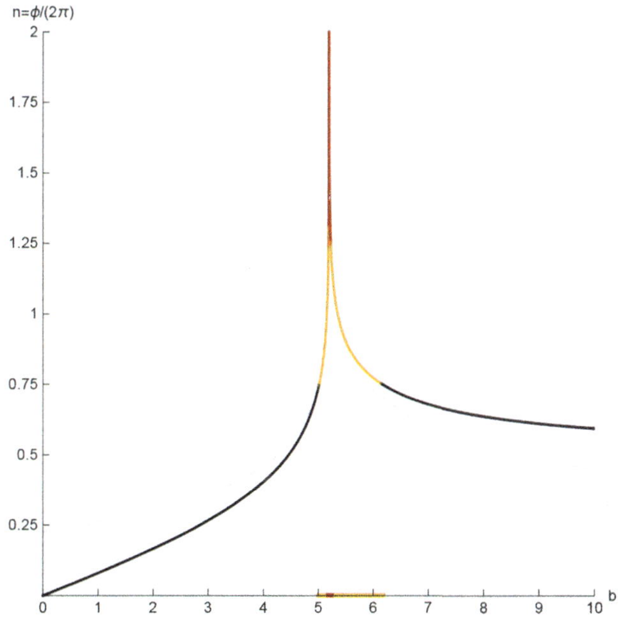

Fig. 2 Deflection angle for the null geodesics for the Schwarzschild black hole as a function of the impact parameter. The dashed line represents the exact solution of the integral, while the solid line corresponds to the approximation given by Eq. (13). The region denoted in red already provides a good approximation for the photon sphere. Reprinted figure with permission from [43]. Copyright 2019 by the American Physical Society

3 Shadow of Stationary and Axisymmetric Compact Objects

In this section we will reduce the symmetry assumptions and consider the shadow cast by stationary and axisymmetric compact objects when illuminated by a uniform distribution of distant light sources. In this setting we have two Killing vectors associated with the conservation of the photon's specific energy E and angular momentum L on the geodesics and the constraint given by Eq. (2) which ensures that the geodesic's type is preserved. These integrals of motion are not sufficient for the complete integrability of the geodesic equations and in general solutions are obtained numerically. However, there exist particular spacetimes possessing additional symmetries called hidden symmetries. They are associated with the existence of irreducible Killing and Killing–Yano tensors which do not generate isometries but lead to additional integrals of motion along the geodesics. This property allows to separate the variables in the geodesic equations and reduce formally the problem of light propagation to a one-dimensional motion in the field of an effective potential. Then, we can obtain the shadow boundary analytically by analysing the radial motion similar to the spherically symmetric case.

Black hole solutions in general relativity belong to this type of integrable spacetimes since it was proven that the Kerr–Newman-NUT-(A)dS family possesses a Killing tensor. Its explicit form was derived initially by Carter [20] and the corresponding integral of motion was called a Carter constant. Penrose and Walker obtained this result simultaneously in an independent way relating the presence of a Killing tensor to the algebraic type of the spacetime [91].

Let us demonstrate how we can obtain the shadow boundary in stationary axisymmetric spacetimes possessing hidden symmetries. In this case it is convenient to consider the Hamilton–Jacobi formulation of the geodesic problem. Due to the presence of a Killing tensor the Hamilton–Jacobi equation

$$\frac{\partial S}{\partial \lambda} = -\frac{1}{2} g^{\mu\nu} \frac{\partial S}{\partial x^\mu} \frac{\partial S}{\partial x^\nu}, \tag{14}$$

is separable, and introducing a spheroidal coordinate system $x^\mu = \{t, r, \theta, \varphi\}$ there exists a solution in the form

$$S = \frac{1}{2}\mu^2\lambda - Et + L\varphi + S_r(r) + S_\theta(\theta). \tag{15}$$

In this expression we denote by μ the mass of the test particle, λ is the affine parameter on the geodesics, while the functions $S_r(r)$ and $S_\theta(\theta)$ depend only on the specified coordinates.The geodesic equations are obtained by the standard procedure using the fact that the partial derivatives of the Jacobi action S with respect to the constants of motion should vanish. Setting $\mu = 0$ for null geodesics, we obtain two decoupled equations for the radial and polar motion in the form

$$\frac{dr}{d\lambda} = R(r, E, L, Q), \tag{16}$$

$$\frac{d\theta}{d\lambda} = T(\theta, E, L, Q), \tag{17}$$

where Q is the integral of motion associated with the hidden symmetries. These equations combined with the conservation laws associated with the Killing vectors determine the photon propagation.

In Eqs. (16)–(17) we see that the photon trajectories depend on the conserved charges $\{E, L, Q\}$. However, the number of independent parameters can be reduced by recognizing that the specific energy E is merely a scale parameter, and can be eliminated by rescaling the affine parameter as $\lambda \to E\lambda$. Then, the geodesic motion will depend only on the ratios $\xi = L/E$ and $\eta = Q/E^2$, which play the role of impact parameters.

We can obtain the shadow boundary by analysing the radial equation similar to the spherically symmetric case. In analogy we can transform Eq. (16) into an energy-like equation for a certain effective potential $V(r, \eta, \xi)$

$$\left(\frac{dr}{d\lambda}\right)^2 + V_{eff} = 1. \tag{18}$$

Performing the qualitative analysis which we described in Sect. 2 we can determine the trajectories which scatter away from the gravitational potential and those which overcome the potential barrier and plunge into the black hole. The critical orbits which separate gravitational scattering from gravitational capture correspond to the highest maximum of V_{eff}. Thus, they satisfy the conditions

$$V_{eff} = 1, \qquad V'_{eff} = 0, \qquad V''_{eff} < 0. \tag{19}$$

An important distinction from the spherically symmetric case is that the critical orbits are in general not circular. They represent unstable spherical orbits, i.e. they lie on a sphere with a certain radius, but the θ-motion can be very complicated.[2] Due to this distinction we cannot define a photon sphere as in the spherically symmetric case. However, we can introduce an appropriate generalization. The collection of all the unstable spherical orbits is now called a photon region and the shadow boundary arises as its lensed image on the observer's sky.

We can outline the following general procedure for constructing the shadow boundary for stationary axisymmetric spacetimes with integrable geodesic equations.

- From the definition of the unstable spherical orbits given by Eq. (19) we obtain two algebraic equations for the impact parameters ξ and η. They determine a curve in the impact parameter space $\eta = \eta(\xi)$, which can be also represented in a parametric form using the radial coordinate, i.e. $\xi = \xi(r)$, $\eta = \eta(r)$. This curve corresponds to the shadow boundary in the impact parameter space.
- The range of the radial coordinate on the curve $\{\xi(r), \eta(r)\}$ is constrained by the geodesic equation for the polar motion. The impact parameters $\xi = \xi(r)$ and $\eta = \eta(r)$ should be such that the function $T(r, \xi, \eta)$ on the right-hand side of Eq. (17) is well-defined. This ensures the existence of the unstable spherical orbits determining the photon region.
- Similar to the spherically symmetric case, for every radial coordinate $r = r_0$ belonging to the photon region there exist ingoing geodesics with the same impact parameters $\xi(r_0)$ and $\eta(r_0)$ which spiral towards the corresponding spherical orbit as a limit curve. Then, the observable image of the photon region is determined by the projection of these geodesics on the observer's sky.

The last step of the procedure is constructing explicitly the projected image. The projection on the observer's sky is not defined uniquely and there are different types of celestial coordinates proposed in the literature. In our approach we will follow the Bardeen procedure which was applied initially to obtain the Kerr black hole shadow [16] (see also [28]). For the purpose we should first define the observer's frame by introducing a local orthonormal frame (tetrad) at the observer's position. Considering the general form of a stationary axisymmetric metric in the spheroidal

[2] The spherical orbits for the Kerr black hole are discussed in detail in [84] and illustrated with particular examples.

coordinates $x^\mu = \{t, r, \theta, \varphi\}$ a natural choice of the orthonormal tetrad is given by

$$e_{(r)} = \frac{1}{\sqrt{g_{rr}}}\partial_r, \quad e_{(\theta)} = \frac{1}{\sqrt{g_{\theta\theta}}}\partial_\theta, \quad e_{(\phi)} = \frac{1}{\sqrt{g_{\varphi\varphi}}}\partial_\varphi, \tag{20}$$
$$e_{(t)} = \zeta\,\partial_t + \gamma\,\partial_\varphi,$$

where the quantities ζ and γ are defined as

$$\zeta = \sqrt{\frac{g_{\varphi\varphi}}{g_{t\varphi}^2 - g_{tt}g_{\varphi\varphi}}}, \qquad \gamma = -\frac{g_{t\varphi}}{g_{\varphi\varphi}}\zeta. \tag{21}$$

Then, we can obtain the locally measured 4-momentum $p^{(\mu)}$ by projecting the tangent vector to the geodesics, i.e. the photon's 4-momentum p_μ, in the observer's frame

$$p^{(r)} = \frac{p_r}{\sqrt{g_{rr}}}, \qquad p^{(\theta)} = \frac{p_\theta}{\sqrt{g_{\theta\theta}}}, \qquad p^{(\phi)} = \frac{L}{\sqrt{g_{\phi\phi}}}, \tag{22}$$
$$p^{(t)} = \zeta E - \gamma L.$$

The projection is determined by two celestial angles α and β which serve as coordinates on the observer's sky. They are expressed explicitly by the components of the local momentum as

$$\alpha = \arcsin\frac{p^{(\theta)}}{p^{(t)}}, \qquad \beta = \arctan\frac{p^{(\phi)}}{p^{(r)}}. \tag{23}$$

We can further set $p^{(t)} = 1$ by rescaling the affine parameter along the geodesics.

In order to obtain the image of the shadow boundary on the observer's sky we should select the celestial angles α and β, which correspond to the photon region. Since the 4-momentum p_μ is determined from the geodesic equations, we should simply substitute in these expressions the relevant values of the impact parameters ξ and η. Then, using Eq. (23) we can obtain the celestial angles describing the shadow boundary

$$\alpha = \alpha(\xi, \eta, r_{\text{obs}}, \theta_{\text{obs}}), \qquad \beta = \beta(\xi, \eta, r_{\text{obs}}, \theta_{\text{obs}}), \tag{24}$$

for a given position of the observer $(r_{\text{obs}}, \theta_{\text{obs}})$. Since α and β decrease with the radial distance of the observer, i.e. $\alpha \sim \frac{1}{r}$, $\beta \sim \frac{1}{r}$, it is convenient to rescale them as $\alpha \to \alpha r_{\text{obs}}$, $\beta \to \beta r_{\text{obs}}$ introducing celestial coordinates with dimension of mass.

By the described procedure we can obtain the image of the shadow boundary for any radial position of the observer. However, if we aim at describing the observational data from the Event Horizon Telescope, we should assume that the observer is located at spacetime infinity. In the limit $r_{\text{obs}} \to \infty$ we further obtain simpler expressions for the celestial coordinates.

3.1 Example: Kerr Black Hole

As an example we will discuss the shadow of the Kerr black hole. The Kerr solution is characterized by two parameters representing its ADM mass M and angular momentum J, which is commonly substituted by the spin parameter $a = J/M$. Following the outlined procedure we can derive the impact parameters which define the photon region [16]

$$\begin{aligned} \xi(r) &= -\frac{r^3 - 3Mr^2 + a^2(M+r)}{a(r-M)}, \\ \eta(r) &= \frac{r^3[4a^2M - r(r-3M)^2]}{a^2(r-M)^2}, \end{aligned} \tag{25}$$

where the radial coordinate satisfies the inequality

$$\eta + \cos^2\theta\left(a^2 - \frac{\xi^2}{\sin^2\theta}\right) \geq 0. \tag{26}$$

In this way we obtain the shadow boundary in the impact parameter space. Assuming an asymptotic observer, i.e. $r_{\rm obs} \to \infty$, these expressions lead to the following celestial coordinates

$$\begin{aligned} \alpha &= -\frac{\xi}{\sin\theta_{\rm obs}}, \\ \beta &= \left[\eta + a^2\cos^2\theta_{\rm obs} - \xi^2\cot^2\theta_{\rm obs}\right]^{1/2}, \end{aligned} \tag{27}$$

which determine the observable shadow image. We see that the shadow depends on the spin of the black hole and the angular position of the observer called also an inclination angle. In Fig. 3 we illustrate the shadow boundary for the Kerr black hole for different spin parameters and inclination angles. For static black holes the shadow is circular as its size depends on the black hole mass. Introducing rotation leads to characteristic deformation of the boundary curve. The particular shape of the shadow boundary contains enough information to estimate the black hole spin if it is restored from the observational data with sufficient resolution [52,58,86]. It can further determine whether the spacetime is characterized by higher multipole moments, in this way serving as a means for detecting black hole hair or existence of more exotic compact objects.

Section summary: We demonstrated how the notion of photon sphere can be generalized for a stationary and axisymmetric spacetime possessing an irreducible Killing tensor. Due to the 'hidden' symmetries the geodesic motion becomes completely integrable. We showed how we can obtain the shadow boundary analytically using the Hamilton–Jacobi formulation of the geodesic problem. We considered the Kerr black hole as a particular example.

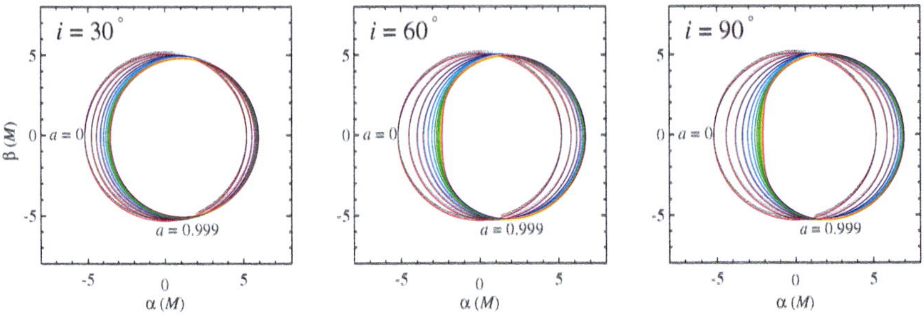

Fig. 3 Shadow of the Kerr black hole for different inclinations angles i and spin parameters a. In each panel the spin parameter grows continuously from $a = 0$ (black curve) to the near-extremal value $a = 0.999$ (red curve). Increasing the spin and the inclination angle leads to stronger deformation of the shadow boundary. Reprinted from [102].

4 Shadow of Traversable Wormholes

One of the most exciting theoretical predictions is that our universe may contain traversable wormholes which we have not detected so far because they mimic the observational features of black holes. As we already explained the shadow is such a strong gravity phenomenon which is common for a range of compact objects of different physical nature. In this section we will construct explicitly the shadow of a certain class of traversable wormholes and discuss how closely its boundary curve resembles the Kerr black hole.

For the purpose we consider the general form of the metric describing stationary axisymmetric wormholes suggested by Teo [83] as a rotating generalization of the Morris–Thorne wormhole

$$ds^2 = -N^2 dt^2 + \left(1 - \frac{b}{r}\right)^{-1} dr^2 + r^2 K^2 \left[d\theta^2 + \sin^2\theta (d\varphi - \omega dt)^2\right]. \tag{28}$$

The metric functions depend only on the spherical coordinates r and θ and under some mild conditions define a completely regular geometry which represents a tunnel connecting two distant regions in spacetime. The wormhole throat corresponds to the minimal surface located at the constant radius $b = r$, where the metric function g_{rr} becomes divergent. This behavior reduces to an apparent singularity if $\partial_\theta b(r, \theta) = 0$ at the throat. The spacetime does not contain any curvature singularities or event horizons, if we require further that the redshift function N is finite and non-zero in all the coordinate range. In addition, satisfying $\frac{db}{dr} < 1$ at the throat provides the characteristic flaring out shape of the embedding of the constant t and θ cross-sections of the wormhole spacetime into three-dimensional Euclidean space.

Since the spherical coordinates break down at the wormhole throat, we can represent only one of the asymptotic regions in this coordinate system. However, we can introduce a global radial coordinate l defined as

$$dl = \pm \frac{dr}{\sqrt{1 - \frac{b}{r}}}, \tag{29}$$

which extends the wormhole metric across the throat. It takes the range $-\infty < l < +\infty$ and describes smoothly the transition between two asymptotic regions for positive and negative values of l connected by the wormhole throat at $l = 0$.

For our purposes we will impose restrictions on the general wormhole geometry (28) by requiring that the metric functions N, b, ω and K depend only on the radial coordinate r. In this way we obtain a family of traversable wormholes with integrable geodesic equations and we can calculate the shadow boundary analytically as described in Sect. 3. Separating the variable in the Hamilton–Jacobi equation we obtain for the null geodesics [64]

$$N\left(1-\frac{b}{r}\right)^{-1/2}\frac{dr}{d\lambda} = \left[(1-\omega\xi)^2 - \eta\frac{N^2}{r^2K^2}\right]^{1/2} = \sqrt{R(r)}, \tag{30}$$
$$r^2K^2\frac{d\theta}{d\lambda} = \left[\eta - \frac{\xi^2}{\sin^2\theta}\right]^{1/2}, \qquad \frac{dt}{d\lambda} = \frac{1-\omega\xi}{N^2},$$
$$\frac{d\varphi}{d\lambda} = \frac{\omega(1-\omega\xi)}{N^2} + \frac{\xi}{r^2\sin^2\theta K^2},$$

where ξ and η are the impact parameters determined by the integrals of motion. The radial equation can be transformed into an energy-like equation

$$\left(\frac{dr}{d\lambda}\right)^2 + V_{eff} = 1,$$

by introducing the effective potential

$$V_{eff} = 1 - \frac{1}{N^2}\left(1-\frac{b}{r}\right)R(r). \tag{31}$$

Analysing its behavior we obtain two families of unstable spherical orbits [48, 78]. The first family corresponds to maxima of the effective potential lying outside the wormhole throat and obeys the conditions

$$R(r) = 0, \qquad \frac{dR}{dr} = 0, \qquad \frac{d^2R}{dr^2} > 0. \tag{32}$$

Solving these equations for the impact parameters ξ and η we obtain the shadow boundary in the impact parameter space [64]

$$\eta = \frac{r^2K^2}{N^2}(1-\omega\xi)^2,$$
$$\xi = \frac{\Sigma}{\Sigma\omega - \omega'}, \qquad \Sigma = \frac{1}{2}\frac{d}{dr}\ln\left(\frac{N^2}{r^2K^2}\right). \tag{33}$$

The corresponding observable image at the asymptotic infinity is determined by the celestial coordinates

$$\alpha = -\frac{\xi}{\sin i}, \qquad \beta = \left(\eta - \frac{\xi^2}{\sin^2 i}\right)^{1/2}, \tag{34}$$

where i is the inclination angle of the observer.

In addition, there exists a second type of unstable spherical orbits located at the wormhole throat. They are described by the conditions

$$1 - \frac{b(r)}{r} = 0, \qquad R(r) = 0, \qquad \frac{dR}{dr} > 0. \tag{35}$$

Using the expressions for the celestial coordinates these conditions reduce to the implicit relation [78]

$$(\omega^2 r_0^2 K^2 \sin^2 i - N^2)\alpha^2 + 2\omega r_0^2 K^2 \sin i \alpha + r_0^2 K^2 - N^2 \beta^2 \mid_{r_0} = 0, \tag{36}$$

where r_0 is the location of the wormhole throat. The observable shadow boundary at large distances and inclination angle i is constructed as the collection of images determined by the relations given by Eqs. (33) and (36). Thus, both families of spherical orbits correspond only to certain portions of the boundary curve.

In the following we will illustrate explicitly the shadow boundary for some particular wormhole geometries. We consider the metric functions

$$b = r_0, \qquad K = 1, \qquad \omega = \frac{2J}{r^3}, \tag{37}$$

where r_0 is the mass of the wormhole, J is its angular momentum, and two different choices for the redshift function

$$N^{(1)} = \exp\left(-\frac{r_0}{r}\right), \qquad N^{(2)} = \exp\left(-\frac{r_0}{r} - \frac{r_0^2}{r^2}\right). \tag{38}$$

The shadow for the first redshift function $N^{(1)}$ is presented in Fig. 4 in comparison with the Kerr black hole shadow for the same spin parameter. The shadow boundary determined by the spherical orbits outside the throat corresponds to positive values of the celestial coordinate α (orange curve), while the blue curve for negative α represents the spherical orbits at the throat. Both portions of the boundary curve join smoothly at $\alpha = 0$.

In Fig. 5 we illustrate the shadow for the second type of redshift function $N^{(2)}$ using the same conventions. A major distinction from the previous case is that the two portions of the shadow boundary defined by the two families of spherical orbits do not merge smoothly but form a cusp at the intersection. Such features were also discovered in the shadow of hairy black holes [31,92] and in general they can

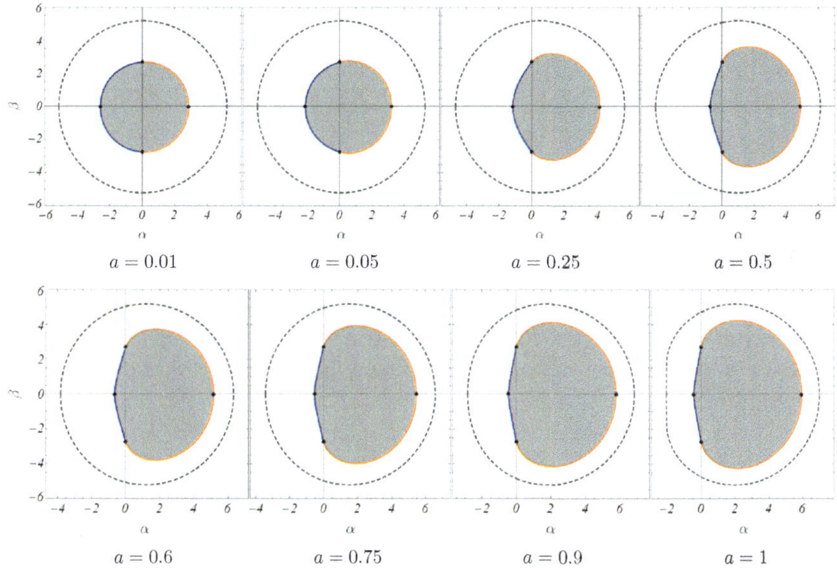

Fig. 4 Wormhole shadow for the redshift function $N^{(1)} = \exp\left(-\frac{r_0}{r}\right)$ and different values of the spin parameter. The shadow of the Kerr black hole is presented with a dashed line for comparison. As the spin increases the shadow of the wormholes grows and approaches the Kerr black hole. Adapted from [48]

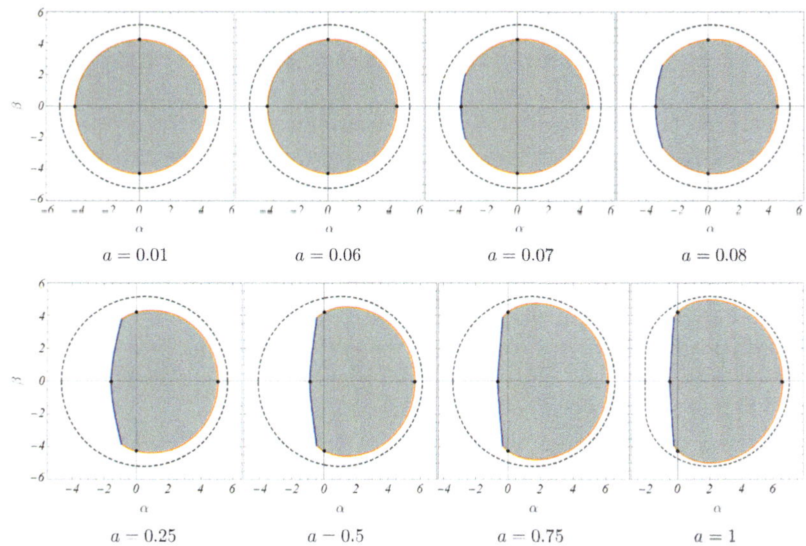

Fig. 5 Wormhole shadow for the redshift function $N^{(2)} = \exp\left(-\frac{r_0}{r} - \frac{r_0^2}{r^2}\right)$ and different values of the spin parameter. The shadow of the Kerr black hole is presented with a dashed line for comparison. This class of wormholes resembles the shadow of the Kerr black hole more closely. Adapted from [48]

serve as observational signatures for refuting the Kerr hypothesis if sufficiently high resolution is reached.

Comparing the wormhole shadow images with the Kerr black hole, we see that for the first redshift function the shadow is significantly smaller than for the Kerr black hole. For the second redshift function the differences decrease and when the spin parameter grows the shadows of the two types of compact objects approach a similar size. However, the deformation of the shadow boundary due to rotation is more pronounced for the wormholes and may be observationally significant.

Despite these deviations the wormholes generally produce qualitatively similar shadows as the Kerr black hole. Keeping in mind that we examined two particular examples we can conjecture that the quantitative discrepancies can be minimized at least for some spin parameters by fine-tuning the wormhole metric. Thus, some of the wormhole solutions will closely mimic the Kerr black hole. For example, in [35] we studied a static wormhole geometry leading to a shadow which cannot be distinguished from the Schwarzschild black hole at the current resolution of EHT.

Section summary: We demonstrated that horizonless spacetimes can cast a shadow by considering a particular class of traversable wormholes. The geodesic equations in these geometries are completely integrable and we calculated explicitly the boundary of the wormhole shadow by using the Hamilton–Jacobi approach. The wormhole shadow is qualitatively similar to the Kerr black hole and by fine-tuning the wormhole geometry we can obtain images which resemble so closely black holes that the two types of compact objects cannot be distinguished at the current resolution of EHT.

5 Image of the Thin Accretion Disk Around Compact Objects

In the previous sections we discussed the images of compact objects when they are illuminated by a uniform distribution of distant light sources. This physical setting is useful for obtaining the photon region and its projection on the observer's sky since it is independent of the light sources location as long as they reside outside the last spherical orbit. However, if we are interested in constructing a realistic image of the compact object's vicinity we should take into account the accretion disk surrounding them. The accretion disk is the major source of electromagnetic radiation in most of the astrophysical scenarios. It produces a bright image on the observer's sky with a characteristic shape and intensity which contain information about the compact object's spacetime and the properties of the accreting plasma.

In this section we will consider one of the simple analytical models of accretion representing geometrically thin and optically thick disk [66,68]. It assumes stationary and axisymmetric fluid distribution such that the disk's height is negligible compared to its radial dimension and the fluid motion is Keplerian. Then, the disk can be approximated by a collection of particles moving on stable circular geodesics in the equatorial plane and the radiation flux can be calculated by means of their kinematic quantities as

$$F(r) = -\frac{\dot{M}_0}{4\pi\sqrt{-g^{(3)}}}\frac{\Omega_{,r}}{(E-\Omega L)^2}\int_{r_{ISCO}}^{r}(E-\Omega L)L_{,r}dr, \tag{39}$$

where the integration starts from the innermost stable circular orbit (ISCO). In this expression we denote the accretion rate by $\dot{M}_0$, $g^{(3)}$ is the determinant of the induced metric in the equatorial plane, while E, L and Ω are the specific energy, angular momentum and angular velocity on the circular orbits. Considering a general stationary and axisymmetric metric we can express the kinematic quantities on the circular orbits in terms of the metric functions as

$$E = -\frac{g_{tt}+g_{t\phi}\Omega}{\sqrt{-g_{tt}-2g_{t\phi}\Omega-g_{\phi\phi}\Omega^2}}, \qquad L = \frac{g_{t\phi}+g_{\phi\phi}\Omega}{\sqrt{-g_{tt}-2g_{t\phi}\Omega-g_{\phi\phi}\Omega^2}},$$

$$\Omega = \frac{d\phi}{dt} = \frac{-g_{t\phi,r}+\sqrt{g_{t\phi,r}^2-g_{tt,r}g_{\phi\phi,r}}}{g_{\phi\phi,r}}. \tag{40}$$

where the terms $(...)_{,r}$ denote differentiation with respect to the radial coordinate.

The observable image of the thin disk is constructed by obtaining the projection of the photon trajectories originating from the disk on the observer's sky. This is achieved by solving numerically the null geodesic equations with the appropriate boundary conditions and associating with each geodesic the appropriate celestial angles α and β according to the procedure described in Sect. 3 (see Eqs. (20)–(23)). Practically, it is more convenient to scan all the celestial angles in the range $\alpha \in [0, \pi]$ and $\beta \in [-\frac{\pi}{2}, \frac{\pi}{2}]$ and integrate the corresponding photon trajectories backwards towards their emission point. Then, we select these trajectories which intersect the disk, i.e. pass through the equatorial plane at a radial distance within the range of stability of the timelike circular geodesics. The corresponding set of celestial angles α and β build up the image of the accretion disk on the observer's sky.

The observable radiation intensity at each point of the image is calculated by modifying the emitted flux given by Eq. (39) by the gravitational redshift z

$$F_{\text{obs}} = \frac{F}{(1+z)^4}, \tag{41}$$

where the gravitational redshift for a trajectory with an impact parameter $b = L/E$ can be expressed as

$$1+z = \frac{1+\Omega b}{\sqrt{-g_{tt}-2g_{t\phi}\Omega-\Omega^2 g_{\phi\phi}}}, \tag{42}$$

for a general stationary and axisymmetric spacetime.

5.1 Thin Accretion Disk Images in Static Spherically Symmetric Spacetime

In static spherically symmetric spacetime the construction of the accretion disk image simplifies due to the integrability of the null geodesic equations. Similar to the calculation of the shadow boundary it reduces to the integration of the radial geodesic equation or equivalently to the estimation of the deflection angle for the photon trajectories which originate from the disk and reach the observer [59,62]. Thus, for the general static spherically symmetric metric

$$ds^2 = -A(r)dt^2 + A^{-1}(r)dr^2 + r^2 D(r)(d\theta^2 + \sin^2\theta d\varphi^2), \tag{43}$$

we should calculate the integral

$$\varphi = \int_{r_{source}}^{r_{obs}} \frac{dr}{r\sqrt{D(r)\,A(r)}\sqrt{\left(\frac{r^2 D(r)}{b^2 A(r)} - 1\right)}}, \tag{44}$$

where $b = L/E$ is the impact parameter on the geodesic, r_{source} denotes the radial coordinate of the photon's emission point, while r_{obs} is the position of the observer. This reduces the construction of the lensed image of the thin accretion disk to a semi-analytic procedure and allows to make predictions about the morphology of the image based on the qualitative behavior of the deflection angle [50].

As we already discussed, if the spacetime possesses a photon sphere, the null geodesics can perform an arbitrary large number of turns around the compact object before reaching the observer resulting in a diverging behaviour of the deflection angle in the vicinity of the photon sphere. Therefore, it is convenient to classify the trajectories according to the number of half-loops k which they perform. Direct trajectories are characterized with $k = 0$ and deflection angles in the range $\phi \in [0, \pi)$, while trajectories of higher order k lead to deflection angles $\phi \in [k\pi, (k+1)\pi)$. The images on the observer's sky are also classified into direct ($k = 0$) and indirect ($k \geq 1$) according to the order of the null geodesics which give rise to them.

In Fig. 6 we illustrate the image of the thin accretion disk around the Schwarzschild black hole for a distant observer located at $r = 5000M$ and inclination angle $i = 80°$. On the left-hand side we present the apparent shape of the disk by constructing its projection on the observer's sky without considering its radiation. On this diagram we can differentiate between the photon trajectories of different order k which build up the image. Direct trajectories with $k = 0$ give rise to the main hat-like image in orange as the images of some particular circular geodesics with radius in the range $r \in [r_{\text{ISCO}}, 30M]$ are highlighted with solid lines. The image of order $k = 1$ is depicted in blue, while the higher order images with $k \geq 2$ correspond to the central black circle. These images are located in a very close neighbourhood of the image of the photon sphere approaching it asymptotically when $k \to \infty$.

In the right panel of Fig. 6 we present the observable radiation associated with each point of the image. We depict the observable flux calculated by Eq. (41) normalized

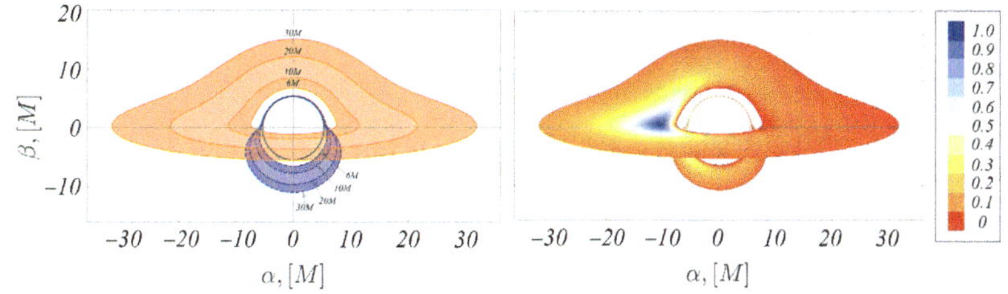

Fig. 6 Image of the thin accretion disk around the Schwarzschild black hole as seen by a distant observer at the inclination angle $i = 80°$. The left panel illustrates the disk morphology by highlighting the images of the particular circular orbits. The right panel represents the observable radiation from the disk. The most intensive flux is presented in blue, while the dimmest regions are in red. Adapted from [49]

by the maximal value F^{max}_{obs} reached in the image. We further map continuously this quantity $F_{obs}/F^{max}_{obs} \in [0, 1]$ to the color spectrum from red to blue, as the highest values correspond to dark blue.

Section summary: We discussed the Novikov-Thorne model of thin accretion disk. We demonstrated how we can obtain its observable image giving as a particular example the Schwarzschild black hole.

6 Images of Thin Accretion Disks Around Naked Singularities

In this section we will discuss three types of static spherically symmetric naked singularities which possess qualitatively different behavior of the null geodesics. The first type is characterized by a single maximum of the effective potential for the null geodesics which corresponds to a photon sphere similar to the Schwarzschild black hole. The second spacetime contains no photon sphere but the gravitational field becomes repulsive in the close vicinity of the singularity which results into reflective behavior of the effective potential in this region. The third example combines the presence of a photon sphere with a reflective potential for the null geodesics near the singularity. These features lead to a different appearance of the thin disk around the corresponding compact objects. While the first type of naked singularities mimic closely black holes, the other two cases show distinctive phenomenology.

6.1 Weakly Naked Janis–Newman–Winicour Singularities

Spherically symmetric naked singularities can be classified into weakly and strongly naked according to their lensing properties. We can consider as an example the Janis–Newman–Winicour naked singularity which arises as a solution to the Einstein-scalar field equations [39,53,90,95]. In this case spherically symmetric black holes do not exist and it was proven that the Janis–Newman–Winicour naked singularity represents the unique static spherically symmetric solution. It is given by the metric

$$ds^2 = -\left(1-\frac{2M}{\gamma r}\right)^{\gamma} dt^2 + \left(1-\frac{2M}{\gamma r}\right)^{-\gamma} dr^2 + \left(1-\frac{2M}{\gamma r}\right)^{1-\gamma} r^2\left(d\theta^2+\sin^2\theta d\phi^2\right), \quad (45)$$

and the scalar field takes the form

$$\varphi = \frac{q\gamma}{2M}\ln\left(1-\frac{2M}{\gamma r}\right), \quad (46)$$

where the parameter γ is determined by the conserved charges of the solution, i.e. its ADM mass M and scalar charge q. It takes the range $\gamma \in [0, 1]$ as $\gamma = 1$ corresponds to vanishing scalar charge and the solution reduces to the Schwarzschild black hole.

The Janis–Newman–Winicour class of solutions describes both weakly and strongly naked singularities depending on the value of the scalar field parameter γ. The weakly naked regime is realized in the range $\gamma \in (0.5, 1)$. Then, the spacetime contains a photon sphere with radial position determined by the expression

$$r_{\mathrm{ph}} = (2\gamma + 1)M/\gamma, \quad (47)$$

while the innermost stable circular orbit for the particle motion corresponds to the solutions of the equation [49]

$$r^2\gamma^2 - 2r\gamma(3\gamma + 1) + 2(2\gamma^2 + 3\gamma + 1) = 0. \quad (48)$$

The effective potential for the null geodesics possesses only a single maximum determining the photon sphere and resembles qualitatively that for the Schwarzschild black hole. Under these conditions the shadow and the accretion disk images are expected to mimic the Schwarzschild black hole with only quantitative deviations. The shadow was calculated in [80] and it was demonstrated in [55] that its apparent size can become by approximately 20% smaller than for the Schwarzschild black hole. Still, for most values of the scalar field parameter, i.e. in the range $\gamma \in [0.53, 1)$, its dimensions are compatible with the EHT observations of the supermassive compact object at the center of the galaxy M87.

The image of the thin accretion disk was obtained in [49] leading to similar conclusions. We observe a qualitatively similar apparent shape as for the Schwarzschild black hole with quantitative deviations in the observable size of the circular orbits since the naked singularity causes a stronger focussing effect (see Fig. 7). The deviation depends on the radius of the circular orbit as it is stronger in the inner part of the disk reaching $\sim$18% for the ISCO for $\gamma = 0.51$. It decreases in the regions with

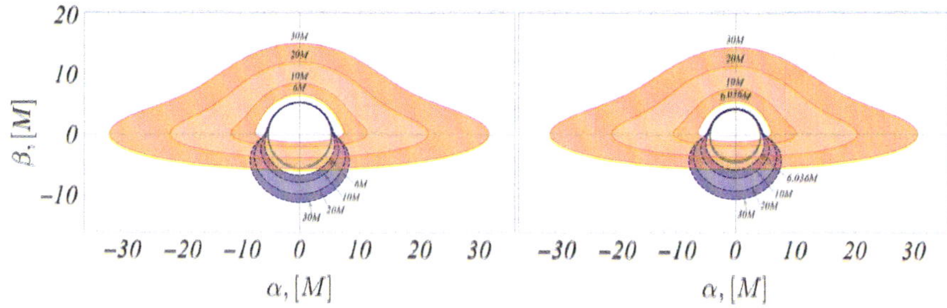

Fig. 7 Image of the thin accretion disk around the weakly naked Janis–Newman–Winicour singularity (right) compared to the Schwarzschild black hole (left) for the scalar field parameter $\gamma = 0.51$. The disk image for the naked singularity closely resembles the black hole. Direct images are presented in orange, indirect images of order $k = 1$ are denoted in blue, while the higher order images are in black. The observer is located at $r = 5000M$, while the inclination angle is $i = 80°$. Adapted from [49]

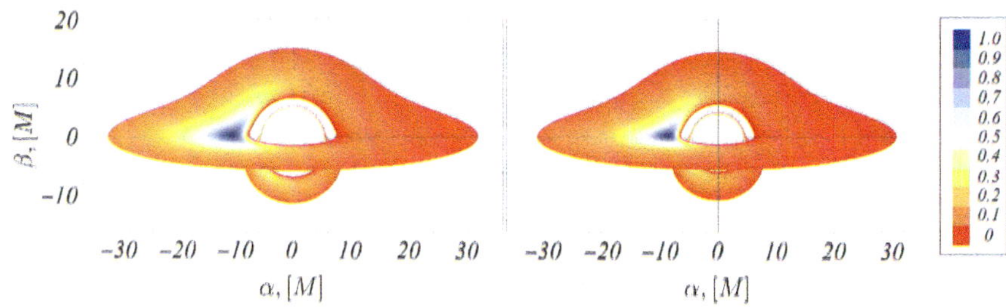

Fig. 8 Observable radiation from the thin accretion disk around the weakly naked Janis–Newman–Winicour singularity for the scalar field parameter $\gamma = 0.51$ (right) compared to the Schwarzschild black hole (left). The flux is normalized to its maximum value for each solution. The observer is located at $r = 5000M$, while the inclination angle is $i = 80°$. Adapted from [49]

weaker gravitational field as for more distant orbits located in outskirts of the disk at $r = 30M$ it becomes $\sim 3\%$ for the same value of the scalar field parameter.[3]

The observable radiation from the disk also resembles qualitatively the distribution for the Schwarzschild black hole with a similar position of its maximum (see Fig. 8). Quantitatively the maximum value of the observable flux for the naked singularity is higher than for the Schwarzschild black hole reaching approximately two times difference for the scalar field parameter $\gamma = 0.51$. Since we plot the normalized flux F_{obs}/F_{obs}^{max} this leads to a dimmer appearance of the outskirts of the naked singularity thin disk.

We discussed the shadow and accretion disk images for a particular type of naked singularities represented by the Janis–Newman–Winicour solution. However, qualitatively similar phenomenology is expected for any solution with the same behaviour of the effective potential for the null geodesics, i.e. type of extrema and asymp-

[3] We adopt as a measure of the deviation of the apparent size of the circular orbits the horizontal dimension of their direct images at the inclination angle $i = 80°$ (see [49]).

totics, and the same structure of the circular geodesics for the massive particles. This includes a broad class of compact objects such as black holes in the modified theories of gravity [51], wormholes [70], and regular back holes [15].

6.2 Strongly Naked Janis–Newman–Winicour Singularities

In the range of the scalar field parameter $\gamma \in (0, 0.5)$ the Janis–Newman–Winicour singularity classifies as strongly naked. The spacetime contains no photon sphere and the effective potential for the null geodesics possesses no further extrema (see Fig. 9). Instead, it diverges in the vicinity of the singularity preventing the null geodesics from reaching it. In this way, if the naked singularity is illuminated by a uniform distribution of light sources all the photon trajectories will scatter away to infinity and no shadow will be observed.

Another important feature of this geometry is that the stable circular geodesics for the massive particles are located into two disconnected regions with a gap in between. Thus, we have an inner disk delimited by two marginally stable orbits and an outer disk spanning from another marginally stable orbit to infinity. The emission from both regions gives contribution to the image of the thin accretion disk. The location of the marginally stable orbits is determined by the roots of Eq. (48). For $\gamma \leq 1/\sqrt{5}$ it has no real solutions and the stable circular orbits extend up to the singularity.

In Fig. 10 we present the optical appearance of the thin disk for $\gamma = 0.48$ without taking into account its radiation. For clarity we provide separate images for the outer and the inner disk as the complete observable image is constructed as their superposition. As in the previous sections, we denote the parts of the image created by photon trajectories of different order k by distinct colors. The direct image of the outer disk only partially resembles the case of the Schwarzschild black hole. On the one hand, it produces the characteristic hat-like image similar to the Schwarzschild black hole. However, it contains also a second disconnected part which represents a

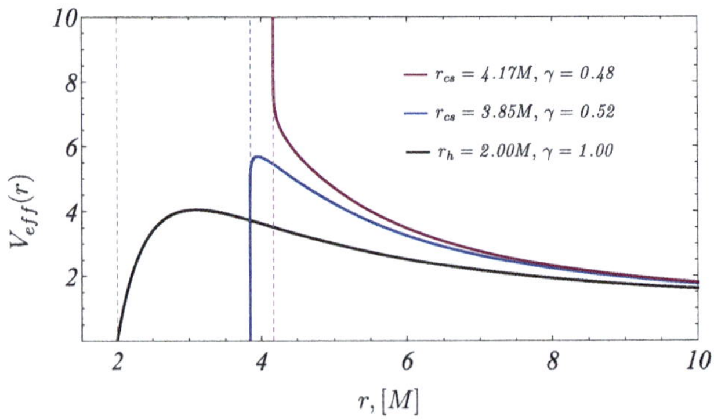

Fig. 9 Behavior of the effective potential for the null geodesics for the strongly naked Janis–Newman–Winicour singularity. The effective potential for the Schwarzschild black hole and the weakly naked Janis–Newman–Winicour singularity are presented for comparison. We specify the values of the scalar field parameter γ for each solution and the location of the singularity r_{cs} or the event horizon r_h in the case of black holes. Adapted from [50]

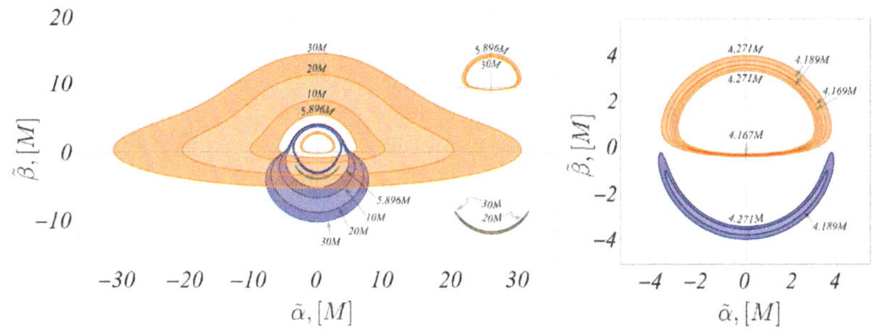

Fig. 10 Image of the outer disk (left) and inner disk (right) for the strongly naked Janis–Newman–Winicour singularity with scalar field parameter $\gamma = 0.48$. Direct images are presented in orange, indirect images of order $k = 1$ are denoted in blue, while the higher order images are in black. The observer is located at $r = 5000M$, and the inclination angle is $i = 80°$. Adapted from [50]

bright ring at the center of the disk. This phenomenon of creating double images of the disk distinguishes observationally the strongly naked singularities. In addition, the inner disk gives rise to another central bright ring. This image is further absent for the Schwarzschild black hole since it does not possess a similar disconnected region of stable circular orbits.

The properties of the accretion disk image can be predicted by examining the behavior of the deflection angle on the photon trajectories [50]. In Fig. 11 we consider the photon trajectories emitted by the ISCO for the outer disk for naked singularities with $\gamma = 0.48$. We assume that the observer is located at the radial coordinate $r = 5000M$ and the inclination angle is $i = 80°$. Then, we plot the deflection angle of the trajectories which reach the observer's position as a function of their impact parameter b as the solutions for the different order k are outlined in distinct color strips. The intersection of the curve $\phi(b)$ with each color strip corresponds to an image of the ISCO of the same order. If there exist two disconnected intersections of the deflection angle with a certain color strip they will produce a couple of disconnected images of the corresponding order on the observer's sky.

Since the strongly naked singularities do not possess a photon sphere, the deflection angle cannot grow arbitrary large. Instead, it is a bounded function in contrast to the Schwarzschild black hole. We see that it reaches a maximum which corresponds to photon trajectories of order $k = 1$ for the ISCO. As a result of this behavior we obtain two branches of solutions for the observable direct trajectories with $k = 0$ which correspond to two disconnected intervals for the impact parameter b. Since the impact parameter is directly related to the celestial coordinates, these solutions give rise to two disconnected images in the observer's sky. The solution for larger b results in the hat-like disk image which is present also in the Schwarzschild case, while the smaller impact parameters produce the central ring.

Although in Fig. 11 we considered a particular example, the described behavior of the deflection angle is representative for any circular orbit from the accretion disk. The deflection angle always possesses a maximum which can be located at most in the range of photon trajectories of order $k = 2$. Thus, we obtain an upper limit for the number of half-turns which the null geodesics originating from the disk can perform

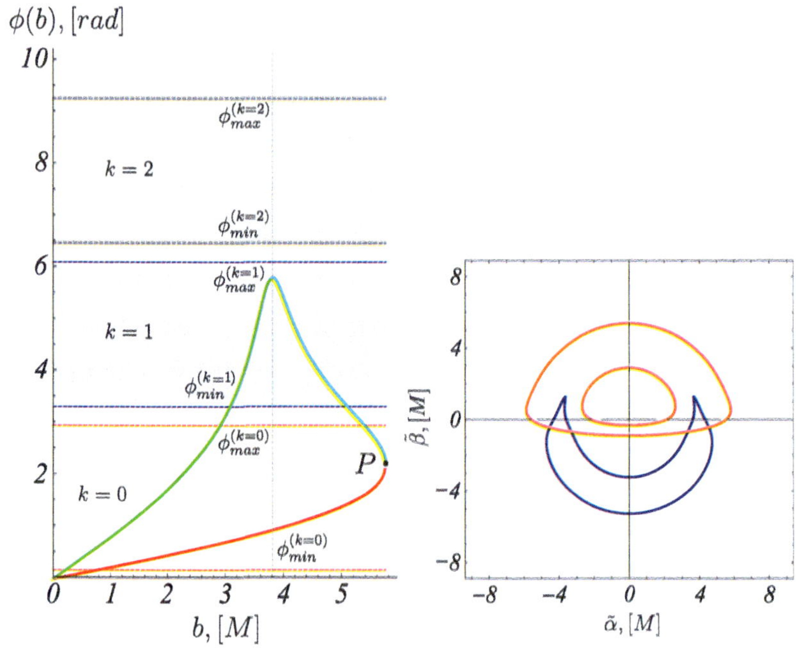

Fig. 11 Behavior of the deflection angle for photon trajectories originating from the ISCO and reaching an asymptotic observer at the inclination angle $i = 80°$ (left panel). The color strips correspond to images of different order k. Disconnected intersections of the deflection angle with a certain color strip give rise to a couple of disconnected images of the corresponding order. The ISCO possesses a double direct image with $k = 0$ and an indirect image of order $k = 1$ presented in the right panel in orange and blue respectively. We consider the scalar field parameter $\gamma = 0.48$. Adapted from [50]

before reaching infinity. In this case, in addition to the double direct image of the disk we observe also double image of order $k = 1$. It is produced by the two disconnected solutions for the trajectories of order $k = 1$ which now become possible.

In Fig. 12 we present the observable radiation from the thin disk around the strongly naked singularity. The central rings are observationally significant within this model of accretion since the maximum of the radiation flux is reached in the image of the inner disk. On the other hand, the central rings which result from the double image of the outer disk emit with around 30% of the maximum radiation.

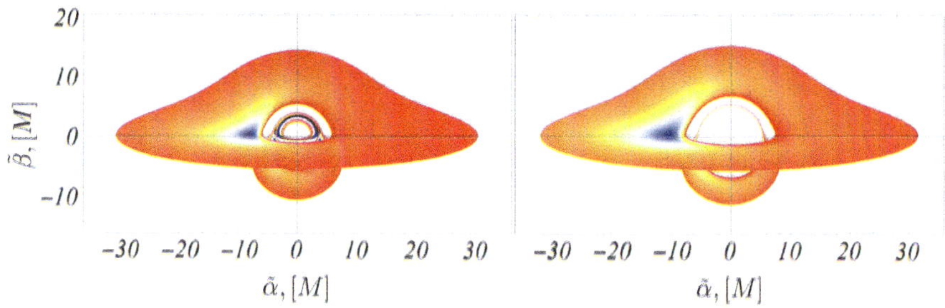

Fig. 12 Observable radiation from the thin accretion disk around the strongly naked Janis–Newman–Winicour singularity with scalar field parameter $\gamma = 0.48$ (left panel). The Schwarzschild black hole is presented for comparison in the right panel. The observable flux is normalized to its maximum value for each solution. The observer is located at $r = 5000M$, while the inclination angle is $i = 80°$. Adapted from [50]

6.3 Einstein–Gauss–Bonnet Naked Singularities

The four-dimensional Einstein–Gauss–Bonnet gravity admits a static spherically symmetric solution in the form [18,25,42,85]

$$ds^2 = -f(r)dt^2 + \frac{1}{f(r)}dr^2 + r^2(d\theta^2 + \sin^2\theta\phi^2),$$

$$f(r) = 1 + \frac{r^2}{2\hat{\gamma}M^2}\left(1 - \sqrt{1 + \frac{8\hat{\gamma}M^3}{r^3}}\right),$$

where M is the ADM mass of the solution, while $\hat{\gamma}$ is the Gauss–Bonnet coupling constant. In the range $\hat{\gamma} \in [0, 1]$ the solution describes black holes while for $\gamma > 1$ it represents naked singularities. The naked singularities with coupling constant $1 < \hat{\gamma} < 3\sqrt{3}/4$ lead to particularly interesting lensing properties [51]. The effective potential for the null geodesics possesses a stable and unstable photon ring and diverges in the vicinity of the singularity (see Fig. 13). Thus, it reflects infalling photon trajectories preventing them from reaching the singularity. Such spacetimes cannot cast a shadow although they possess a photon sphere. The photon trajectories which pass through the photon sphere will still be reflected back to infinity due to the potential barrier in the vicinity of the singularity leaving no dark directions on the observer's sky.

The stable circular timelike orbits are located in two disconnected regions consisting of an inner annulus and an outer disk. The limits of these regions are determined by the marginally stable orbits which are solutions to the equation

$$r^3 - 9M^2r + 8M^3\hat{\gamma} = 0. \tag{49}$$

Thus, the thin accretion disk consists of two disconnected portions, i.e. inner and outer disk, similar to the strongly naked Janis–Newman–Winicour singularity.

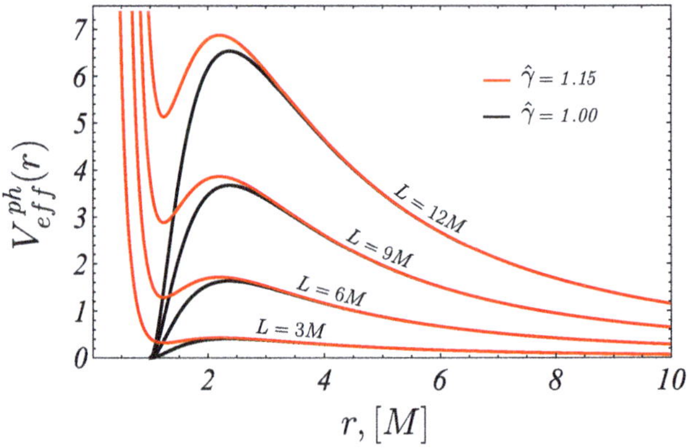

Fig. 13 Behavior of the effective potential for the null geodesics for the Gauss–Bonnet naked singularities with coupling constant in the range $1 < \hat{\gamma} < 3\sqrt{3}/4$. We plot the effective potential for the representative value $\hat{\gamma} = 1.15$ for several values of the specific angular momentum (red line). The effective potential for the Schwarzschild black hole is presented for comparison (black line). Adapted from [51]

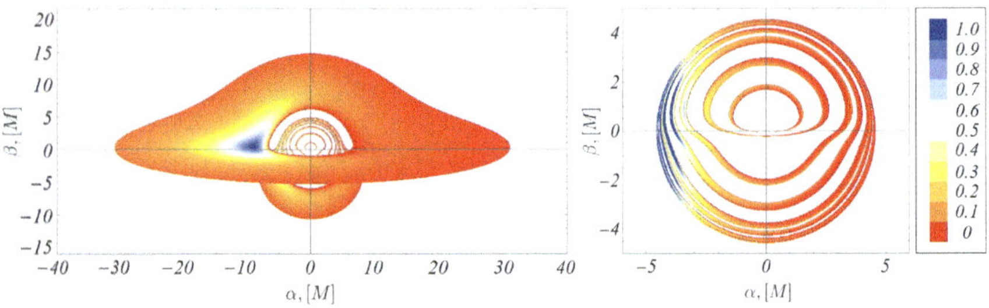

Fig. 14 Image of the outer disk (left) and inner disk (right) for the Gauss–Bonnet naked singularity with coupling constant $\hat{\gamma} = 1.15$. The observable flux is normalized to its maximum value for each image. The observer is located at $r = 5000M$, while the inclination angle is $i = 80°$. Adapted from [51]

These properties determine the structure of the disk images presented in Fig. 14. The images of the inner and the outer disks are constructed separately as the observable image represents their superposition. The outer disk possesses a double image of any order $k \geq 1$. One of the images produces an observable structure similar to the thin disk for the Schwarzschild black hole. The second image represents a nested multi-ring structure at small inclination angles which is absent for black holes. The inner disk also leads to double images of any order, however both of them correspond to central bright rings. The central rings represent the brightest part of the image radiating with intensity $\sim 10^3$ higher than the hat-like disk image.

Examining the behavior of the deflection angle we can predict the formation of the qualitatively different morphology of the disk images [51]. In Fig. 15 we present the deflection angle for photon trajectories originating at the ISCO of the outer disk and reaching an observer located at $r = 5000M$ and inclination angle $i = 80°$. We see that the deflection angle diverges at the location of the photon sphere. However,

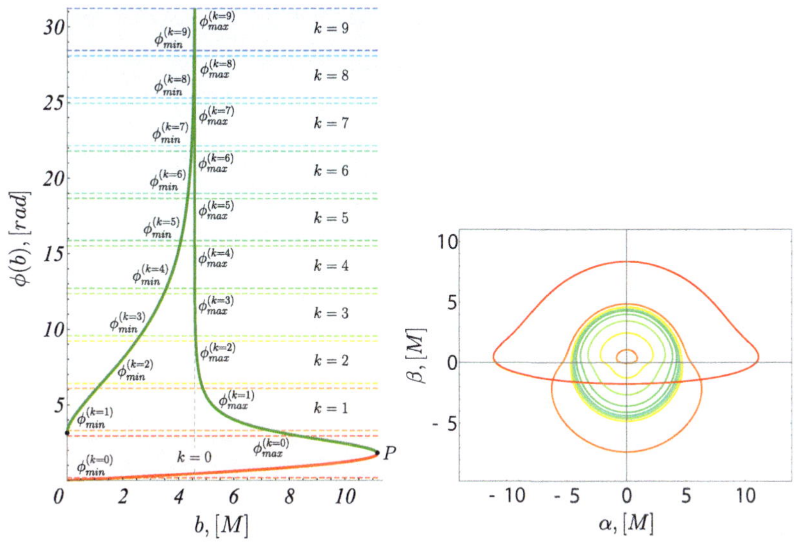

Fig. 15 Behavior of the deflection angle for photon trajectories originating from the ISCO and reaching an asymptotic observer at inclination angle $i = 80°$ (left panel). The color strips correspond to images of different order k. Disconnected intersections of the deflection angle with a certain color strip give rise to a couple of disconnected images of the corresponding order. In the right panel we present the corresponding images of the ISCO up to order $k = 6$. We consider the coupling constant $\hat{\gamma} = 1.15$. Adapted from [51]

in contrast to the Schwarzschild black hole, there exist two families of null geodesics scattering away to infinity which approach the photon sphere as a limit surface with higher and lower values of the impact parameters, respectively. The first family gets reflected from the maximum of the effective potential and gives rise to a similar image as for the Schwarzschild black hole. The second family gets reflected from the potential barrier in the vicinity of the singularity and produces the multi ring structure at small celestial angles. As a result the ISCO possesses double images of any order $k \geq 1$. The images of lower order are distinguishable but when k grows they converge towards the images of the photon sphere. This behavior is representative for any orbit from the outer disk, thus demonstrating how the disk image is created.

The multi-ring structure which we described in the thin disk images for the strongly naked JNW singularities and the Gauss–Bonnet singularities appears in much more general cases than these particular spacetimes. As we demonstrated it is governed by the behavior of the deflection angle for the scattering photon trajectories, which on the other hand depends on the form of the effective potential for the null geodesics. This implies that any spacetime with a qualitatively similar behavior of the effective potential will give rise to a qualitatively similar morphology of the disk images encompassing a large class of compact objects of diverse physical nature. According to a general theorem [29] any regular compact object with trivial topology which possesses a photon sphere should further possess a stable light ring.[4] The thin disk images of these spacetimes would be characterised by the same morphology as for the Gauss–Bonnet naked singularity although they can describe very different self-gravitating objects like regular black holes or exotic stars. Some examples of such compact objects were presented in [19,36,37,47].

Section summary: We discussed the thin accretion disk images for three types of spherically symmetric naked singularities. The first spacetime produces a very similar image to the Schwarzschild black hole, while the other two give rise to a characteristic multi-ring structure at the center of the image. The key concept which determines the morphology of the image is the behavior of the deflection angle on the scattering photon trajectories which reach the observer. Since the deflection angle depends directly on the properties of the effective potential for the null geodesics, compact objects with similar effective potentials will lead to qualitatively similar accretion disk images despite their different physical nature.

Further Reading

- **On the calculation of the black hole shadow**:
  The analytical calculation of the shadow boundary for the Kerr black hole is described in detail in the books [24,40], as well as in the recent reviews [32,74]. The procedure is generalized in the presence of a cosmological constant [73], NUT charge [5,45], and plasma environment [71,72]. Shadows of black holes interacting with an external gravitational source are obtained numerically in [3, 4,46,65,81].
- **Black hole shadow in the modified theories of gravity**:
  These exist a large number of works studying the black hole shadow in various modifications of general relativity. Analytical results include [12–14,56,61,69], while [26,27,30,87] investigate the problem numerically.
- **On the calculation of the thin disk image**:
  The construction of observable image of the thin accretion disk for the Schwarzschild black hole is developed in the classical works [33,34,41,59,62, 63]. The procedure can be generalized straightforwardly for a general static spherically symmetric compact objects as demonstrated in [49–51].
- **Accretion disk images of exotic compact objects**:
  Exotic compact objects such as wormholes, naked singularities, boson stars and regular black holes are considered as viable alternatives of the Kerr black hole and simulations of their images are used in the interpretation of the EHT results [8,11,55]. Accretion disk images of exotic compact objects considering different models of accretion include [57,67,76,79,88,89,101].

[4] This light ring structure could be extended to a further number of light rings, but they should always come in pairs of stable and unstable ones.

Acknowledgements I would like to thank Galin Gyulchev and Stoytcho Yazadjiev for the fruitful and inspiring collaboration which lead to the original results reviewed in this chapter. The financial support by the Bulgarian NSF Grant KP-06-H38/2 is gratefully acknowledged.

References

1. B. Abbott et al. (LIGO Scientific Collaboration and Virgo Collaboration), Phys. Rev. Lett. **116**, 061102 (2016); Phys. Rev. Lett. **116**, 241103 (2016); Phys. Rev. Lett. **118**, 221101 (2017); Phys. Rev. Lett. **119**, 141101 (2017); Phys. Rev. Lett. **119**, 161101 (2017); Astrophys. J. Lett. **851**, L35 (2017); Phys. Rev. Lett. **119**, 161101 (2017); Astrophys. J. Lett. **851**, L35 (2017); Phys. Rev. X **9**, 031040 (2019)
2. B. Abbott et al. (LIGO Scientific Collaboration and Virgo Collaboration), Astrophys. J. Lett. **892**, L3 (2020); Phys. Rev. D **102**, 043015 (2020); Astrophys. J. Lett. **896**, L44 (2020); Phys. Rev. Lett. **125**, 101102 (2020); Phys. Rev. X **11**, 021053 (2021); Astrophys. J. Lett. **915**, L5 (2021)
3. S. Abdolrahimi, R. Mann, C. Tzounis, Phys. Rev. D **91**, 084052 (2015)
4. S. Abdolrahimi, R. Mann, C. Tzounis, Phys. Rev. D **92**, 124011 (2015)
5. A. Abdujabbarov, F. Atamurotov, Y. Kucukakça, B. Ahmedov, U. Camci, Astrophys. Space Sci. **344**, 429 (2012)
6. A.A. Abdujabbarov, L. Rezzolla, B.J. Ahmedov, Mon. Not. Roy. Astron. Soc. **454**, 2423 (2015)
7. K. Akiyama et al., Event Horizon Telescope. Astrophys. J. Lett. **875**, L1–L6 (2019)
8. K. Akiyama et al., Event Horizon Telescope. Astrophys. J. Lett. **875**, L5 (2019)
9. K. Akiyama et al., Event Horizon Telescope. Astrophys. J. Lett. **910**, L12–L13 (2021)
10. K. Akiyama et al., Event Horizon Telescope. Astrophys. J. Lett. **930**, L12–L16 (2022)
11. K. Akiyama et al., Event Horizon Telescope. Astrophys. J. Lett. **930**, L17 (2022)
12. L. Amarilla, E.F. Eiroa, G. Giribet, Phys. Rev. D **81**, 124045 (2010)
13. L. Amarilla, E.F. Eiroa, Phys. Rev. D **85**, 064019 (2012)
14. L. Amarilla, E.F. Eiroa, Phys. Rev. D **87**, 044057 (2013)
15. P. Bambhaniya, K. Saurabh, K. Jusufi, P. Joshi, Phys. Rev. D **105**, 023021 (2022)
16. J.M. Bardeen, in *Black Holes*, ed. by C. DeWitt, B. DeWitt (Gordon and Breach, New York, 1973)
17. A. Broderick, T. Johannsen, A. Loeb, D. Psaltis, Astrophys. J. **784**, 7 (2014)
18. R. Cai, L. Cao, N. Ohta, JHEP **1004**, 082 (2010)
19. R. Carballo-Rubio, V. Cardoso, Z. Younsi, Phys. Rev. D **106**, 084038 (2022)
20. B. Carter, Phys. Rev. **174**, 1559 (1968)
21. C. Cederbaum, Contemp. Math. **667**, 86 (2015)
22. C. Cederbaum, G.J. Galloway, Class. Quantum Grav. **33**, 075006 (2016)
23. G. Munoz, Orbits of massless particles in the Schwarzschild metric: Exact solutions. Am. J. Phys. **82**, 564 (2014)
24. S. Chandrasekhar, *The Mathematical Theory of Black Holes* (Oxford University Press, Oxford, 1983)
25. G. Cognola, R. Myrzakulov, L. Sebastiani, S. Zerbini, Phys. Rev. D **88**, 024006 (2013)
26. P. Cunha, C. Herdeiro, E. Radu, H. Runarsson, Phys. Rev. Lett. **115**, 211102 (2015)
27. P. Cunha, J. Grover, C. Herdeiro, E. Radu, H. Runarsson, A. Wittig, Phys. Rev. D **94**, 104023 (2016)
28. P. Cunha, C. Herdeiro, E. Radu, H. Runarsson, Int. J. Mod. Phys. D **25**, 1641021 (2016)
29. P. Cunha, E. Berti, C. Herdeiro, Phys. Rev. Lett. **119**, 251102 (2017)
30. P. Cunha, C. Herdeiro, B. Kleihaus, J. Kunz, E. Radu, Phys. Lett. B **768**, 373 (2017)
31. P. Cunha, C. Herdeiro, E. Radu, Phys. Rev. D **96**, 024039 (2017)
32. P. Cunha, C. Herdeiro, Gen. Rel. Grav. **50**, 42 (2018)
33. C. Cunningham, J. Bardeen, ApJ **173**, L137 (1972)
34. C. Cunningham, Phys. Rev. D **12**, 323 (1975)

35. V. Deliyski, G. Gyulchev, P. Nedkova, S. Yazadjiev, Phys. Rev. D **106**, 104024 (2022)
36. A. Eichhorn, R. Gold, A. Held, arxiv:2205.14883 [astro-ph.HE]
37. A. Eichhorn, A. Held, JCAP **01**, 032 (2023)
38. H. Falcke, F. Melia, E. Agol, Astrophys. J. Lett. **528**, L13 (2000)
39. I. Fisher, Zh. Exp, Teor. Fiz. **18**, 636 (1948)
40. V. Frolov, A. Zelnikov, *Introduction to Black Hole Physics* (Oxford University Press, New York, 2011)
41. J. Fukue, T. Yokoyama, Publ. Astron. Soc. Japan **40**, 15 (1988)
42. D. Glavan, C. Lin, Phys. Rev. Lett. **124**, 081301 (2020)
43. S. Gralla, D. Holz, R. Wald, Black hole shadows, photon rings, and lensing rings. Phys. Rev. D **100**, 024018 (2019)
44. S. Gralla, A. Lupsasca, Phys. Rev. D **102**, 124003 (2020)
45. A. Grenzebach, V. Perlick, C. Lämmerzahl, Phys. Rev. D **89**, 124004 (2014)
46. J. Grover, J. Kunz, P. Nedkova, A. Wittig, S. Yazadjiev, Phys. Rev. D **97**, 084024 (2018)
47. M. Guerrero, G.J. Olmo, D. Rubiera-Garcia, D. Sáez-Chillón Gómez, Phys. Rev. D **106**, 044070 (2022)
48. G. Gyulchev, P. Nedkova, V. Tinchev, S. Yazadjiev, EPJC **78**, 544 (2018)
49. G. Gyulchev, P. Nedkova, T. Vetsov, S. Yazadjiev, Phys. Rev. D **100**, 024055 (2019)
50. G. Gyulchev, J. Kunz, P. Nedkova, T. Vetsov, S. Yazadjiev, EPJC **80**, 1017 (2020)
51. G. Gyulchev, P. Nedkova, T. Vetsov, S. Yazadjiev, Eur. Phys. J. C **81**, 885 (2021)
52. K. Hioki, K. Maeda, Phys. Rev. D **80**, 024042 (2009)
53. A. Janis, E. Newman, J. Winicour, Phys. Rev. Lett. **20**, 878 (1968)
54. T. Johannsen, D. Psaltis, Astrophys. J. **718**, 446 (2010)
55. P. Kocherlakota et al., EHT Collaboration. Phys. Rev. D **103**, 104047 (2021)
56. M. Kuniyasu, K. Nanri, N. Sakai, T. Ohgami, R. Fukushige, S. Koumura, Phys. Rev. D **97**, 104063 (2018)
57. F. Lamy, E. Gourgoulhon, T. Paumard, F. Vincent, Class. Quant. Grav. **35**, 115009 (2018)
58. Z. Li, C. Bambi, J. Cosm. Astropart. Phys. **01**, 041 (2014)
59. J.-P. Luminet, Astron. Astrophys. **75**, 228 (1979)
60. Y. Mizuno, Z. Younsi, C. Fromm et al., NatAs **2**, 585 (2018)
61. J.W. Moffat, Eur. Phys. J. C **75**, 130 (2015)
62. T. Müller, Gen. Rel. Grav. **41**, 541 (2009)
63. T. Müller, J. Frauendiener, Eur. J. Phys. **33**, 955 (2012)
64. P. Nedkova, V. Tinchev, S. Yazadjiev, Phys. Rev. D **88**, 124019 (2013)
65. D. Nitta, T. Chiba, N. Sugiyama, Phys. Rev. D **84**, 063008 (2011)
66. I. Novikov, K. Thorne, in *Black Holes*, ed. by C. DeWitt, B. DeWitt (Gordon and Breach, New York, 1973)
67. H. Olivares, Z. Younsi, C. Fromm, M. De Laurentis, O. Porth, Y. Mizuno, H. Falcke, M. Kramer, L. Rezzolla, MNRAS **497**, 521 (2020)
68. D. Page, K. Thorne, Astrophys. J. **191**, 499 (1974)
69. U. Papnoi, F. Atamurotov, S. Ghosh, B. Ahmedov, Phys. Rev. D **90**, 024073 (2014)
70. S. Paul, R. Shaikh, P. Banerjee, T. Sarkar, JCAP **03**, 055 (2020)
71. V. Perlick, O. Tsupko, G. Bisnovatyi-Kogan, Phys. Rev. D **92**, 104031 (2015)
72. V. Perlick, O. Tsupko, Phys. Rev. D **95**, 104003 (2017)
73. V. Perlick, O. Tsupko, G. Bisnovatyi-Kogan, Phys. Rev. D **97**, 104062 (2018)
74. V. Perlick, OYu. Tsupko, Phys. Rept. **947**, 1 (2022)
75. D. Psaltis, Gen. Rel. Grav. **51**, 137 (2019)
76. J. Rosa, D. Rubiera-Garcia, Phys. Rev. D **106**, 084004 (2022)
77. N. Sakai, H. Saida, T. Tamaki, Phys. Rev. D **90**, 104013 (2014)
78. R. Shaikh, Phys. Rev. D **98**, 024044 (2018)
79. R. Shaikh, P. Joshi, JCAP **1910**, 064 (2019)
80. R. Shaikh, P. Kocherlakota, R. Narayan, P. Joshi, MNRAS **482**, 52 (2019)
81. J. Shipley, S. Dolan, Class. Quant. Grav. **33**, 175001 (2016)
82. J. Synge, Mon. Not. Roy. Astron. Soc. **131**, 463 (1966)

83. E. Teo, Phys. Rev. D **58**, 024014 (1998)
84. E. Teo, Gen. Rel. Grav. **35**, 1909 (2003)
85. Y. Tomozawa, arxiv:1107.1424 [gr-qc]
86. N. Tsukamoto, Z. Li, C. Bambi, J. Cosm. Astropart. Phys. **06**, 043 (2014)
87. F. Vincent, E. Gourgoulhon, C. Herdeiro, E. Radu, Phys. Rev. D **94**, 084045 (2016)
88. F. Vincent, Z. Meliani, P. Grandclément, E. Gourgoulhon, O. Straub, Class. Quant. Grav. **33**, 105015 (2016)
89. F. Vincent, M. Wielgus, M. Abramowicz, E. Gourgoulhon, J.-P. Lasota, T. Paumard, G. Perrin, A&A **646**, A37 (2021)
90. K. Virbhadra, Int. J. Mod. Phys. A **12**, 4831 (1997)
91. M. Walker, R. Penrose, Commun. Math. Phys. **18**, 265 (1970)
92. M. Wang, S. Chen, J. Jing, JCAP **1710**, 051 (2017)
93. C. Will, *Theory and Experiment in Gravitational Physics* (Cambridge University Press, Cambridge; New York, 1993)
94. C. Will, Living Rev. Relativ. **17**, 4 (2014)
95. M. Wyman, Phys. Rev. D **24**, 839 (1981)
96. S. Yazadjiev, Phys. Rev. D **91**, 123013 (2015)
97. S. Yazadjiev, B. Lazov, Class. Quantum Grav. **32**, 165021 (2015)
98. S. Yazadjiev, B. Lazov, Classification of the static and asymptotically flat Einstein-Maxwell-dilaton spacetimes with a photon sphere. Phys. Rev. D **93**, 083002 (2016)
99. S. Yazadjiev, Phys. Rev. D **104**, 124070 (2021)
100. Ya. Zeldovich, I. Novikov, Sov. Phys. Usp. **8**, 522 (1966); Usp. Fiz. Nauk **86**, 447 (1965) (in Russian)
101. M. Zhou, A. Cardenas-Avendano, C. Bambi, B. Kleihaus, J. Kunz, Phys. Rev. D **94**, 024036 (2016)
102. C. Chan, D. Psaltis, F. Özel, GRay: a Massively parallel GPU-based code for ray tracing in relativistic spacetimes. Astrophys. J. **777**, 13 (2013)

The Optical Appearance of Compact Stars: Shadows and Luminous Rings

Merce Guerrero

Abstract

In 2019, the direct imaging of M87* [1] by the Event Horizon Telescope opened the door to understand the nature of the central object and the underlying theory of gravity, since it explore the regions where the gravitational field is extremely strong. In the images obtained by the collaboration, we can observe two distinct regions: a dark circular center called shadow and an enveloping luminous ring produced by the hot accretion disk surrounding the astrophysical object. In this chapter, we want to explain the basic tools to analyse the optical appearance of a compact object. Starting by the light bending near a massive body and the expected detected images when a star illuminates a black hole. Finally, we assume different models of accretions disks as well as the technique to obtain the similar images as the one obtained by the Einstein telescope.

1 Introduction

The optical appearance of a black hole or any other ultra-compact object (UCO) is the image one would expect to obtain after processing the data measured by several telescopes around the world. This technique for acquiring such an image is called *imaging* and can be done using different methodology [2,6,24,29]. Indeed this is currently done by the *Event Horizon Telescope* (EHT) collaboration, whose first result of M87* on April 10, 2019 [1,2] was a game changer in this field. This collaboration uses radio telescopes spread in few groups around the globe to collect the data. Since every measurement needs to be done by a pair of telescopes, which is linked to the distance between them, there are many gaps on the resulting image as a consequence of the small number of telescopes. The missing data is filled by an

M. Guerrero (✉)
Physics Department, Autonomus University of Barcelona, Barcelona, Spain
e-mail: merguerr@ucm.es

B. Hartmann and J. Kunz (eds.), *Gravity, Cosmology, and Astrophysics*,
Lecture Notes in Physics 1022, https://doi.org/10.1007/978-3-031-42096-2_4

Fig. 1 Representation of a shadow and a wide luminous ring using the AI Dall-e

algorithm that generates possible 'realistic' solutions. The possible resulting images that can fit reasonable the small collected data are huge. Nevertheless, by increasing the number of telescopes, the resolution is going to be better since there are going to have more image 'covered with data' and not by the algorithm.

In the corresponding images obtained by the EHT collaboration, which are similar to Fig. 1, we can identify two distinct regions: a nearly circular dark center called shadow and a wide surrounding luminous ring produced by the extremely hot accretion disk surrounding the astrophysical object [13].

These types of observations are extraordinarily important, since they explore the most extreme conditions of matter and spacetime. In fact, they detect photons that have traveled very close to the horizon of a black hole, where the gravitational field is extremely strong. It is therefore quite possible that these measurements will highlight any small discrepancies between theories, since gravitational effects are more prominent there. Not only black hole images, but also other observations of strong fields, such as gravitational wave signals, are essential for understanding gravity. For example, both allow testing the Kerr hypothesis, according to which the only physically acceptable solution for a rotating, uncharged black hole in General Relativity (GR) is the Kerr geometry. Therefore, if any signal is detected coming from a compact rotating and uncharged object inexplicable by Kerr phenomenology, this would be the signature of the New Physics. This New Physics may come from considering exotic matter (violating energy conditions) or theories beyond GR.

Now that we know why it is relevant to study the optical appearance of UCOs, let us explain the phenomenon. First, black holes have a so-called event horizon, for which if something enters in the region below it, they will never be able to escape, even photons. Therefore, these objects cannot emit light like other astrophysical objects such as stars or any kind of dwarfs. The only way we can directly observe such an object is because it is illuminated by one of the following luminous sources,

- A distant point-like source, for example, a star that is loosely attached to the compact object.
- An accretion disk emitting around it. Such a possibility could occur in a tighter binary system or in a dense center of a galaxy.

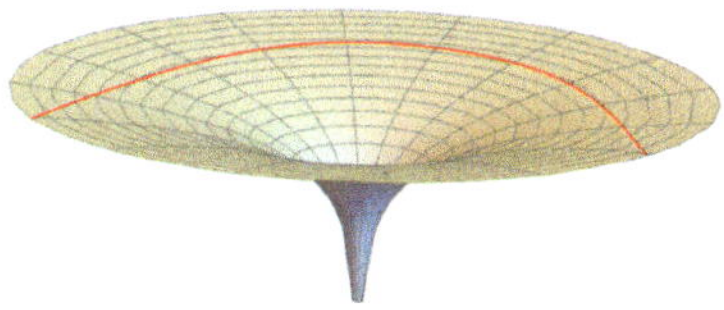

Fig. 2 3D plot of the light bending phenomenon in a spacetime created by a static, chargless (Schwarzschild) black hole in GR

Although both are likely to reveal the presence of a black hole, the expected brightness of the ring provided by both sources will not be the same, since the number of photons deflected by the gravitational field differs due to the proximity of the light sources and the emitting surface. The first case is simpler to analyze, since we expect the light rays to concentrate on the critical curves. However, the brightness of the corresponding luminous rings may be fainter than those coming from a surrounding accretion disk. This later scenario is actually the case we have already observed, but it involves much more arduous machinery to model and study the corresponding luminous rings and shadow.

We shall also understand the trajectories followed by the observed photons. These trajectories deviate as they approach a massive body (see Fig. 2 for descriptive image), giving rise to curved trajectories called geodesics. This phenomenon was predicted by GR, and was used as a test of this theory. Sir Arthur Eddington first observed this effect during a solar eclipse in 1919.[1] He measured the displacement between the apparent and real positions of some stars as a consequence of the light deflection produced by the Sun's gravitational field. This feature is also the main responsible for the optical appearance, since the photons emitted or traveling near the object are deflected as a consequence of the strong gravitational field. Therefore, it is necessary to analyze the geodesics to know what the image of the object will look like.

The theories of gravity predict the angular size and the shape of the shadow, which depends on the geometry of the spacetime. For example, in GR, the shadows are almost circular, but its size and shape depend on the mass and not so strongly on the spin of the Kerr black hole. However, these and other features in the image are sensible to astrophysical properties of the plasma near the black hole as it is going to be shown in Sect. 4.2.

2 Geodesics in GR

Geodesics are those curved trajectories described by the functions $x^\mu = x^\mu(\lambda)$, with λ being the affine parameter, whose tangent vector, $t^\mu = dx^\mu/d\lambda$, is invariant under parallel transport (autoparallels) defined by the connection of a metric. Thus, they

[1] Even though there had been two unfruitful expeditions before, in 1912 in Brazil spoiled by the thick clouds covering the sky and, in 1914, in Crimea for the start of World War I [9].

are described by the following equations

$$t^\mu \nabla_\mu t^\nu = \ddot{x}^\alpha + \Gamma^\mu_{\alpha\beta}\dot{x}^\alpha \dot{x}^\beta = 0\,, \tag{1}$$

where $\Gamma^\mu_{\alpha\beta}$ are the Christoffel symbols of the metric, $g_{\mu\nu}$, and dots denote derivatives with respect to an affine parameter, λ. As one can already imagine, solving the above equation to find the geodesics can be quite demanding as one has to calculate all the components of the connection. Thankfully, such an equation can also be obtained through the variational procedure by considering the Lagrangian,

$$\mathcal{L} = \frac{1}{2} g_{\mu\nu}\dot{x}^\mu \dot{x}^\nu\,, \tag{2}$$

where $g_{\mu\nu}$ is a general metric solution of the field equations. Substituting the above Lagrangian into the Euler-Lagrange equations, one can explicitly check that it leads to Eq. (1). For any spherically symmetric spacetime with line element defined as

$$ds^2 = -A(r)dt^2 + B(r)dr^2 + C(r)(d\theta^2 + \sin^2\theta d\varphi^2)\,, \tag{3}$$

one can assume that the motion takes place in the plane $\theta = \pi/2$ without loss of generality, because of the symmetry of the geometry we can always redefine the coordinates so that the geodesics happen in such a plane. Then, the geodesic equations become

$$A(r)\,\dot{t} = E\,, \tag{4}$$

$$2B(r)\ddot{r} + B'(r)\dot{r}^2 + A'(r)\dot{t}^2 - C'(r)\dot{\phi}^2 = 0\,, \tag{5}$$

$$C(r)\,\dot{\phi} = L\,. \tag{6}$$

Since the Lagrangian does not depend on the coordinates t and ϕ, the first and third equations are constants of motion, also known as Killing symmetries, where E is the total energy and L the angular momentum of a particle per unit of mass. Indeed, if we compute the Hamiltonian, $H = p\dot{q} - \mathcal{L}$, where $\dot{q}$ are the derivatives of the coordinates with respect to the affine parameter,

$$2H = -\frac{E^2}{A(r)} + \dot{r}^2\, B(r) + \frac{L^2}{C(r)}\,. \tag{7}$$

We can also check that it is a constant of motion as it does not explicitly depend on the affine parameter. Actually, we can redefine $2H = k$, where $k = -1, 0$ for timelike and null observers, respectively. For timelike observers we mean those that travel slower than light, linked with massive particles. On the contrary, null observers are those that travel at the speed of light or that are massless. With the redefinition of the Hamiltonian, we can rewrite the above equation as

$$AB\dot{r}^2 = E^2 - V(r)\,, \tag{8}$$

where we have introduced the effective potential

$$V(r) = A(r)\left(-k + \frac{L^2}{C(r)}\right) . \tag{9}$$

Remember that geodesics are related with the motion of particles on a curved spacetime. In the following subsections, we are going to turn our attention to the main important aspects of null and timelike geodesics for the study of the optical appearance.

2.1 Null Geodesics and Gravitational Lensing

As we mentioned in the introduction, particles' trajectories are bent as a consequence of an ample curvature of spacetime generated by a massive body. This is similar to what happens to the light when passes through an optical lens, however in this case the lens is a gravitational source and therefore such a process is called *gravitational lensing*. Here, we want to explain the mathematical framework that describes such an effect by using the geodesic equation for null (massless) observers.

Consider a light ray starting from spatial infinity and approaching to a gravitational lens. As the photon gets sufficiently close to the gravitational source, due to the spacetime geometry, it begins to deviate from their initial direction until they get to the closest radius and subsequently turn back to spatial infinity again.

To comprehend this effect, we should rewrite Eq. (8) in terms of the impact parameter[2] defined as $b = L/E$,

$$\frac{A(r)B(r)}{L^2}\left(\frac{dr}{d\lambda}\right)^2 = \frac{1}{b^2} - V_{eff}(r) \geq 0 , \tag{10}$$

where now the effective potential is

$$V_{eff}(r) = \frac{A(r)}{C(r)} . \tag{11}$$

Equation (10) describes a one-dimensional trajectory of a photon with impact parameter b governed by a potential V_{eff}. This equation gives us an idea of how close to the object we are and helps us to classify the trajectories depending on the number of turns around the center. In order to understand this issue check Fig. 3, where we have depicted a potential for a static, chargeless black hole (blue curve) and a trajectory of a photon (black line) with an impact parameter $b = 6.5$. Note that a light ray only propagates in those regions fulfilling $1/b^2 \geq V_{eff}$. Additionally, there is a particular radius where both functions intersect. This is the turning point or the radius of closest approach, r_0, as $dr/d\lambda = 0$ there. After reaching this point, the photon is going to go from reducing the distance with the black hole to grow it again.

[2] In the next section we explain why do we need to do it.

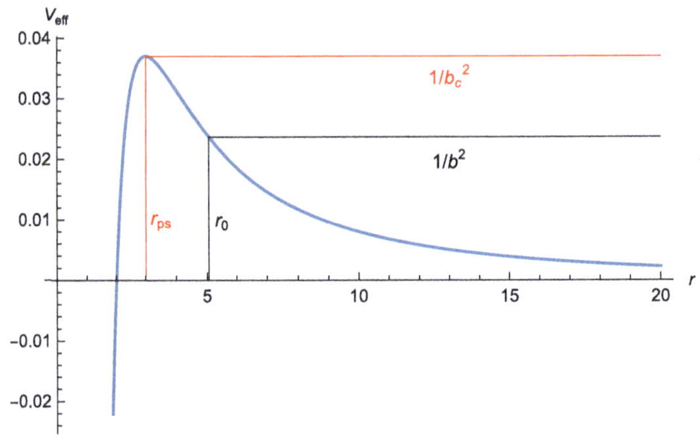

Fig. 3 Schwarzschild effective potential (for a unitary mass, $M = 1$), a photon trajectory with impact parameter $b = 6.5$ and b_c depicted in blue black and red, respectively. With this plot, it is clear why Eq. (10) hints on how close a photon can be from the object's center and which the radius of closest approach (given by the intersection between the blue and black lines) depending on the impact parameter

Going back to the mathematics, one can obtain the impact parameter value for a particular r_0 as

$$b = V_{eff}^{-1/2}(r_0) = \sqrt{\frac{C_0}{A_0}}, \tag{12}$$

where the subscript $_0$ means evaluated at r_0. If the effective potential has a maximum, there is a radius of closest approach corresponding to the unstable photon orbit or photon sphere radius, r_{ps} and the impact parameter leading to such curve is

$$b_c = \sqrt{\frac{C_{ps}}{A_{ps}}}, \tag{13}$$

called critical impact parameter. Since this corresponds to a maximum of the potential, it effectively splits the space of light rays issued from the observer's screen into two classes: those with $b > b_c$ are deflected at r_0 back towards asymptotic infinity, while those with $b < b_c$ will inspiral down towards the center of the object (thus meeting the event horizon in a black hole case). Since this orbit is unstable, any small perturbation will make the photon to eventually fall into the black hole horizon or escape to the asymptotic infinity. Thus, a photon with an impact parameter arbitrarily close to $b \gtrsim b_c$ will turn a large number of times around the compact object. To calculate r_{ps}, we should find the maximum of the potential defined in Eq. (11)

$$V'_{eff}(r_{ps}) = -\frac{A(r)}{C(r)}D(r)\bigg|_{r=r_{ps}} = 0 \quad \text{with} \quad D(r) = \left(\frac{C'(r)}{C(r)} - \frac{A'(r)}{A(r)}\right), \tag{14}$$

where primes denote derivative respect to radial coordinate. Thus, a photon sphere exists (critical curve) if $D(r) = 0$. Since the maxima of the effective potential are the responsible of the photon spheres, they are a useful tool to study the black hole shadow. However, when we analyse the non-spherical static case, the concept of the photon sphere is generalized [8,30].

If one wants to analyze the optical appearance of a compact object illuminated by the light rays passing close by, one has to suitably rewrite the geodesic Eq. (10) in order to be able to calculate the deflection angle. Thus, the equation must be expressed in terms of the variation of the azimuth angle ϕ with respect to the radial coordinate. Using Eq. (6), we obtain

$$\frac{A(r)B(r)}{C(r)^2}\left(\frac{dr}{d\phi}\right)^2 = \frac{1}{b^2} - V_{eff}(r)\,. \tag{15}$$

The above equation is the one we are going to use when we want to get the optical appearance of a UCO.

2.2 Timelike Geodesics

Remember that timelike geodesics describe the motion of a massive particles. This will play a role when we want to model the accretion disk composed by plasma, an ionized gas formed by ions and free electrons, *i.e.* massive particles. Recall once again the geodesic Eq. (8), which in this case reduces to

$$AB\left(\frac{dr}{d\lambda}\right)^2 = E^2 - A\left(1 + \frac{L^2}{C}\right), \tag{16}$$

where the last term of the equation is the effective potential for timelike observers. In this case, the potential usually has a minimum, that is

$$\frac{dV}{dr} = -\frac{A}{C}D(r) - A' = 0\,. \tag{17}$$

Such a position is known as the Inermost Stable Circular Orbit (ISCO). Conversely to the photon sphere, the fact that this radius corresponds to a minimum of the potential instead of a maximum translates to a stable orbit for timelike observers. As a consequence, one would expect that the inner edge of the orbit is placed here, since it is stable. However, for supermassive black holes this is not the case, and typically the inner edge is going to be even closer to the event horizon.

Let us now apply this knowledge to the optical appearance.

3 Illumination from a Loosely Bounded Star

Remember that the shadow of an UCO can be observed by two different means; the first one is by a distant orbiting star while the second is by an accretion disk. We begin by the former case, which is simpler and sets the ground of the main phenomenology happening in this framework. In this situation, we assume the star to be a punctual isotropic light emitter.

Figure 4 represent the system star-object with the main photon trajectories coming out of the star that are able to reach us. Another assumption made here is that the UCO corresponds to a Schwarzschild black hole with unitary mass ($M = 1$) placed at the origin of coordinates. The star is located at $r = -5$ in two different configurations with respect to the central object and we have also considered our observatory far away on the right hand side of the plot. The light rays depicted in both figures are divided into three different colors: blue, red and green. The first one is the typical example of gravitational lensing since its trajectory is slightly deviated, while the second and third colors can barely be distinguishable as the initial and final parts of the trajectories overlap. The red one only does one turn around the massive body, whereas the green does two. Thus, we can note that, as the radius of the turning point reduces, the deflection angle increases until reaching the critical distance in which light is not able to escape from the object. At such a distance, the trajectory yields a circular orbit around the center depicted in the figures as the yellow dashed circle. This orbit is the so-called photon sphere which is circular because its radius corresponds to the maximum of the potential and, consequently, it will not change if there is no perturbation. If the radius keep reducing, then the light ray will be dragged into the center of the object.[3]

After these loops, photons depart from the object recovering its almost straight paths when they are far away, since the gravitational field of the black hole decreases with distance. Thus, when they reach us, their trajectories are practically parallel to each other (see the above figures). Furthermore, we can see that the more turns the light does around the black hole, the closer they get when they leave. Indeed, the distance between them gets exponentially small on each turn, so after some loops, they are going to essentially lead to the critical impact parameter. Let us pay attention to Fig. 4b. This image corresponds to a case where the system is not exactly aligned, *i.e.* the star, black hole and us do not form a straight line. Normally when the system is slightly unaligned, what we would see is a 'cross': four identical images of the star. This can be seen in the Universe, although the gravitational lens is a galaxy instead of a black hole (look for Einstein's cross, for example). For the extreme case of Fig. 4b, what we would expect to observe is only two identical reproductions of the source at opposite sites, corresponding to the light rays reaching us from above the abscissa axis and from below.

[3] Note that this is the case for the case of Schwarzschild solution, which corresponds to a rather simple effective potential. However, if we consider different potentials, this might be different.

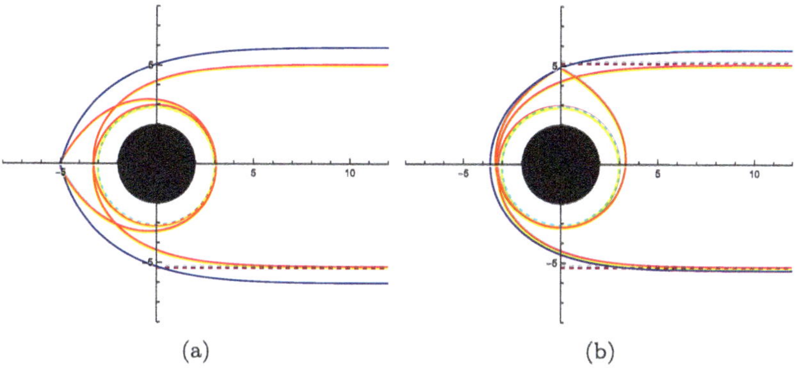

Fig. 4 Representation of the Schwarzschild black hole-star system and photon trajectories in two different configurations: in **a** the star his 'hidden' behind the black hole while in **b** the star is above. We assume that we are at asymptotic infinity to the right of each plot. The black disk symbolizes the interior of the event horizon, the dashed yellow circle is the photon sphere, while the purple dashed line is the value of the critical impact parameter for Schwarzschild. The red, green and blue curves correspond to the trajectories of the photons that completely turn around the center one, two and zero times, respectively

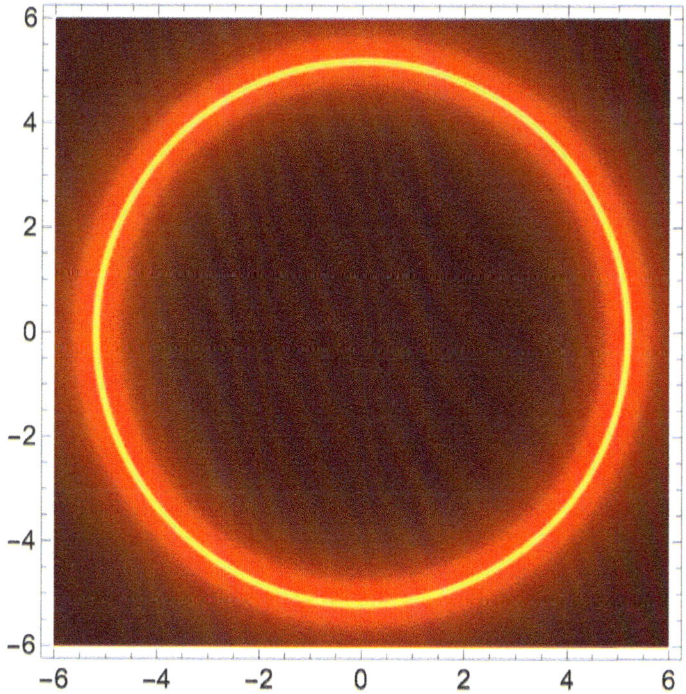

Fig. 5 Representation of the shadow and the luminous ring surrounding it for the configuration seen in Fig. 4a

On the contrary, for the case depicted in Fig. 4a, where the star is completely aligned with the object and us, the image is going to be completely different. In this case, if we perform a revolution around the horizontal axis, we will get the 3D version of the trajectories. Once we have this picture in mind, let's go one step further. Imagine yourself located at the end of the horizontal axis; what you would see is shown in Fig. 5, where we can see a luminous ring surrounding a black circular area. This thick yellowish ring represents the regions where a large number of photons accumulate and is the one that an observer is expected to see. The radius of these luminous rings is given by the critical impact parameter since the observed photons will reach us with a certain impact parameter and, therefore, the radius of the luminous ring is given by this parameter. This means that the radius of the shadow, the central black region, in this case is given by the critical impact parameter and not by the event horizon. However, if the star would have been placed inside the photon sphere, then a smaller shadow would appear, since photons emitted in that region can still arrive to us if they are emitter as 'direct' as possible, that is without turning around the object. Therefore, even though the photon sphere and the event horizon given by the geometry of the spacetime play a main role on determining the size of the shadow, the position of the photon's emission is also a meaningful factor on the optical appearance.

4 Illumination from an Accretion Disk

Let us now move to the second case of illumination: the one produced by an accretion disk, an extremely hot disk formed by plasma. The first question that may arise is: how is it formed? At the beginning, the black hole is surrounded by a gas that orbits around the object far from the event horizon. Contrary to the common idea we have of a black hole, they can have other bodies rotating at a sufficient distance in the same way as the planets of our Solar System do around the Sun, instead of being a huge astrophysical object that swallows everything that comes near it. Therefore, the surrounding gas would be orbiting around it without falling into it. The reason why we all have the image of a black hole attracting and eating everything is because there is something else besides gravity that makes the gas approach the black hole: friction.

Friction heats up the surrounding gas, which means that the gravitational energy of the system has to be transformed to thermal, causing the gas to fall into the black hole. As a consequence, the gas becomes a hot disk around the black hole. The hotter the accreting material becomes, the more energetic light it emits, as happens, for example, with stars or incandescent light bulbs.

In this scenario, we have to face a more challenging situation. On the one hand, the accretion disk can certainly be well inside the photon sphere and its shape allows to emit all along its surface instead of a point-like seen before, requiring a precise knowledge of the trajectories of light in spacetime. On the other hand, the received luminosity depends strongly on the emission profile of the disk. This demands a realistic model of the disk to establish the luminosity profile, which is obtained by simulations combining GR and Magnetohydrodynamics. In this section, we intend to explain in detail both needed ingredients, the geodesic structure analysis as well as the modelling of the disk itself, and the assumptions we use to simplify the study of the problem.

4.1 Ray Tracing

The first step is to get under control the trajectories followed by photons passing near the compact object. Before getting into the mathematics, let's start by explaining the fundamental idea of the ray tracing. Its main objective is to classify light rays according to the number of turns they make around the object. This information is necessary because, after all, when an accretion disk is added, the light rays will intersect with it at most twice per turn around the object. Depending on the characteristics of the accretion disk, these could mean additional luminous 'enhancements'.

Moreover, as we mentioned before, the optical appearance of a compact object is closely related to the impact parameter. This is so because we are so distant of the gravitational source that the light rays come almost parallel among themselves and with a different impact parameters, as we saw in Fig. 4. Using this fact, we assume that each pixel of the image corresponds to a wave detector that received a light with an impact parameter b. The responsible geodesic obtained from Eq. (15) is traced backwards towards the black hole, ending either close to the horizon or when the geodesic escapes again to a large distance from the black hole. With this, we know the total number of revolutions around the compact object. This procedure is known as *ray tracing*.

To define how we count the number of turns, we take into account the configuration of the object-observer system introduced in the previous section. Recall that we place the observer at asymptotic infinity on the right-hand side, so a photon that was not deflected at all by the compact object would have turned $n = 1/2$ times, i.e., it would have gone directly from the left-hand side to the right-hand side of the plot. In fact, the total number of orbits made by a single light ray is the (normalized) change of the azimuth angle, $n(b) \equiv \frac{\phi}{2}$. Consequently, the number of intersections with the equatorial plane of a given line is $[2n]$. Finally, recall that as we approach the critical impact parameter, $b \gtrsim b_c$, a light ray will have a longer trajectory around the neighborhood of a black hole until it is formally there forever at the critical value (or until a perturbation causes it to fall into or out of the object). The number of orbits will obviously depend on how close the impact parameter is to the critical one.

Under the above conditions, typical relevant contributions to the total luminosity on the observer's screen will be given by three types of trajectories indexed by an integer m which counts the number of intersections of a particular light ray with the vertical axis, i.e.

$$\frac{m}{4} - \frac{1}{4} \leq n < \frac{m}{2} + \frac{1}{4}, \tag{18}$$

except for the first case, $m = 1$, for which the lower limit corresponds to $n = 1/2$ and remember that n is the normalized change of the azimuth angle. We will use this number to classify the different types of emission[4] :

- Direct ($m = 1$): represents trajectories that intersect just $m = 1$ times the vertical axis ($1/2 \leq n < 3/4$), meaning that the light rays emitted by the disk go directly to the observer. This is the dominant contribution to the optical appearance of the object, in terms of luminosity and width of the associated radiation ring. However, it essentially reproduces the characteristics of the accretion disk rather than those of the background geometry and its critical curve.
- Lensed ($m = 2$): corresponds to the light rays crossing the equatorial plane for a second time, and it is defined by $3/4 < n \leq 5/4$, being the subdominant contribution to the luminosity.
- Photon ring ($m = 3$): composed by light rays intersecting the equatorial plane at least three times, and is defined by $n > 5/4$.
- Higher order ($m > 3$): typically contribute negligibly to the total luminosity (see [15] for a general discussion), as a consequence of the reduction of their impact parameter range. For this reason, they are usually integrated into the photon ring emission. These modes are much more sensitive to the characteristics of the background geometry than the rest of the emission.

The fact that higher order emissions can be neglected was already depicted in Fig. 4, where the blue curve corresponds to the direct emission, the red to the photon ring (has three cuts with the vertical axis) and the green to a high order emission of $m = 5$, where these two later contributions are indistinguishable when they reach the observer.

Note that a light ray with lensed emission contains the direct emission and the photon ring contains both. As mentioned before, the contribution of higher order emissions is negligible. Indeed, their contribution to the total luminosity can be dismissed. Indeed, its contribution exponentially decrease as they approach the critical curve [3] such that beyond $m = 3$ all additional emissions are typically accumulated in the $m = 3$ mode and thus giving the position of the critical curve [11, 15]. Although we have said that higher order emission is usually omitted, the shape of the effective potential plays an important role in the contribution of these lower order trajectories, for example the geometry studied in [16] is richer compared to the Schwarzschild case, allowing higher-order emission to contribute significantly to the total luminosity.

Last but not least, it should be noted that for impact parameters $b < b_c$ the light ray will also perform a series of half turns. These light beams will be emitted near

[4] This notation is slightly different from the more canonical one in the community, where $m = 0$ is reserved for direct emission and $m = 1, 2, \ldots$ for photon ring images, see e.g. [35]. Given the correspondence between m numbers and photon ring images in our case, we find it clearer to use m for the number of intersections with the disk, which means that our m is always one unit larger than the usual convention.

the central region of the object within the photon sphere. However, when calculating their inner ray-traced trajectories we will see two different cases: if the object has an event horizon like a black hole, the last trajectory we can detect is emitted very close to the event horizon, while for those that they do not have, such trajectories continue their path to the center of the solution. For the case of a black hole, there are light rays that do not intersect the equatorial plane because they cross the event horizon before without encountering the accretion disk on their trajectory. These trajectories form the **inner shadow** [7], $b = b_{is}$, and defines the brightness depression of a black hole independently of the emission properties of the geometrically thin accretion disk, since we will never detect those photons. However, such an inner shadow may be missing for a compact object without a horizon.

4.1.1 Schwarzschild Black Hole

Let us start by considering a static, spherically symmetric solution: the Schwarzschild black hole. Even though we know black holes are rotating objects, this assumption turns out to be a good approximation since the size and shape of the shadow, as seen by an asymptotic observer, depend very weekly on the spin of the black hole in combination with the inclination with respect to the line of sight, with deviations from circularity lying within $\sim 7\%$ for ultra-fast spinning black holes [31].

As we have already seen, for this case, one needs to calculate first the location of the horizon, since depending on the position of the horizon, the impact parameters that belong to the inner shadow are going to be different,

$$r_h = 2M \ , \tag{19}$$

with M being the mass of the black hole. Apart from the position of horizon, there is another main radial distance which plays a role when analyzing the optical appearance of a compact object: the critical curve (for which we shall also reserve the word 'photon sphere'), which for Schwarzschild is

$$r_{ps} = 3M \ . \tag{20}$$

The next step in our analysis is to integrate the geodesic equation for a bunch of light rays spanning the whole region of impact parameter values. The corresponding trajectories can be therefore classified according to the number of (half-)orbits around the solution as follows (Table 1):

We have ordered them from the outermost to the innermost emission. To illustrate this general discussion, the trajectories of a bunch of photons are depicted in Fig. 6 for $b \in (0, 10)$. We point out that the observer's screen is located at the far right side of this plot in all these cases. In these figures one can see the direct (green), lensed (orange) and photon ring (red) trajectories outside the photon sphere (dashed yellow). In addition, we have plotted the photon ring (blue), lensed (purple), and direct (cyan) emission originated from inside the photon sphere, $b < b_c$, while the black trajectories correspond to the inner shadow, those light rays that do not cross at any time the vertical axis.

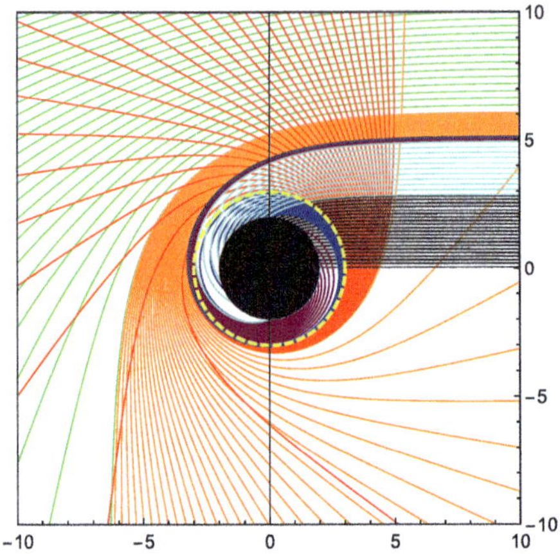

Fig. 6 Ray tracing of the Schwarzshild black hole of unitary mass (M=1). The observer's screen is located in the far right side of this plot and the type of emission is defined with respect to the number of intersections with the equatorial plane (vertical line): for $b > b_c$ we have direct (green), lensed (orange) and photon ring (red) emissions reaching to a minimum distance from the photon sphere (dashed yellow circumference) before running away, while for $b < b_c$ we also have direct (cyan), lensed (purple) and photon ring (blue) emissions. The latter three trajectories intersect the BH horizon (black central circle) after crossing the photon sphere. The bunch of black curves do not intersect the equatorial plane and therefore no emission can come out on them no matter the accretion disk model, therefore corresponding to the inner shadow of the solutions

At this point, we already know which is the impact parameter range for each type of photon trajectory. Despite with Fig. 6 we can visually reason why we only classify the null geodesics into these three groups, as we can barely see the photon ring emissions (red and blue), there is a better way to understand it graphically. This is Fig. 7 which depict the transfer functions, r_m, in terms of the impact parameter. The *transfer functions*, r_m, account for the location of the m-th intersection between the light ray and the vertical axis (i.e., the future disk). Therefore, the information one can subtract about this plot is how demagnified the light ring will be by the slope of the transfer function; the steeper it is, the lesser the contribution. This is so because the ring's width will be continuously shrinking since its thickness depends on the impact parameter range. Bearing this in mind, the direct emission is the largest contribution to the total luminosity by far, and the lensed and photon ring are highly diminished as we expected from the previous section.

Table 1 Impact parameter range for direct, lensed and photon ring emissions for Schwarzschild black hole

	Direct	Lensed	Photon ring	Inner shadow
Above the critical curve	$b > 6.15$	$b \in (5.23, 6.15)$	$b \in (5.19, 5.23)$	
Below the critical curve	$b \in (5.19, 5.23)$	$b \in (5.02, 5.19)$	$b \in (2.85, 5.02)$	$b < 2.85$

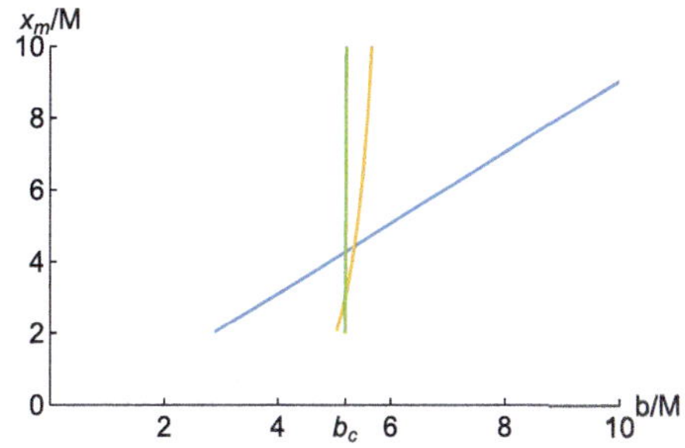

Fig. 7 The first three transfer functions for the direct (blue), lensed (orange) and photon ring (green) emissions. b_c denotes the location of the corresponding photon sphere. The slope of each curve is interpreted as the demagnification factor of the corresponding emission

This plot of the transfer is also very useful to understand how many rings and how they are going to be distributed depending on the inner edge of the accretion disk. Imagine that the inner edge is placed at $r = 6M$, this means that the accretion disk emits from $r \geq 6M$. If we draw an horizontal line in the plot at such a radius,[5] we are only going to receive those light rays above the horizontal line. Thus, for such a case, we will be able to distinguish three rings, one for each type of emission. The 'direct' ring would go from more or less $b = 7$ to infinity, whereas the lensed would go from $b \sim [5.5, 6.15]$ and the photon ring near the critical value, so the shadow is going to extend up to there. But if we move the inner edge down to $r = 2M$, we can see how now all the curves intersect and, therefore, all the rings are going to be overlapped at some points. For this case, the shadow is going to be the smallest, the one we have defined as inner shadow. Now that we completely know the trajectories of the photons and what we can expect of the luminous rings and shadows, we are able to move to the modelling of the accretion disk.

4.2 Accretion Disk Model

Treating the disk can become a real ordeal very easily, as its modeling requires the use of General Relativistic Magneto-Hydrodynamic (GRMHD) simulations, requiring the models fully account for relativistic effects, matter dynamics and photon propagation. Nonetheless, let us begin with a toy model proposed in [15] that will already show the effects of the gravitational lensing and redshift of the emitted photons. Therefore, let us itemize the several considerations we have assumed for this first example:

[5] Caution: note that in this section we talk about two different types of radius, one of them is r, the physical radial distance between the central object and, in this case, the inner edge of the accretion disk. The second one is the optical radius of the shadow and the luminous rings, which are given by the impact parameter.

- Placed on the equatorial plane of the object which is perpendicular to us: the image seen from the observer will be face-on.
- *Optically thin*: the disk does not re-absorb the photons. On each intersection with the equatorial plane the light ray will 'pick up' additional brightness, in the sense that when the photon crosses, a new photon can be emitted. This strongly depends on the particular assumed emission intensity profile of the disk.
- *Geometrically thin*: the width of the disk is negligible as compared to the radial extension of the disk, which means that most of the matter lies close to the radial plane.[6] This property produces an infinite sequence of concentric rings from photons that have completed n half-orbits in their approach to the critical curve.
- The specific luminosity only depends on the radial coordinate, $I^{em} = I(r)$.
- Isotropic emission in the rest frame of matter: the intensity does not depend on the frequency, ν, in the static frame.
- Monochromatic emission: the emissivity, j_ν, depends on the frequency as $j_\nu \sim \nu^2$.
- The intensity profile is higher close to the black hole, where the deflection of the light is strongest and the emitting plasma velocity is close to those of light.

Note that the last assumptions assume a static frame, this can be done because the disk is assumed to be face-on, so the effects between the dynamics of the disk are degenerated by the choice of the radial profile. Additionally, significant progress can also be made by using analytical models of static accretion disks with a localized emission starting from a finite-size region of the disk. Here we have assumed the disk thin, but if it is spherically symmetric, the luminous rings would converge to the critical curve itself and delimit the outer edge of the shadow [25], instead of the infinite sequence of concentric rings proper of an infinitely thin disk (or even thicker [35]).

Once we have defined the properties of emission and absorption to the disk, we can produce a model total intensity image by solving the unpractised radiative transport equation, governed by the Boltzmann equation for photons. The relativistic Boltzmann equation, which is written in terms of invariant quantities or frame-independent, reduces to

$$\frac{d}{d\lambda}\left(\frac{dI_\nu}{d\nu^3}\right) = \left(\frac{j_\nu}{\nu^2}\right) - (\nu\alpha_\nu)\left(\frac{I_\nu}{\nu^3}\right) . \tag{21}$$

where I_ν is the intensity for a given frequency ν, j_ν is the emissivity, α_ν the absorptivity, and quantities inside parenthesis are frame-independent. The resolution of such an equation demands precise knowledge of the fluid forming the disk (*i.e.* number density, angular momentum, emissivity and absorptivity).

[6] This is a good approximation for small compact objects or the rate is sub-Eddington [28], but for supermassive ones we know that the accreting material is very massive, leading to a geometrically thick disk whose inner edge can extend inside the Innermost Stable Circular Orbit for timelike observers [32].

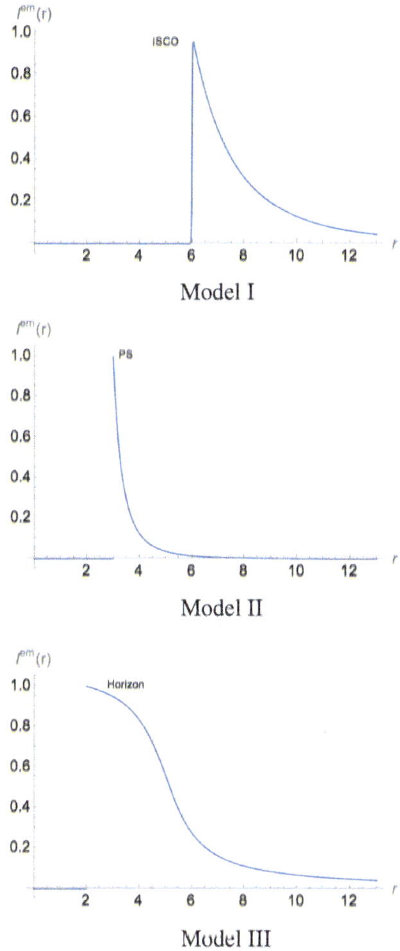

Fig. 8 Plots of the three different normalized intensity profiles emitted by the accretion disk

For the purpose of simulating different stages in the temporal evolution of such an accretion disk, we are modelling such a profile by truncating the inner edge of the disk, r_{ie}, at different radius. Also, we assume that there the intensity actually takes its maximum value, and smoothly falls off outwards until asymptotic infinity (so that the outer edge of the disk is assumed to be infinitely far away) with a given radial decay. To simplify the analysis of this aspect, typically in the literature different decay profiles for the emission are taken *ad hoc* depending on how close to the innermost region of the geometry the inner edge of the disk is. Indeed, there are different possibilities to write the expressions that describe the profiles, but here we are interested in their shape:

- Model I: The emission starts at the ISCO for timelike observers, while vanishing in the region internal to it and falling off asymptotically to zero beyond of it, see Fig. 8
- Model II: The emission has a sharp peak at the critical curve also known as photon ring, (20), having a qualitatively similar central and asymptotic behaviour as Model I, see Fig. 8
- Model III: The emission starts right off the event horizon (in the black hole case[7]) or the innermost region for horizonless UCOs. This profile decays more smoothly to zero than in the previous two cases, see Fig. 8

Once the emission profile of the accretion disk is set, we can turn our attention to the observed intensity. If we assume that the photons emitted reach us without interacting with anything, then the observed intensity is the emitted but altered due to both gravitational redshift and the optical properties of the accretion disk. The former phenomena happens when a light ray is emitted close to a gravitational object and escapes from it, the emitted frequency is going to be affected; in particular, if the frequency of the photon in the rest frame of the plasma in the disk is given by ν_e with associated intensity I_{ν_e}, then the photon frequency measured by the distant observer will be ν_o with intensity I^{ob}. To relate both intensities we use the assumption of a geometrically (infinitesimally) thin accretion disk, for which Eq. (21) implies that I_ν/ν^3 is conserved along a photon's trajectory. Since I_ν/ν^3 is conserved along a photon's trajectory, radiation emitted from a radius r and received at any frequency ν' has specific intensity [15]

$$I^{obs} = g^4 I_\nu(r) \,, \tag{22}$$

where g is the square root of the time metric component; for spherically symmetric spacetimes this would be $g = A(r)^{1/2}$ with A defined in Eq. (3). Thus, in the spherically symmetric geometry considered in this work $I^{ob}_{\nu'} = A^{3/2}(r)I(r)$. On the contrary, the implications of an optically thin disk are less known. The raw idea is that each additional intersection of the trajectories with the accretion disk will contribute to pick up additional luminosities according to the emission profile of the disk. Therefore, the total observed intensity will be

$$I^{obs}(b) = \sum_m A^2 I_{|r=rm(b)} \,, \tag{23}$$

where remember that the transfer function, $r_m(b)$, contains the information about the radius of the disk where a given light ray with impact parameter b will have its mth-intersection with the disk (in the coordinate r).

[7] From the point of view of the GRMHD simulations relevant for the EHT observations this is the most suitable scenario [14].

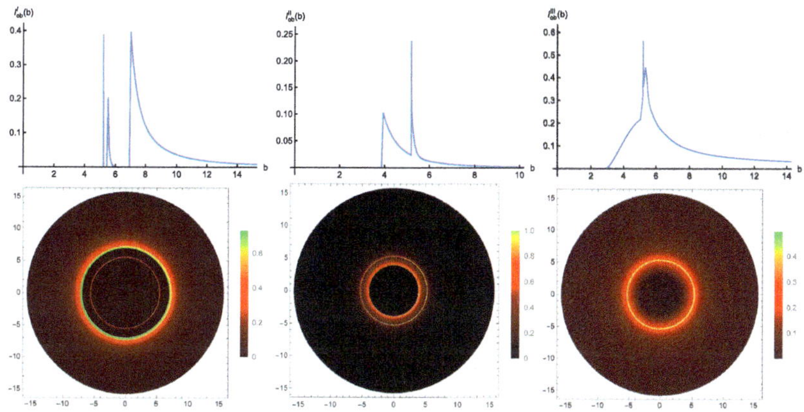

Fig. 9 The observed luminosity (top) and the optical appearance (bottom) for the Schwarzschild BH with an accretion disk based on the Model I (left), Model II (middle) and Model III (right panel) depicted in Fig. 8, viewed from a face-on orientation

4.3 The Optical Appearance of an Schwarszchild Black Hole

The first model of the accretion disk is extended up to the ISCO for timelike observers. Its emitted intensity profile is plotted in Fig. 8, and the observed intensity together with the optical appearance with its intensity legend is depicted in left panel of Fig. 9. The fact that the emission starts at the ISCO allows to clearly identify the impact parameter regions on the observed intensity corresponding to (from larger to smaller b's) a small reproduction of the emission profile and two spikes representing the direct, lensing and photon ring emissions, respectively. This is translated into a clean view of the three kinds of light rings in the optical appearances image (after zooming in a little bit). The direct emission is largely dominating the total luminosity with a broad ring very bright at the inner edge and smoothly fading out for larger impact parameters. This ring encloses a thinner and dimmer ring (the lensed emission) and inside this latter an even thinner photon ring which is barely visible at naked eye.

In Model II, depicted in the middle panel of Fig. 9, the direct, lensed, and photon ring types are overlapped in the observed intensity as a consequence of the inner edge location of the accretion disk being on the critical curve itself, which enables the direct emission via the gravitational redshift correction to pierce well inside the critical impact factor region and become the dominant contribution there, while for larger impact parameter values the combined lensed and photon ring emissions occurring roughly at the same location produce a large but narrow spike in the observed emission. Indeed, if we zoom in, we see a split between the photon ring (being fainter and closer to the direct peak) and the lensed spikes. After this luminosity boosts, the direct emission dominates again in a fainter way. The net result is that in the optical appearance the lensing and photon rings are superimposed with the direct emission. The lensing ring contribution can be appreciated in this figure, though the one of the photon ring is highly diluted and barely visible.Lastly, Model III is depicted in right panel of Fig. 9. Since the inner edge of the disk extends all the way down to the event horizon this translates into a much wider region of luminosity in the observed emission, thanks to the stretching of the direct emission to a larger distance. As a consequence, the photon ring and lensed emissions appear now as two separated but superimposed spikes with the direct emission. Another meaningful feature is the enlargement of the range of lensed emission. At the same time, this discussion is reflected in the optical appearance which shows a much wider region of luminosity with the contributions of the direct, lensed and photon ring emission. However, the second type of light rays encloses a wide ring right on the middle of it, whereas another (dimmer) one right on the inner boundary comes from the photon ring emission.

With this we have the basic tools to analyze and understand the techniques used to obtain the shadows and photon rings of a static and spherically symmetric UCO, for example [4, 16–18, 26, 27, 33, 34].

5 Discussion

The simplifications assumed here make the employed methods more accessible, but in order to compare their results with real data, the models must be improved. The possibilities for doing so are numerous. On the one hand, in terms of geometry, we can extend the above framework to rotating case [11, 12, 36]. In the cases of considering non-Schawzschild or Kerr geometries, one has to be careful when studies these kind of objects, since if they present an anti-photon sphere, it can produce a non-perturbative instability [5, 10, 19, 20, 23]. On the other hand, one can also improve the modeling of the accretion disk. For example, the disk can be considered to be tilted (*e.g.* the disk of M87* is likely to have an inclination of 17°), as well as a larger profile [35]. For a more realistic model used by the EHT collaboration, see [14], for example.

Acknowledgements M.G. wants to thank her collaborators, G. Olmo, D. Rubiera-Garcia and D. Saez-Chillon for all their discussions and fruitful collaborations that yield to Refs. [16–18]. Finally, thanks to D. Reyes Nozaleda and I. Gonzalo Fonrodona for their useful comments to make this contribution the most accessible for a broader audience as possible.

References

1. K. Akiyama et al., Event horizon telescope. Astrophys. J. Lett. **875**, L4 (2019)
2. K. Akiyama et al., Event horizon telescope. Astrophys. J. Lett. **875**, L6 (2019)
3. G.S. Bisnovatyi-Kogan, O.Y. Tsupko, Phys. Rev. D **105**, 064040 (2022)
4. R. Carballo-Rubio, V. Cardoso, Z. Younsi, Phys. Rev. D **106**, 084038 (2022)
5. V. Cardoso, L.C.B. Crispino, C.F.B. Macedo, H. Okawa, P. Pani, Phys. Rev. D **90**, 044069 (2014)
6. C.L. Carilli, N. Thyagarajan, Astrophys. J. **924**, 125 (2022)
7. A. Chael, M.D. Johnson, A. Lupsasca, Astrophys. J. **918**, 6 (2021)
8. C.M. Claudel, K.S. Virbhadra, G.F.R. Ellis, J. Math. Phys. **42**, 818 (2001)

9. L.C.B. Crispino, S. Paolantonio, Nature Astron. **4**, 6 (2020)
10. P.V.P. Cunha, E. Berti, C.A.R. Herdeiro, Phys. Rev. Lett. **119**, 251102 (2017)
11. P.V.P. Cunha, C.A.R. Herdeiro, Gen. Rel. Grav. **50**, 42 (2018)
12. P.V.P. Cunha, C.A.R. Herdeiro, E. Radu, H.F. Runarsson, Phys. Rev. Lett. **115**, 211102 (2015)
13. H. Falcke, F. Melia, E. Agol, Astrophys. J. Lett. **528**, L13 (2000)
14. R. Gold, A.E. Broderick, Z. Younsi, C.M. Fromm, C.F. Gammie, M. Mościbrodzka, H.Y. Pu, T. Bronzwaer, J. Davelaar, J. Dexter et al., Astrophys. J. **897**, 148 (2020)
15. S.E. Gralla, D.E. Holz, R.M. Wald, Phys. Rev. D **100**, 024018 (2019)
16. M. Guerrero, G.J. Olmo, D. Rubiera-Garcia, D. Sáez-Chillón Gómez, Phys. Rev. D **106**, 044070 (2022)
17. M. Guerrero, G.J. Olmo, D. Rubiera-Garcia, D.S.C. Gómez, JCAP **08**, 036 (2021)
18. M. Guerrero, G.J. Olmo, D. Rubiera-Garcia, D. Gómez Sáez-Chillón, Phys. Rev. D **105**, 084057 (2022)
19. M. Guo, Z. Zhong, J. Wang, S. Gao, Phys. Rev. D **105**, 024049 (2022)
20. G. Guo, P. Wang, H. Wu, H. Yang, JHEP **06**, 060 (2022)
21. C.A.R. Herdeiro, A.M. Pombo, E. Radu, P.V.P. Cunha, N. Sanchis-Gual, JCAP **04**, 051 (2021)
22. H. Huang, J. Kunz, J. Yang, C. Zhang, Phys. Rev. D **107**, 104060 (2023)
23. J. Keir, Class. Quant. Grav. **33**, 135009 (2016)
24. W. Lockhart, S.E. Gralla, Mon. Not. Roy. Astron. Soc. **509**, 3643 (2021)
25. R. Narayan, M.D. Johnson, C.F. Gammie, Astrophys. J. Lett. **885**, L33 (2019)
26. G.J. Olmo, J.L. Rosa, D. Rubiera-Garcia, D. Saez-Chillon Gomez, arXiv:2302.12064 [gr-qc]
27. G.J. Olmo, D. Rubiera-Garcia, D.S.C. Gómez, Phys. Lett. B **829**, 137045 (2022)
28. D.N. Page, K.S. Thorne, Astrophys. J. **191**, 499 (1974)
29. R. Patel, B. Roachell, S. Caino-Lores, R. Ketron, J. Leonard, N. Tan, K. Vahi, D.A. Brown, E. Deelman, M. Taufer, Comput. Sci. Eng. **24**, 42 (2022)
30. V. Perlick, O.Y. Tsupko, Phys. Rept. **947**, 1 (2022)
31. D. Psaltis, Gen. Rel. Grav. **51**, 137 (2019)
32. S. Riaz, D. Ayzenberg, C. Bambi, S. Nampalliwar, Mon. Not. Roy. Astron. Soc. **491**, 417 (2020)
33. J.L. Rosa, D. Rubiera-Garcia, Phys. Rev. D **106**, 084004 (2022)
34. S. Vagnozzi, R. Roy, Y.D. Tsai, L. Visinelli, M. Afrin, A. Allahyari, P. Bambhaniya, D. Dey, S.G. Ghosh, P.S. Joshi, et al. arXiv:2205.07787 [gr-qc]
35. F.H. Vincent, S.E. Gralla, A. Lupsasca, M. Wielgus, Astron. Astrophys. **667**, A170 (2022)
36. S.W. Wei, Y.X. Liu, Eur. Phys. J. Plus **136**, 436 (2021)

Beyond Newtonian Dynamics of Planar CRTBP with Kerr—Like Primaries

Suparna Roychowdhury and Roopkatha Banerjee

Abstract

In this article, we shall discuss the dynamics of the classic planar circular restricted three-body problem (CRTBP) with spinning primaries in the context of a beyond-Newtonian approximation [27]. We shall begin by discussing the construction of the beyond-Newtonian potential in the so-called Fodor–Hoenselaers–Perjés procedure where we keep first order non-Newtonian contributions in both the mass and spin. Using this potential, we shall then discuss our model for a test particle of infinitesimal mass orbiting in the equatorial plane of the two primaries. The talk shall then discuss the dynamics as the system transitions from the Newtonian to the beyond-Newtonian regime. We shall then study the evolution and stability of the fixed points of the system as a function of the parameter ϵ with the dynamics of the particle analyzed using the Poincaré map of section and the Maximal Lyapunov Exponent as indicators of chaos. We shall establish that the intermediate values of ϵ seem to be the most chaotic for the two cases of primary mass ratios ($= 0.001, 0.5$) examined. We also conclude that the amount of chaos in the system remains higher than the Newtonian system as well as for the planar circular restricted three-body problem with Schwarzschild-like primaries for all non-zero values of ϵ.

S. Roychowdhury (✉)
Department of Physics, St. Xavier's College, 30, Mother Teresa Sarani, Kolkata, India
e-mail: suparna@sxccal.edu

R. Banerjee
Department of Computational and Data Sciences, Indian Institute of Science, Bangalore, India
e-mail: roopkathab@iisc.ac.in

B. Hartmann and J. Kunz (eds.), *Gravity, Cosmology, and Astrophysics*,
Lecture Notes in Physics 1022, https://doi.org/10.1007/978-3-031-42096-2_5

1 Introduction

In the field of modern celestial mechanics and dynamical astronomy, one of the most intriguing and important problems is the dynamics of few bodies, in particular being the case of a circularly restricted three body problem [113]. This problem has been applied in various fields in astronomy, like planetary dynamics, galactic and stellar cluster dynamics and even molecular dynamics. From Euler first formulating the CRTBP in a synodic (or rotating) coordinate system to Lagrange demonstrating the existence of five equilibrium points at which the gravitational forces of the bodies cancelled out, the CRTBP has been an intensive field of research for almost three centuries now. Soon after Lagrange, Jacobi introduced his integral of motion, which was then extensively used by Hill to determine the motion of an asteroid in the three-body problem and to introduce the so-called zero velocity curves (ZVC), which establish regions in space where the bodies are allowed to move. In the 19th century, Poincaré notably studied Hill's problem and was able to choose initial conditions that resulted in periodic orbits in CRTBP. This led to systematic searches for periodic orbits in the three-body problem which boomed with the advent of modern-day computers. Currently, with the advent of LIGO and the detection of gravitational waves from binary black hole mergers [1,2], the investigation of such systems in strong gravitational fields have become a field of intense research once again. The black holes involved in these discoveries span a mass range of 10–100 $M_{\odot}$, and are all consistent to have initially formed from the death of massive stars. For an excellent review on the history of CRTBP and its development through the centuries, the reader is referred to [83].

There is also strong observational evidence that a different class of super massive black holes (SMBHs), with masses ranging from 10^5 to 10^{10} $M_{\odot}$, are residing in almost all centres of galaxies [7]. It is expected that some of these SMBHs will pair up as binaries as their host galaxies merge [8]. In fact, there is ample evidence of several active galaxies with double nucleus [59,81]. It is also speculated that the eventual inspiral and merger of some of these SMBH binaries constitutes a prime gravitational wave source for the planned LISA observatory [3]. In addition, there is also increasing evidence that there are Kerr black hole binaries which are merging [7,48,93] and are sources of gravitational radiation.

In such binary black hole mergers which also accrete, the investigation of the chaotic dynamics of test particles within accretion discs or inside the halo surrounding these compact objects has become a subject of prime importance [23,24,46,50, 53,55,56,70,72,74,77,95,119,129–134,137]. Some authors have also studied the numerical schemes and techniques which can be used for such non-linear, chaotic problems along with the dynamics of these systems [74,134,138]. Investigations of such dynamics of charged particles moving under the influence of magnetic and strong gravitational fields of a single compact object have already been studied in some detail within the general relativistic framework [60–64,115]. Such studies have been extended to the motion of test particles under the influence of the relativistic gravitational field of accreting black holes [98–100,118,125] and also for motion under the influence of gravity produced by an extended body [30,68,117,127].

On the other hand, escaping particles from dynamical systems has also been a subject under focus for decades. Especially the issue of escape in Hamiltonian systems is directly related to the problem of chaotic scattering which has been an active field of research over the last decade and it still remains an open area [9, 10, 12–14, 17, 20, 21, 36, 79, 88, 96]. It is well known that some types of Hamiltonian systems have a finite energy of escape. For lower values of the particle energy, the equipotential surfaces of these systems are closed and therefore escape is impossible. For energies above the escape energy, these surfaces open and exit channels emerge through which particles can escape to infinity. There is a comprehensive body of work on such "open" or "leaking" Hamiltonian systems [6, 22, 38, 57, 67, 87, 102–104, 139–144]. However, it is needless to say that this list of citations is neither complete nor exhaustive. It is just indicative of the body of work that has happened in these fields and is still continuing.

The restricted three-body problem (RTBP) is an excellent example of such open Hamiltonian systems with escape [123, 124]. Over the last few decades, a large number of studies have been devoted to the classification of orbits in the RTBP. It all started with the pioneering works of [85, 86] where initial conditions of orbits were classified as bounded, escaping or collisional. Moreover, bounded orbits were further classified into orbital families by taking into account the type of motion of the test particle around the primary sources. Such classifications have also been done in the context of planetary systems, Earth-Moon system and Saturn-Titan system [29, 143]. In this context, it is important to mention that a simplified modification of the RTBP is the Hill approximation which focuses on the vicinity of the secondary source [52, 91, 92, 108, 109]. This facilitates for the study of the motion of test particles in the neighborhood of the Lagrange (equilibrium) points L_1 and L_2. At this point it should be mentioned that the Hill approximation is valid only when the mass of the secondary is much smaller than the mass of the primary body. One can directly obtain the Hill model from the classical RTBP by translating the origin to the center of the secondary body and also by re-scaling the coordinates suitably. The Hill problem was proved to be non-integrable by [78], and is chaotic, as shown by [101]. Subsequently, thorough numerical investigations of this problem were performed by carrying out a systematic classification of the initial conditions of the orbits [144]. More precisely, the initial conditions of the orbits were classified into four categories: (i) non escaping regular orbits; (ii) trapped chaotic orbits; (iii) escaping orbits; and (iv) collisional orbits. In addition, the issue of equilibrium points in circular restricted three body problem (CRTBP) has also been studied widely and in great detail (see [51] and references there in). The discovery of the Trojan asteroids around the Lagrangian points L_4 and L_5 in the Sun-Jupiter system [82], and the recent observations of asteroids around L_4 for the Sun-Earth system [18], has added a great impetus to theoretical studies on the subject. Moreover, the dynamics of non-conservative RTBP have also been investigated extensively, like the case of CRTBP with gravitational radiation [94], an elliptic restricted three-body problem [121] and that of a dissipative CRTBP with drag forces [122].

One of the first attempts at studying the relativistic CRTBP under the assumptions of low velocities and weak gravity was made by [65] in the year 1967. He looked at the post-Newtonian equations for the first time using the Einstein-Infeld-Hoffmann (EIH) formalism [37]. Since then this problem has been studied by several authors where they have attempted to present the Lagrangian explicitly [19]. Some authors have also tried to explore the deviations to the Lagrangian points due to the post-Newtonian corrections [75]. In addition, analytical solutions were also attempted in the GR regime using the EIH approximation up-to the first order [135]. Recently, as one of the first studies of chaotic orbits in the post-Newtonian CRTBP, [53] explored the influence of the distance of separation between the two primaries. They observed that if the primary bodies are close enough, the post-Newtonian dynamics is qualitatively quite different, particularly where some Newtonian bounded orbits become unstable.

In more recent studies, several authors have formulated this problem using pseudo-Newtonian potentials developed for non-rotating Schwarzschild-like (Paczyńsky-Witta potential) [89] and rotating Kerr-like primaries [4, 80, 97] to avoid the complications of a post-Newtonian formulation. Subsequently, detailed studies of orbits and the dynamics of test particles around a single Schwarzschild primary and a binary system, as well as Kerr like primaries have been made in recent years with the idea of investigating the chaotic and unstable nature of orbits in the relativistic regime. In a very recent study, [33] used the Fodor-Hoenselaers-Perjés (FHP) procedure [41] (taking into account the corrections made by [107]) to derive an approximate potential for the gravitational field of two uncharged spin-less particles modeled as sources with multi-pole moment, m. In this work, they have explored the dynamics of a massless test particle using the Poincaré section and the Lyapunov exponent as indicators of chaos. As they have mentioned, this potential is not ad-hoc as other pseudo-Newtonian potentials but rather it is exactly derived from the multipolar structure of the sources. In our current study, we also follow a similar route and use the FHP procedure to derive the multipolar structure of a spinning binary system. Subsequently, we construct a beyond-Newtonian potential to imitate the gravitational effects of this system on a test particle in the CRTBP scheme.

The article is organized as follows. In the next section, we present the formulation of the gravitational beyond-Newtonian potential of each Kerr-like source using the FHP procedure. Next, we present the Lagrangian and the equations of motion of a test particle in context to CRTBP. In the subsequent section, we present a detailed analysis of the Hill curves or the zero velocity surfaces as the system makes a gradual transition from the FHP beyond-Newtonian approximation to the classical regime through a parameter ϵ in the beyond-Newtonian potential. Here we also present a detailed analysis of the orbits and a discussion on the fixed points of this system along with their stability as a function of the parameter ϵ. The classification of the nature of orbits is made using Poincaré surfaces of section and the variational method for the calculation of the largest Lyapunov exponent, as done by several previous authors. In the next section, we present a comparison between the dynamics of a test particle around a binary system of Schwarzchild and spinning primaries. Finally, in the last section we conclude with a summary of our main results and present certain new directions that we intend to investigate in the near future.

2 Formulation of Beyond-Newtonian Potential for Kerr Binary

The version of CRTBP we consider consists of two massive, spinning primaries with masses $\mathcal{M}_1$ and $\mathcal{M}_2$ and intrinsic angular momenta a_1 and a_2, at positions X_1 and X_2, respectively, describing a circular orbit in the $x - y$ plane about their common centre of mass (taken to be the origin $\mathcal{O}$). The centre-to-centre distance remains fixed and remains sufficiently far apart, while the orbital angular velocity is ω_0. The aim is to set up the beyond-Newtonian potential (up to the first non-Newtonian term) for this CRTBP system and consequently write down the Euler-Lagrange equations of motion of a test particle under the influence of this potential. The schematic of the configuration is illustrated in Fig. 1.

To simulate the dynamics of the CRTBP at hand, we employ the Fodor-Hoenselaers-Perjés (FHP) procedure to perform an expansion in the mass and rotation potential of each primary up to the first non-Newtonian term. This essentially generates first-order general relativistic effects, the dynamics of which is analyzed at length in the following section. The beyond-Newtonian potential for the system is then constructed by virtue of a superposition of the potentials corresponding to the two primaries, modelled to describe circular orbits around their common centre of mass. We then write down the Lagrangian and consequently the equations of motion for a test particle under the influence of such a potential.

2.1 Beyond-Newtonian Potential

We shall now briefly outline the steps involved in the FHP procedure leading to the construction of the beyond-Newtonian potential for the problem at hand. The FHP algorithm involves the decomposition of the Einstein field equation in the so-called

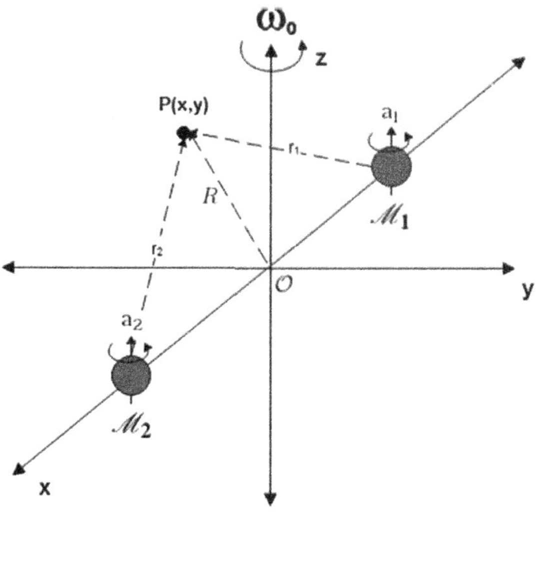

Fig. 1 The configuration of the two primaries, $\mathcal{M}_1$ and $\mathcal{M}_2$, in the centre-of-mass frame which is rotating about the z-axis with angular frequency ω_0 (= 1). A test particle with infinitesimal mass **P** is placed at an arbitrary position in the equatorial plane

Ernst formalism. In this formalism, the field equations of GR are reduced to a pair of complex equations by virtue of introducing the complex potentials ζ and Ψ. These complex potentials are further defined in terms of two new potentials ξ and ς through the relations

$$\zeta = \frac{1-\xi}{1+\xi}, \quad \Psi = \frac{\varsigma}{1+\xi}. \tag{1}$$

The field potentials satisfy [39,40]

$$(\xi\xi^* - \varsigma\varsigma^* - 1)\,\nabla^2\xi = 2(\xi^*\nabla\xi - \varsigma^*\nabla\varsigma)\cdot\nabla\xi\,, \tag{2}$$

$$(\xi\xi^* - \varsigma\varsigma^* - 1)\,\nabla^2\varsigma = 2(\xi^*\nabla\xi - \varsigma^*\nabla\varsigma)\cdot\nabla\varsigma\,. \tag{3}$$

The above set of equations are an alternative representation of the Einstein-Maxwell field equations. As a matter of fact, they could be interpreted as the generalization of Laplace's equation for the Papapetrou's metric describing the space-time around a stationary and axisymmetric source

$$ds^2 = -F(dt - \omega d\phi)^2 + F^{-1}[e^{2\gamma}(d\rho^2 + dz^2) + \rho^2 d\phi^2]\,, \tag{4}$$

where the metric coefficients F, ω, and γ depend only on the Weyl-Papapetrou coordinates ρ and z. These metric functions can be reformulated in terms of the Ernst complex potentials [107] ζ and Ψ and described by the associated Einstein-Maxwell field equations (2) and (3).

The new set of field potentials ξ and ς are related to the classical gravitational and electromagnetic potentials in the following way

$$\xi = \Phi_M + i\,\Phi_J\,, \quad \varsigma = \Phi_E + i\,\Phi_H\,, \tag{5}$$

where Φ_M, Φ_J, Φ_E, and Φ_H represent the mass, angular momentum, electrostatic and magnetic potentials, respectively. As our massive, spinning primaries do not possess electromagnetic fields, we set $\Phi_E = \Phi_H = 0$, which from (1) implies $\varsigma = \Psi = 0$. The seminal work of [44,49] allows us to determine the multipolar moments of asymptotically flat spacetimes. In this prescription, the induced 3-metric h_{ij} is mapped by virtue of a conformal transformation $h_{ij} \to \tilde{h}_{ij} = \Omega^2(x)h_{ij}$ onto a conformal metric $\tilde{h}_{ij}$. This conformal factor Ω satisfies the conditions

$$\Omega\big|_\Lambda = \tilde{D}_i\Omega\big|_\Lambda = 0, \quad \tilde{D}_i\tilde{D}_j\Omega\big|_\Lambda = 2h_{ij}\big|_\Lambda, \tag{6}$$

where $\tilde{D}$ denotes the covariant derivative on the induced surface and Λ denotes the point added due to conformal compactification. Essentially, Ω transforms the potential ξ into $\tilde{\xi} = \Omega^{-1/2}\xi$ with the explicit transformation being $\Omega = r'^2 = \rho'^2 + z'^2$. The relation between the primed and unprimed Weyl-Papapetrou coordinates are

$$\rho' = \frac{\rho}{\rho^2+z^2}, \quad z' = \frac{z}{\rho^2+z^2}, \tag{7}$$

with ϕ remaining unchanged. This helps in mapping the infinity to the origin of the primed coordinates $(\rho', z') = (0, 0)$. Besides, the potential $\tilde{\xi}$ can be expressed as a power series expansion in ρ' and z' as

$$\tilde{\xi} = \sum_{i,j=0}^{\infty} a_{ij} \rho'^i z'^j \tag{8}$$

with the coefficients a_{ij} determined by recursive relations presented explicitly in [107]. Following this procedure, one can deduce approximate relations for the gravitational potential ξ, in terms of the parameters of the primary once its gravitational multiple moments P_i are known. Thus, we apply this outlined prescription to a massive, spinning primary whose multipolar structure we take to be:

$$P_0 = m\,, \quad P_1 = ima\,, \quad P_i = 0 \quad \text{for } i \geq 2\,, \tag{9}$$

such that m and a denote the mass and angular momentum of the source, respectively.

We now aim to set up the beyond-Newtonian potential (up-to the first non-Newtonian term) for the CRTBP system at hand and consequently write down the Euler-Lagrange equations of motion of a test particle under the influence of this potential. For clarity, we restate the conditions and assumptions of the CRTBP model we are trying to construct:

- The two primaries, with masses $\mathcal{M}_1$ and $\mathcal{M}_2$ and intrinsic angular momenta a_1 and a_2, at positions X_1 and X_2, respectively, describe a circular orbit about their common centre of mass (taken to be the origin $\mathcal{O}$). The centre-to-centre distance r remains fixed and remains sufficiently far apart, while the orbital angular velocity is ω_0.
- A beyond-Newtonian potential describing the primaries is constructed assuming that the principle of superposition holds: that the total gravitational potential of the system is a linear sum of the mass and rotation potentials (up to first order effects) of the individual sources.
- A test particle of mass $\mathcal{M}$, that is very small compared to the primaries, now moves under the effect of this beyond-Newtonian potential in the $z = 0$ orbital plane of the primaries. The motion of this test particle has no effect on the primaries whatsoever.
- The convention $G = \mathcal{M} = \omega_0 = r = 1$ is used throughout the analysis hereon (further details on this choice of units has been discussed extensively in Sect. 3).

In accordance with the above conditions and following the preceding discussion on the FHP formalism, we now construct the beyond-Newtonian potential Ω describing the primaries of our CRTBP model from the reconstructed potential ξ describing a single source. Keeping explicitly the factors of c to show the corresponding order-wise contributions, we have the beyond-Newtonian potential for our system:

$$\Omega(x, y) = -\sum_{i=1}^{2} \frac{\mathcal{M}_i}{r_i} + \frac{1}{2c^4} \sum_{i=1}^{2} \frac{\mathcal{M}_i^3}{r_i^3} + \frac{1}{c^2} \sum_{i=1}^{2} \frac{\mathcal{M}_i a_i}{r_i^2} \cos\theta_i + \frac{1}{2c^4} \sum_{i=1}^{2} \frac{\mathcal{M}_i a_i^2}{r_i^3} (3\cos^2\theta_i - 1) \tag{10}$$

where the primaries are stationed at positions $X_1 = (x_1, 0)$ and $X_2 = (x_2, 0)$ respectively, and $r_{1,2} = \sqrt{(x - x_{1,2})^2 + y^2}$. We note that the first two terms of Eq. (10) describe the mass potential and the next two terms represent the rotation potential of the binary system upto first order corrections respectively. Also, following the FHP procedure, we see the potential that is constructed is written in terms of powers of $1/c^2$. The 1st order corrections to the Newtonian potentials, both in mass as well as for rotation, are retained and the higher order terms are dropped since their contribution is smaller compared to the leading order (by appropriate factors of $1/c^2$). The term 'beyond-Newtonian' is designated to these 1st order corrections to the Newtonian potentials that arise in our final form of the potential, as seen in Eq. (10).

Moreover, in order to observe the transition of the system from the Newtonian regime to a beyond-Newtonian one, we introduce a parameter ϵ, such that,

$$\frac{1}{c^2} \to \frac{1}{c^2}\epsilon$$

with $\epsilon \in [0, 1]$ using the fact that $\frac{1}{c^2} \to 0$ reduces Eq. (10) to the Newtonian case. That is, the $\epsilon = 0$ classical limit is essentially the Newtonian problem that models non-spinning binaries composed of weak gravitational sources as found in say, planetary systems and binary stars which are not in close contact with each other. On the other hand, the $\epsilon = 1$ beyond-Newtonian case models departures from Newtonian behaviour that can be found in compact spinning binaries constituted of strong gravitational sources, for example black-hole and compact binaries. The parameter ϵ can thereby be thought of as a knob that slowly "turns on" corrections (both in the mass and rotation potentials as seen from Eq. (10)) to the Newtonian potential as we gradually go from the classical limit $\epsilon = 0$ to the beyond-Newtonian regime $\epsilon = 1$.

3 Dynamics of a Test Particle

In order to simplify the numerical simulation of the three dimensional system described in the section above, we confine ourselves to the plane of the two primaries. We adopt a modified version of the Szebehely convention to de-dimensionalize the problem. Numerous types of scaling transformations have had applications in literature [53,54,111]. For example, studies of chaotic dynamics of asteroids in planetary systems scale primaries to the solar mass. However, for our problem, the absolute masses of the two primaries are irrelevant and do not reveal any new physical information about the system. Therefore, with $\mathcal{M}_1 + \mathcal{M}_2 = \mathcal{M}$ and $a_1 + a_2 = a$, we define a dimensionless mass $\mu_1 = \mathcal{M}_2/\mathcal{M}$ and dimensionless spin $\mu_2 = a_2/a$. Applying the scaling relations described above, we enforce the sum of the masses of the two primaries and the distance between the two to be unity. This has been enforced by adopting geometrized units, $G = 1$ and $c = 1$, with distance and time now having the dimension of mass (this choice of units has been discussed in detail in the next paragraph). Additionally, this scaling also ensures that the sum of the spins of the primaries be unity. Thus, applying the above discussed scaling relations we obtain:

$$\begin{aligned} \mathcal{M}_1 &= 1 - \mu_1 \;;\; \mathcal{M}_2 = \mu_1 \\ a_1 &= 1 - \mu_2 \;\;\,;\; a_2 = \mu_2 \end{aligned} \tag{11}$$

At this point, it is worthwhile to note that different system of units have been used in literature for simplifying the respective problem, both analytically and numerically. The choice of units always mostly depend on the length scales, masses and the timescales involved. As a result of this, the speed of light c can assume different values. For example, in planetary systems, setting $G = 1$, the unit of mass to be the sum of masses, the unit of distance to be the semi-major axis of the secondary body (which is set to unity) and using Kepler's second law, the speed of light assumes different values like c = 22946.5 for the case of Sun-Jupiter, and c = 10065.3 in the case of Sun-Earth [71]. However, while studying the dynamics of test particles around compact objects under the circular restricted three body scheme (CRTBP) in post-Newtonian (PN) treatments [37], the speed of light c surfaces as a parameter which measures the order of the PN contributions. For ease in numerical simulations, $c = 1$ is later enforced and a, which is the separation between the parent bodies, becomes an important parameter for the first post-Newtonian (1-PN) order effect. Thus, this choice of unit and relevant scaling transformations facilitates the study of how the separation between the primaries affect the dynamics of this system [53]. Another variation to this post-Newtonian three body scheme was recently studied by [34] who used $c = 10000$ in his calculations. To show this, one can use the Sun-Earth system as an example (for details refer to [58]). It was shown here that this value of c, the choice of units and relevant scaling transformations, as opposed to $c = 1$ in an earlier work by [53], facilitates a better conservation of the Jacobi integral of motion numerically. This is due to the fact that the contributions of the higher order PN terms vary depending on the formulation and thus a truncation brings about a non-conservation of the Jacobi integral (discussed in detail later). Recently, [31] used different values of c to indicate perturbations from the PN contribution, which were used to find an optimal method for the calculation of eccentric anomaly.

However, in our study, the Jacobi integral of motion is a constant. Thus, our choice of the value of c is to just facilitate the simplification of the system, both algebraically and numerically. As we had noted earlier, the beyond-Newtonian effects are scaled by a factor of $1/c^2$ which is taken care of by the introduction of the parameter ϵ in our system of units. Hence, $c = 10000$ will scale down the beyond-Newtonian terms by a factor of 10^{-8}, which can be compensated by suitably adjusting the range of ϵ, since it is a free parameter in our system. Thus it can be concluded that the nature of the dynamics of the system will not be affected by the choice of the value of c, as has been verified by our simulations too.

The separation between the two primaries is then scaled as,

$$x_1 = -\mu_1 \ ; \ x_2 = 1 - \mu_1 \tag{12}$$

Thus, $\mu_1, \mu_2 \in [0, \frac{1}{2}]$ are the only two control parameters for the system. Applying the earlier described scaling and putting $c = 1$, the potential becomes:

$$\begin{aligned} \Omega(x,y) = &-\left(\frac{1-\mu_1}{r_1} + \frac{\mu_1}{r_2}\right) + \frac{1}{2}\epsilon^2\left(\frac{(1-\mu_1)^3}{r_1^3} + \frac{\mu_1^3}{r_2^3}\right) \\ &+ \epsilon\left(\frac{(1-\mu_1)(1-\mu_2)}{r_1^2}\cos\theta_1 + \frac{\mu_1\mu_2}{r_2^2}\cos\theta_2\right) \\ &+ \frac{1}{2}\epsilon^2\left\{\frac{(1-\mu_1)(1-\mu_2)^2}{r_1^3}\left(3\cos^2\theta_1 - 1\right)\right. \\ &\left. + \frac{\mu_1\mu_2^2}{r_2^3}\left(3\cos^2\theta_2 - 1\right)\right\} \end{aligned} \tag{13}$$

The Lagrangian for the system may be constructed as follows:

$$\mathcal{L} = \frac{V^2 + 2A + R^2}{2} - \Omega(x,y) \tag{14}$$

where $V = \sqrt{\dot{x}^2 + \dot{y}^2}$ represents the magnitude of the velocity of the test particle, $R = \sqrt{x^2 + y^2}$ the position of the test particle with respect to the centre of mass in the non-inertial rotating frame and $A = \dot{y}x - \dot{x}y$. Thus, the Euler-Lagrange equations of motion are:

$$\begin{aligned} \ddot{x} = 2\dot{y} + x &- \left[\frac{(1-\mu_1)}{r_1^3}\left(x+\mu_1\right) + \frac{\mu_1}{r_2^3}\left(x+\mu_1-1\right)\right] \\ &- \epsilon\left\{\frac{(1-\mu_1)(1-\mu_2)}{r_1^4}\left[y\sin\theta_1 - 2\cos\theta_1\left(x+\mu_1\right)\right]\right. \\ &\left. + \frac{\mu_1\mu_2}{r_2^4}\left[y\sin\theta_2 - 2\cos\theta_2\left(x+\mu_1-1\right)\right]\right\} \\ &- \frac{3}{2}\epsilon^2\left\{\frac{(1-\mu_1)(1-\mu_2)^2}{r_1^5}\left[y\sin 2\theta_1 - \left(3\cos^2\theta_1 - 1\right)\right.\right. \\ &\left.\left(x+\mu_1\right)\right] - \frac{(1-\mu_1)^3}{r_1^5}\left(x+\mu_1\right) + \frac{\mu_1\mu_2^2}{r_2^5}\left[y\sin 2\theta_2\right. \\ &\left.\left. - \left(3\cos^2\theta_2 - 1\right)\left(x+\mu_1-1\right)\right] - \frac{\mu_1^3}{r_2^5}\left(x+\mu_1-1\right)\right\} \end{aligned} \tag{15}$$

$$\ddot{y} = -2\dot{x} + y\left[1 - \left(\frac{1-\mu_1}{r_1^3} + \frac{\mu_1}{r_2^3}\right)\right] + \epsilon\left\{\frac{(1-\mu_1)(1-\mu_2)}{r_1^4}\right.$$
$$\left[\left(x+\mu_1\right)\sin\theta_1 + 2y\cos\theta_1\right] + \frac{\mu_1\mu_2}{r_2^4}\left[\left(x+\mu_1-1\right)\right.$$
$$\left.\left.\sin\theta_2 + 2y\cos\theta_2\right]\right\} + \frac{3}{2}\epsilon^2\left\{\frac{(1-\mu_1)(1-\mu_2)^2}{r_1^5}\left[\left(x+\mu_1\right)\right.\right.$$
$$\left.\sin 2\theta_1 + \left(3\cos^2\theta_1 - 1\right)y\right] + \frac{(1-\mu_1)^3}{r_1^5}y + \frac{\mu_1\mu_2^2}{r_2^5}$$
$$\left.\left[\left(x+\mu_1-1\right)\sin 2\theta_2 + \left(3\cos^2\theta_2 - 1\right)y\right] + \frac{\mu_1^3}{r_2^5}y\right\} \tag{16}$$

where,

$$r_1 = \sqrt{(x+\mu_1)^2 + y^2}$$
$$r_2 = \sqrt{(x+\mu_1-1)^2 + y^2}$$
$$\theta_1 = \tan^{-1}[y/(x+\mu_1)]$$
$$\theta_2 = \tan^{-1}[y/(x+\mu_1-1)]$$

The Jacobi integral for the above system is given by,

$$J(x, y, \dot{x}, \dot{y}) = (x^2 + y^2) - 2\Omega(x, y) - (\dot{x}^2 + \dot{y}^2) = C_j \tag{17}$$

where C_j is a constant of motion for the given system and is called the Jacobian constant.

Here we note that the Lagrangian for our system, as stated in Eq. (14), has terms only up to the quadratic order in velocity V of the test particle as a result of which the Jacobian constant (Eq. (17)) is exactly derived. This is in contrast to the post-Newtonian (PN) framework where the Jacobian does not remain conserved and consequently limits the extent of dynamical studies. The reasoning behind this has to do with the relations between the PN Lagrangian and Hamiltonian approaches at the same PN order. Additionally, it also depends on the relations between the approximately truncated as well as the exactly non-truncated Euler-Lagrange equations of motion for this PN Lagrangian approach. The equivalence between the Lagrangian and Hamiltonian approaches at the same PN order was established in [25,26,28,69]. However, recent contradictions of the same have been discussed in [16,56,120,133,134]. It has been shown by [72,73] that the approximately truncated Euler-Lagrange equations of motion for this PN Lagrangian approach have different dynamical behaviours of order and chaos than its exactly non-truncated counterpart. As a result, the reasons why the Jacobian constant cannot be conserved in the PN approach is because (a) some higher-order PN terms are truncated when the Euler-Lagrange equations of motion are derived from this PN Lagrangian approach, and (b) some higher-order PN terms are still truncated when the Hamiltonian (corresponding to the Jacobian constant) is derived from this PN Lagrangian approach. If the truncated higher-order PN terms are large, as in the case of strong gravitational fields of compact objects, the Jacobian constant shows a poor accuracy; while it shows a better accuracy if the same truncated terms are comparatively smaller, as in the case of weak gravitational fields found in our Solar system. It should be expected that for our potential (Eq. (10)), the Lagrangian and Hamiltonian approaches at the same beyond-Newtonian order are not equivalent in general. This in-equivalence should also be true for the approximately truncated as well as the exactly non-truncated Euler-Lagrangian equations of motion for our beyond-Newtonian Lagrangian approach. However, the equations of motion (15) and (16), the corresponding Hamiltonian and the Jacobian constant (17) are exactly derived and have no terms truncated from the beyond-Newtonian Lagrangian (Eq. (14)) because it has no higher-order terms with respect to the test particle velocity V. As a result, the Jacobian constant, given by Eq. (17), is said to be exactly derived.

3.1 Hill Curves

The Hill curves or the zero-velocity curves for the system, for a set of chosen values of C_j, μ_1, μ_2 and ϵ, divide the equatorial plane into regions where the motion of the body is energetically allowed and regions where the motion is energetically disallowed (for a discussion on zero-velocity curves refer to [112] for a Newtonian CRTBP system and [148] for a pseudo-Newtonian CRTBP with Schwarzschild like primaries). All points, where $(x^2 + y^2) - 2\Omega(x, y) - (\dot{x} + \dot{y}) > C_j$, are energetically allowed for the test particle, while all points, where $(x^2 + y^2) - 2\Omega(x, y) - (\dot{x} + \dot{y}) < C_j$ are energetically disallowed. The velocity of the test particle (as we shall calculate from Eq. (19)) in the disallowed region is imaginary (will be calculated explicitly in the next subsection). The Hill curves of the system have an equation,

$$(x^2 + y^2) - 2\Omega(x, y) = C_j \,. \tag{18}$$

Figures 2 and 3 show the evolution of the Hill curves with the introduction of beyond-Newtonian effects for $\mu_1 = \mu_2 = 0.001$ (or the biased-mass system) and $\mu_1 = \mu_2 = 0.5$ (or the Copenhagen system) respectively. The beyond-Newtonian effects are introduced by increasing ϵ from 0.0 to 1.0 in steps of 0.1. The equatorial plane is divided into three regions by the Hill curves—a central region where the particle is energetically allowed but is bounded by the Hill curves, an unbounded energetically allowed region, and a disallowed region in-between them. Test particles with initial conditions in the unbound region may execute stable orbits around both the primaries or may escape to infinity, while test particles with initial positions in the enclosed and energetically allowed regions are 'trapped' and cannot escape to infinity since they cannot cross the Hill curves. The energetically allowed regions are represented by white in Figs. 2 and 3, while the dotted regions are energetically disallowed for the test particle. The two black dots represent the positions of the primaries $\mathcal{M}_1$ and $\mathcal{M}_2$ respectively.

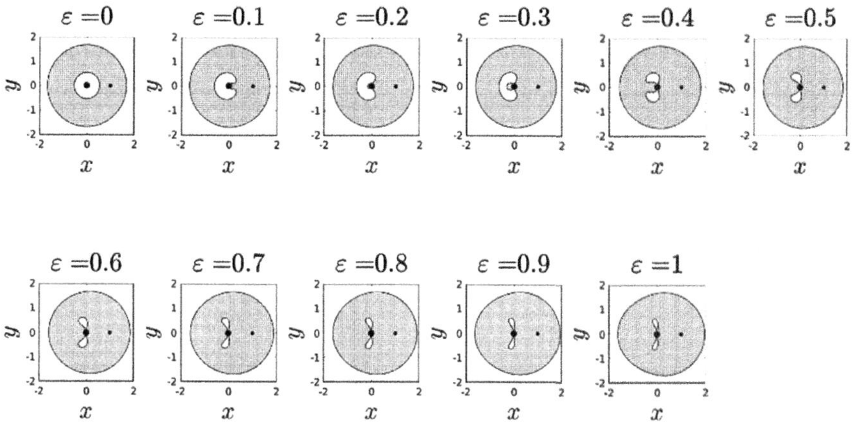

Fig. 2 Evolution of the Hill curves for $\mu_1 = \mu_2 = 0.001$ and $C_j = 4.0$ with the parameter ϵ. The white regions of the plot represents the points in the X-Y plane are energetically allowed, while the dotted regions are energetically disallowed, for the test particle whose Jacobian $C_j = 4.0$. The larger black dot on the left represents the position of the mass $\mathcal{M}_1$ and the smaller black dot on the right represents the position of mass $\mathcal{M}_2$ in each of the plots

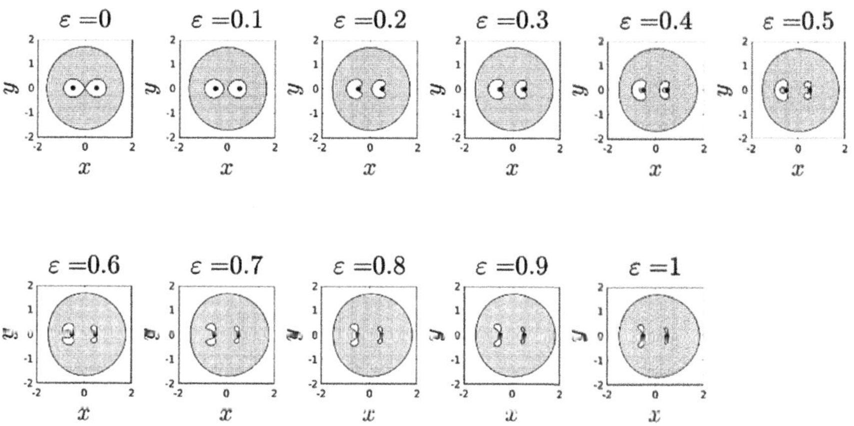

Fig. 3 Evolution of the Hill curves for $\mu_1 = \mu_2 = 0.5$ and $C_j = 4.0$ with the parameter ϵ. The white regions of the plot represents the points in the X-Y plane are energetically allowed, while the dotted regions are energetically disallowed, for the test particle whose Jacobian $C_j = 4.0$. The black dot on the left represents the position of the mass $\mathcal{M}_1$ and the black dot on the right represents the position of mass $\mathcal{M}_2$ in each of the plots

For the biased-mass system, the potential due to mass $\mathcal{M}_1$ dominates the Hill curves. The introduction of beyond-Newtonian effects distorts the curves of the Newtonian system, such that for all values of $\epsilon \gtrsim 0.0865$, no trapped circular orbits exist. For the Copenhagen system, the chosen value of C_j corresponds to the energy at the first Lagrange point L_1. As ϵ increases, the contribution of the spin becomes apparent and the enclosed allowed region becomes smaller. Circular trapped orbits around both the primaries exist for small values of ϵ. For $\epsilon > 0.1248$, circular orbits no longer exist around the primary $\mathcal{M}_2$ while for $\epsilon > 0.134$, circular orbits no longer exist around the primary $\mathcal{M}_1$. Thus, for both systems, we choose our initial conditions in the unbounded energetically allowed region for the sake of consistency of initial conditions for all values of ϵ, μ_1 and μ_2.

3.2 Orbits

Using the six stepped, fifth-order Runge-Kutta method implemented with the Dortmund-Prince algorithm, the equations of motion equations (15, 16) are integrated using time step $\tau = 10$ for $n = 3000$ iterations. For a preliminary investigation of the system, the following initial conditions are considered (similar to [33] which investigates orbits for a system with Schwarzschild like primaries): $x_0 = [3.0, 3.5, 3.75, 4.0, 4.25, 4.5, 4.75, 5.0, 5.25, 5.5, 6.0]$, $y_0 = 0.0$ and $\dot{x}_0 = 0.0$, with $C_j = 4$. The value of $\dot{y}_0(x_0, y_0, \dot{x}_0)$ is calculated from the following equation:

$$\begin{aligned} \dot{x}_0 &= \frac{y_0}{r_0}\sqrt{(x^2+y^2) - 2\Omega(x,y) - C_j} \\ \dot{y}_0 &= -\frac{x_0}{r_0}\sqrt{(x^2+y^2) - 2\Omega(x,y) - C_j} \end{aligned} \tag{19}$$

where $r_0 = \sqrt{x_0^2 + y_0^2}$. The orbits for a test particle for the biased mass and Copenhagen systems are investigated for $\epsilon \in [0, 1]$ and the set of initial conditions mentioned in the paragraph above. Since the system is conservative, the Jacobi constant C_j has to remain constant as the equations of motion are integrated.

The integrator used, being non-symplectic in nature, usually does not conserve the Jacobian. The use of such integrators for conservative systems have been well studied and multiple corrective methods, such as the velocity correction method [31,76,121,122], have been developed for better accuracy. In Fig. 4, we have shown a comparison of the relative error in the Jacobi constant C_j with time for both the non-corrected and velocity corrected integrators. It is observed that the accuracy in the conservation of C_j for the velocity corrected method ranges from 10^{-16}–10^{-14} for stable orbits and goes up to 10^{-8} for chaotic and sticky orbits at large times ($>5 \times 10^3$ years), as has been pointed out in [121,122]. We also observe that our non-corrected integrator has a fairly similar accuracy at the start. However, the growth in error is faster at late times and reaches values of 10^{-10} for stable orbits and goes up to 10^{-8} for chaotic and sticky orbits. Hence, we conclude that the non-corrected fifth-order Runge-Kutta method is also of reasonable accuracy for the relevant time-periods of our investigation.

By observing their evolution, the orbits may be categorized as regular, sticky or escaping. Orbits are said to be sticky if they show regular behavior for a long period of time before their chaotic nature manifest [35] and escaping if the particle directly escapes from the system without executing any regular orbits [20,21]. We classify the stability of the initial conditions based on the number of iterations for which the orbit of the particle is stable. If the test particle executes stable orbits for 3000

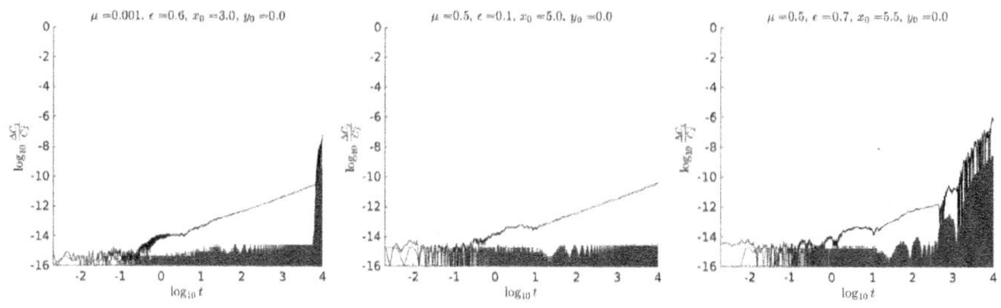

Fig. 4 Plot of the log of the relative error in the Jacobi constant C_j with log of time for the non-corrected Runge-Kutta (4,5) integrator using the Dormand-Prince algorithm (black) and velocity-corrected 4th order Runge-Kutta integrator (grey). The first plot from the left ($\mu = 0.001$, $\epsilon = 0.6$, $x_0 = 3.0$, $y_0 = 0.0$) shows the evolution of C_j with time for a sticky initial condition, the centre plot ($\mu = 0.5$, $\epsilon = 0.1$, $x_0 = 5.0$, $y_0 = 0.0$) shows the evolution of C_j with time for a stable initial condition, and the plot on the right ($\mu = 0.5$, $\epsilon = 0.7$, $x_0 = 5.5$, $y_0 = 0.0$) shows the evolution of C_j with time for a chaotic initial condition

iterations, it is classified as regular. If the orbits are stable for at-least 100 iterations before they escape from the system, they are classified as sticky. If the test particle reaches a distance of 50 times the separation between the two primaries within 1000 iterations, they are said to be escaping.

For the biased mass system, among the initial conditions considered, orbits for $x_0 = [3.5, 3.75, 4.0, 4.25, 4.5]$ are stable for all values of ϵ. Most initial conditions are either sticky or escaping for non-zero values of ϵ. But the interesting initial conditions are the ones where the intermediate values of ϵ are the most chaotic. The initial conditions $x_0 = [5.0, 5.25, 5.5, 6.0]$ show such behavior. For the Copenhagen system, orbits for $x_0 = [3.5, 3.75, 4.0, 4.5, 4.75]$ are stable for all values of ϵ. The initial condition $x_0 = 4.25$ destabilizes for $\epsilon > 0.4$, implying a region of chaotic initial conditions interjects stable initial conditions in the phase space. Orbits for $x_0 = [5.0, 5.25]$ are either sticky or escaping for all values of ϵ except $\epsilon = [0.0, 0.1]$. A stable orbit for $x_0 - 5.5$ exists only for $\epsilon = 0.0$, while no stable orbits exist for $x_0 = 6.0$. for any value of ϵ. This implies that regions of initial conditions allowing stable orbits shrink as ϵ increases for the Copenhagen system.

Contrary to expectation, $\epsilon = 1.0$ does not result in the maximum number of sticky and escaping initial conditions in either of the systems. Instead, the intermediate values of ϵ have the most number of unstable initial conditions. For the biased mass system, $\epsilon = [0.2, 0.4, 0.6, 0.7, 0.8]$ have the least number of stable initial conditions, namely 6 out of the 11 investigated. $\epsilon = 0.9$ has the least number of stable initial conditions for the Copenhagen system, namely 5 out of the 11 investigated. In contrast, $\epsilon = 1.0$ has 9 and 6 initial conditions out of 11 for the biased mass and Copenhagen systems, respectively.

3.3 Fixed Points and Their Stability

For a system having equilibrium points (fixed points), the necessary and sufficient conditions are:

$$\dot{x} = \dot{y} = \ddot{x} = \ddot{y} = 0 \tag{20}$$

Thus, the co-ordinates of the co-planar fixed points are determined by solving the following pair of partial differential equations (refer to Eq. (10) for the detailed expression of $\Omega(x, y)$) [110]:

$$\frac{\partial \Omega(x, y)}{\partial x} = \frac{\partial \Omega(x, y)}{\partial y} = 0 \tag{21}$$

The intersection of the curves for Eq. (21) for a set of values of μ_1, μ_2 and ϵ gives us a set of fixed points for the system. Figures 5 and 6 show the positions of the fixed points for $\epsilon = [0.0, 0.3, 0.5, 0.7, 1.0]$ for the biased-mass and Copenhagen systems respectively. It is evident that for both systems, the number of fixed points is highly dependent on the value of ϵ. A summary of the salient features of the fixed points with respect to ϵ is presented below:

- For $\epsilon = 0$, both the biased-mass and Copenhagen systems reduce to their Newtonian counterparts. These systems have five fixed-points each, as expected.
- For $\epsilon = 0.3$, the biased-mass system has five fixed points while the Copenhagen system has nine.
- For $\epsilon = 0.5$, the biased mass system has nine fixed points while the Copenhagen system has thirteen. The less massive primary in the biased system has three collinear fixed points.
- For both $\epsilon = 0.7$ and $\epsilon = 1.0$, the biased mass system has five fixed points. Only the collinear fixed points in either of the systems is beyond the less massive primary. However, the Copenhagen system has nine equilibrium points for both $\epsilon = 0.7$ and $\epsilon = 1.0$.
- Finally, the more massive primary in the biased mass system as well as both the primaries in the Copenhagen system have two non-collinear equilibrium points very near to it for values of $\epsilon \geq 0.3$ (not shown in Figs. 5 and 6 since they fall very close to the primaries).

It is thus evident that the number of equilibrium points for both the biased-mass and the Copenhagen systems become maximum at intermediate values of ϵ.

Now, moving on to the issue of stability of these fixed points, their linear stability may be determined by Taylor expanding the system's equations of motion around the fixed point (x_0 ,y_0) upto first order. In the perturbation equations, the time-independent coefficient matrix of variations is identified as

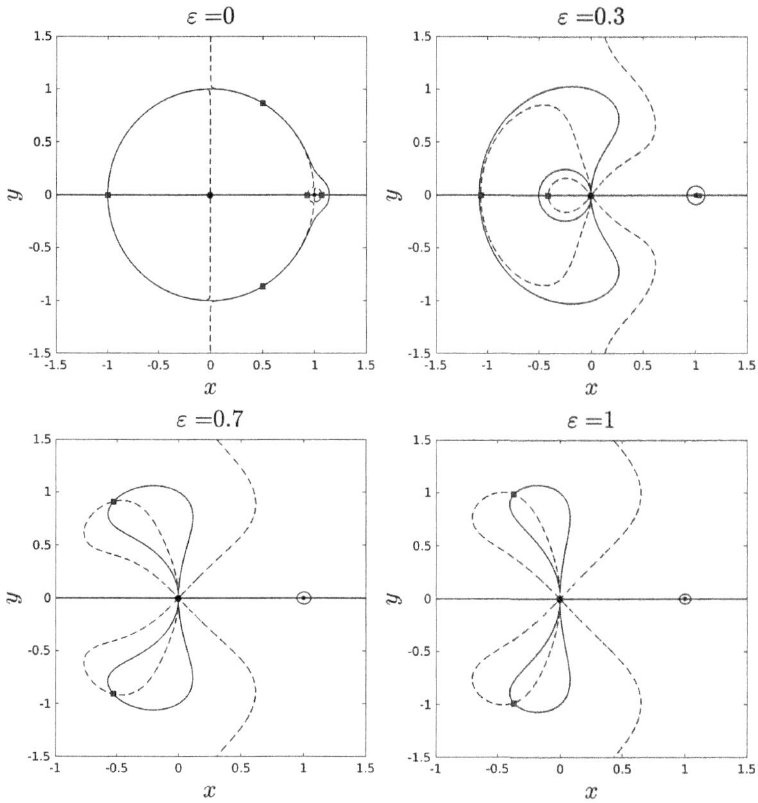

Fig. 5 Locations of some of the equilibrium points of the biased-mass system ($\mu_1 = \mu_2 = 0.001$), marked by grey squares, on the intersection of $\partial\Omega/\partial x = 0$, marked by the dashed line, and $\partial\Omega/\partial y = 0$, marked by the solid line, for $\epsilon = [0.0, 0.3, 0.7, 1.0]$. For $\epsilon \neq 0$, there are two non-collinear equilibrium points on either sides of the more massive primary which could not be shown on the plots due to their close proximity to it. The smaller primary has three more collinear equilibrium points, one of which lies between the two primaries. These too could not be marked on the plots due to their proximity to the primary

$$A = \begin{bmatrix} 0 & 0 & 1 & 0 \\ 0 & 0 & 0 & 1 \\ \frac{\partial^2 \Omega_0}{\partial x^2} & \frac{\partial^2 \Omega_0}{\partial x \partial y} & 0 & 2 \\ \frac{\partial^2 \Omega_0}{\partial y \partial x} & \frac{\partial^2 \Omega_0}{\partial y^2} & -2 & 0 \end{bmatrix} \tag{22}$$

where the subscript 0, attached to the partial derivatives of second order of Ω, denotes evaluation at the position of the equilibrium point (x_0, y_0). The necessary and sufficient condition that a fixed point is stable is that all the eigenvalues of matrix A be purely imaginary. Applying this method to the fixed points, obtained by numerically solving Eq. (21), we can conclude the following:

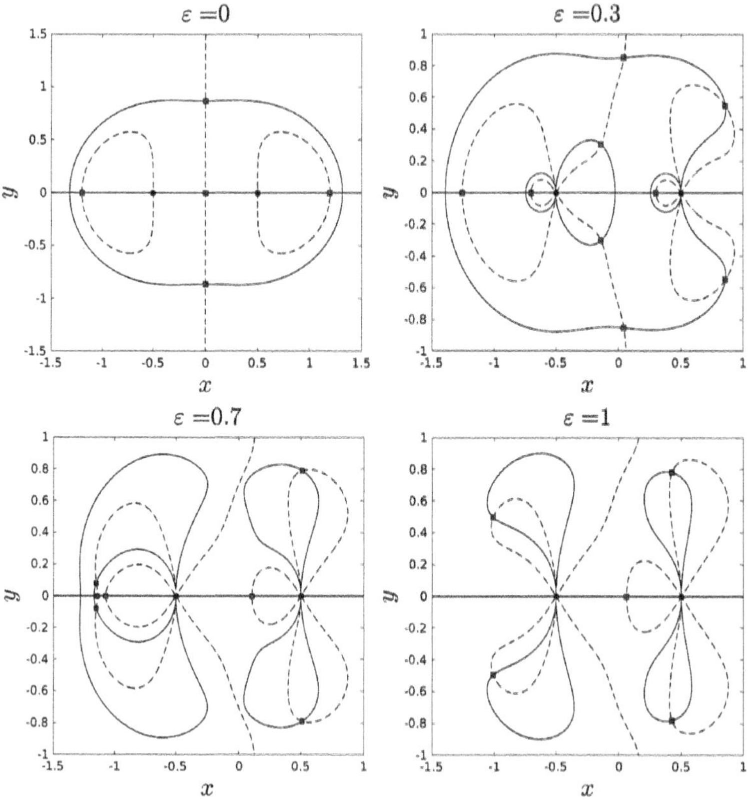

Fig. 6 Locations of some of the equilibrium points of the Copenhagen system ($\mu_1 = \mu_2 = 0.05$), marked by gray squares, on the intersection of $\partial\Omega/\partial x = 0$, marked by the dashed line, and $\partial\Omega/\partial y = 0$, marked by the solid line, for $\epsilon = [0.0, 0.3, 0.7, 1.0]$. For $\epsilon \neq 0$, there are two non-collinear equilibrium points on either sided of both primaries which could not be shown on the plots due to their close proximity to them

- For $\epsilon = 0$, the collinear fixed points for both the biased-mass and the Copenhagen systems are unstable while the triangular fixed points are stable.
- For $\epsilon = 0.3$, none of the fixed points are stable for the biased mass system while two fixed points are stable for the Copenhagen system.
- For $\epsilon = 0.5$, one fixed point is stable for the biased mass system while two are stable for the Copenhagen system.
- For $\epsilon = 0.7$, no fixed point is stable for the biased mass system while one is stable for the Copenhagen system.
- For $\epsilon = 1.0$, no fixed point is stable for either of the systems.

Thus, the evolution and stability of the fixed points of the system under consideration show non-trivial evolution with the parameter ϵ. However, the knowledge about the basins of convergence along with the libration points is of prime importance since the attracting domains reflect some of the most intrinsic properties of the dynamical system. This has been a topic of intense research in recent years for many different dynamical systems such as the Hill problem [32], the four-body problem [5,66,146] and the pseudo-Newtonian planar circular restricted three body problem [145,147, 149]. We plan to investigate these aspects for our beyond-Newtonian potential in detail as part of our future work.

4 Poincaré Map of Section

The Poincaré map, or the first return map, is a powerful and conventional tool for examining the motion of dynamical systems [33,90,114]. In order to construct the map, we evolve the system for 3000 iterations in time-steps of $\tau = 10$ and plot the section of the orbit for $y = 0.0$, $\dot{y} < 0$. This is done for 11 initial conditions $x_0 = [3.0, 3.5, 3.75, 4.0, 4.25, 4.5, 4.75, 5.0, 5.25, 5.5, 6.0]$ and $y_0 = 0.0$ for $\epsilon =$ 0.0, 0.4, 0.8 and 1.0. Figures 7 and 8 show the evolution of the Poincaré map for the biased-mass and and Copenhagen systems respectively. The Poincaré map of a system primarily shows two types of structures: concentric Kolmogorov-Arnold-Moser (KAM) tori which represent bounded, quasi-periodic motions and a sea of chaotic points surrounding such tori. At the centre of each island of concentric tori is a point which corresponds to a stable, periodic and resonant orbit [15,45,53]. The extent of the sea of scattered points in comparison to islands of the tori provides a visual representation of the extent of chaos in the system. For the biased-mass system, the initial conditions $x_0 = [3.5, 3.75, 4.0, 4.25, 4.5]$ show KAM tori on their Poincaré maps for all values of ϵ, implying quasi-periodic orbits. For $x_0 = [4.75, 5.0, 5.25, 5.5, 6.0]$, the destruction of their KAM tori implies chaotic or sticky orbits, as was observed in Sect. 3.2. For the Copenhagen system, the Poincaré maps $x_0 = [3.5, 3.75, 4.0, 4.5, 4.75]$ show KAM tori for all values of ϵ. The torus for $x_0 = 3.0$ breaks up only for $\epsilon = 0.9$, while the tori for $x_0 = 5.5$ and $x_0 = [5.0, 5.25]$ break up for $\epsilon > 0.0$ and $\epsilon > 0.1$ respectively. No KAM tori appear for $x_0 = 6.0$ for any value of ϵ, implying that the initial condition is chaotic for all values of ϵ. Thus, the Poincaré maps for both the systems corroborate the observations presented in Sect. 3.2.

5 Lyapunov Characteristic Exponents

A very popular indicator of chaos in dynamical systems is the calculation of the Lyapunov Characteristic Exponents (LCE), which has been extensively applied to the study of chaos in celestial dynamics especially in the context of the three-body problem [33,34,47,126,128]. It is a measure of the exponential divergence of two neighbouring trajectories in phase space. The rate of separation of the two trajectories is dependent on the initial separation vector. For a pair of trajectories, the number of exponents for the system is equal to dimension of its phase space. However, the largest exponent dominates in the limit $t \to \infty$. The largest Lyapunov exponent, called the Maximal Lyapunov Exponent (MLE), is defined by,

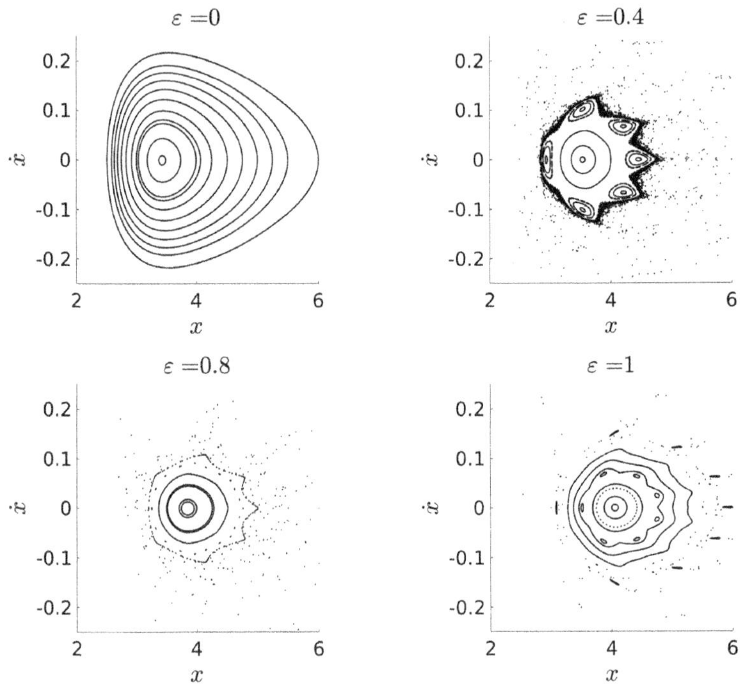

Fig. 7 The Poincaré map for orbits with x_0=[3.0, 3.5, 3.75, 4.0, 4.25, 4.5, 4.75, 5.0, 5.25, 5.5, 6.0] and $y_0 = 0.0$ for $\epsilon = 0.0, 0.4, 0.8, 1.0$. The system is evolved for 3000 iterations in time-steps of $\tau = 10$ and the Poincaré map of section for $y = 0.0$, $\dot{y} > 0.0$ is plotted for all 11 initial conditions for the biased-mass system ($\mu_1 = \mu_2 = 0.001$)

$$\Lambda_{max} = \lim_{t \to \infty} \frac{1}{t} \log \frac{||\Upsilon(t)||}{||\Upsilon(0)||} \tag{23}$$

where $\Upsilon(t)$ is the solution to the variational equations for the potential under consideration [116]. Such a computation mechanism for the MLE is called the variational method and is the most accurate. However, for systems such as the one under consideration where computation of the variational equations are cumbersome, an alternative was introduced in [11]. The Eq. 23 is thus replaced by the following:

$$\Lambda_{max} = \lim_{t \to \infty} \frac{1}{t} \log \frac{||\delta \mathbf{x}(t)||}{||\delta \mathbf{x}(0)||} \tag{24}$$

where, the deviation vector between the two trajectories is $\delta \mathbf{x}(t)$, with $\delta \mathbf{x}(0) \to 0$. The mean rate of deviation of the two trajectories is given by:

$$\Lambda_{max} = \frac{1}{n\tau} \sum_{k=1}^{n} \log \frac{||\delta \mathbf{x}(k\tau)||}{||\delta \mathbf{x}(0)||} \tag{25}$$

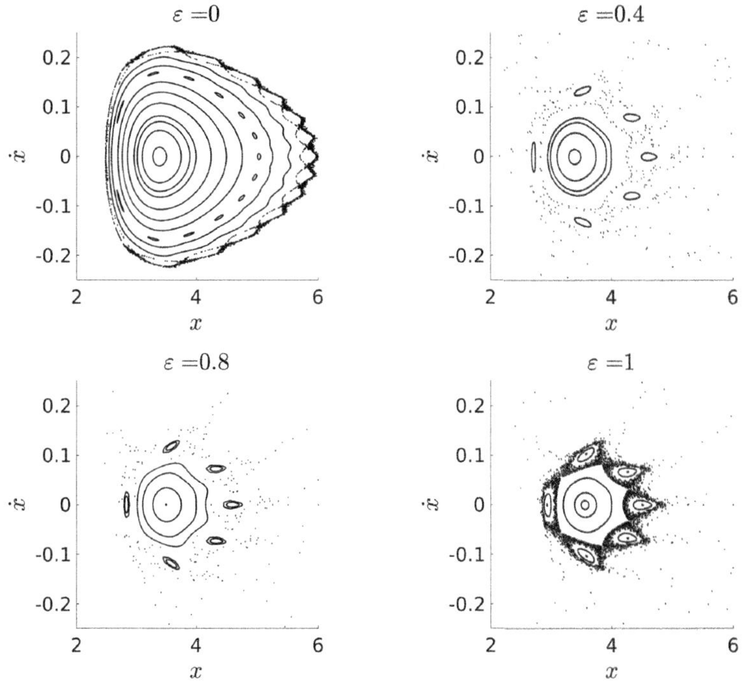

Fig. 8 The Poincaré map for orbits with x_0=[3.0, 3.5, 3.75, 4.0, 4.25, 4.5, 4.75, 5.0, 5.25, 5.5, 6.0] and $y_0 = [0.0]$ for $\epsilon = 0.0, 0.4, 0.8$ and 1.0. The system is evolved for 3000 iterations in time-steps of $\tau = 10$ and the Poincaré map of section for $y = 0, \dot{y} < 0$ is plotted for all 11 initial conditions for the Copenhagen system ($\mu_1 = \mu_2 = 0.5$)

This method is called the two-particle method and is the one we utilize to calculate the MLE for each system. The result is accurate as long as the two trajectories are in the immediate neighbourhood of each other in phase space and the machine used for computation has enough precision. As concluded in the work by [116], we have also taken the initial separation between the two trajectories to be $\delta\mathbf{x}(0) = 10^{-8}$ and have integrated the system in double precision for $n = 10^5$ iterations, each of time step $\tau = 0.1$. Numerical integration diverges rapidly unless the deviation vectors are re-normalized periodically. The two trajectories are evolved separately and the deviation vector is re-normalized using the Gram-Schmidt re-normalization after each time step. To get a quantitative representation of the chaos in the system, the MLE is averaged over the entire phase space. But, as a preliminary investigation of system, we use the initial conditions: $x_0 = [3.0, 3.5, 3.75, 4.0, 4.25, 4.5, 4.75, 5.0, 5.25, 5.5, 6.0]$, $y_0 = 0.0$ and $\dot{x}_0 = 0.0$, with $C_j = 4.0$ for the biased-mass and Copenhagen systems. If trajectories are stable, the value of the MLE remains very small, usually less than 5×10^{-4} (low value of MLE). But for chaotic trajectories, the deviations are exponential and the value of the MLE increases rapidly with time. After 10^5 iterations, its value is usually greater than 5×10^{-4} (high value of MLE). The MLE for initial conditions $x_0 = [3.0, 3.5, 3.75, 4.0, 4.25, 4.5, 4.75, 5.0, 5.25, 5.5, 6.0]$ are calculated and averaged for each value of ϵ and is called the Lyapunov Characteristic

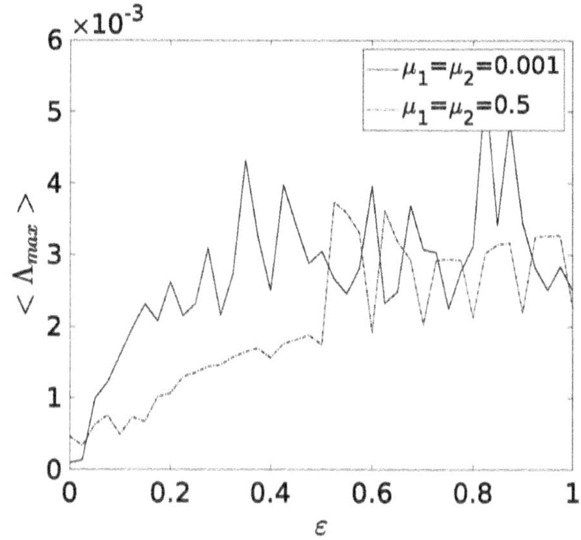

Fig. 9 Plot of the Lyapunov Characteristic Exponent (LCE) versus ϵ

Exponent (LCE) for the particular value of ϵ [84]. The LCE provides a qualitative measure of the amount of chaos in the system, even for the few initial conditions chosen for the study (see [84] as an example). Figure 9 is the plot of the LCE against ϵ for both the biased-mass and Copenhagen systems.

Both for the biased-mass system and the Copenhagen system, the total chaos in the system for small ϵ is low. For the biased-mass system, the LCE for all the initial conditions are $<5 \times 10^{-4}$ for values of $\epsilon = 0.0$ indicating stable orbits. The same is true for the Copenhagen system, except for $x_0 = 6.0$ which gives an LCE of 4.22×10^{-03}. For the biased-mass system, the LCE for all initial conditions are $<5 \times 10^{-4}$ for $x_0 = [3.5, 3.75, 4.0, 4.5]$, implying stable orbits. Some initial conditions, like $x_0 = [4.25, 4.75, 5.0, 5.25, 6.0]$ for the biased mass system and $x_0 = 5.5$ for the Copenhagen system, the system shows high values of LCE for intermediate values of ϵ, but low values of LCE for higher values of ϵ. The most interesting among these is the initial condition $x_0 = 6.0$, which shows low values of LCE only for $\epsilon = 0.0$ and 1.0. This reaffirms the conclusion drawn from the Poincaré maps that the chaos in the system is maximum for intermediate values of ϵ. For the Copenhagen system, values of LCE for $x_0 = [3.5, 3.75, 4.0, 4.5, 4.75]$ are low for all values of ϵ. The initial condition $x_0 = 6.0$ show high values of LCE for values of ϵ. The initial condition $x_0 = 5.5$ shows high values of LCE for all values of ϵ except for $\epsilon = 0.0$.

Figure 9 shows that the chaos in the system is low for both the biased-mass system and the Copenhagen system, as indicated by low values of the averaged LCE. Its value rises rapidly for the biased mass system but much slower for the Copenhagen system. Both the systems show maximum values of the averaged LCE for intermediate values of epsilon, which for the biased mass system is at $\epsilon = 0.825$ and $\epsilon = 0.525$ for the Copenhagen. This re-iterates the observations made from the orbital evolution and the Poincaré maps of the systems.

6 Schwarzschild and Kerr Primaries: A Comparison

In order to examine the effect of the spin of the primaries on the system, we present a comparison to a system with two Schwarzschild like primaries. Using the potential described in [33, 145], we construct a set of Poincaré maps of section for $\epsilon = [0.1, 0.5, 1.0]$. We evolve each orbit for 3000 iterations in time-steps of $\tau = 10$ and plot the section of the orbit for $y = 0.0$ and $\dot{y} < 0.0$. Figures 10 and 11 show Poincaré maps of the biased-mass and Copenhagen systems respectively. For both

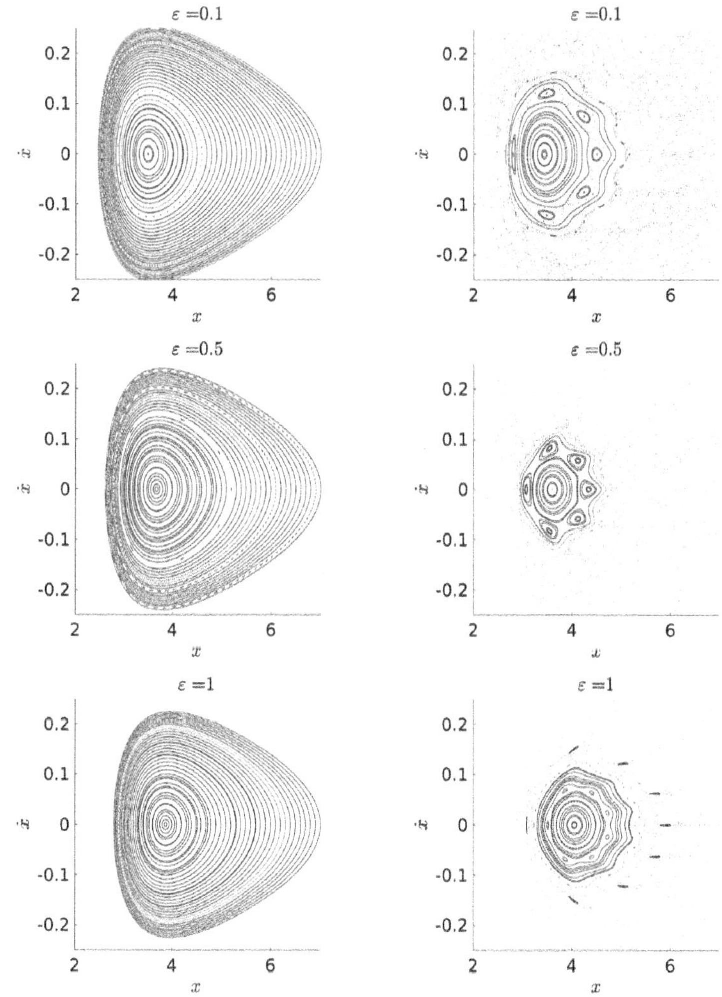

Fig. 10 Poincaré map of section for $y = 0$ and $\dot{y} < 0.0$ for the biased-mass systems (mass ratio of the primaries equals to 0.001) for different values of ϵ. The figures on the left are maps for the system with Schwarzschild-like primaries while those on the right are for the system with Kerr-like primaries

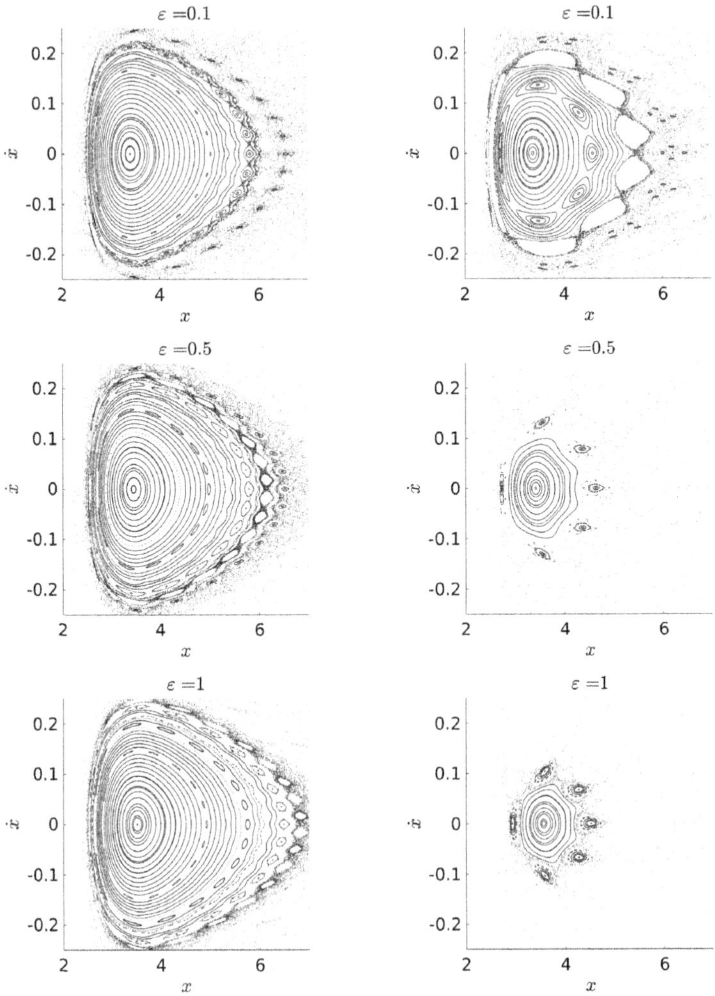

Fig. 11 Poincaré map of section for $y = 0.0$ and $\dot{y} < 0.0$ for the Copenhagen systems (mass ratio of the primaries equals to 0.5) for different values of ϵ. The figures on the left are maps for the system with Schwarzschild-like primaries while those on the right are for the system with Kerr-like primaries

the systems, the plots on the left are for the system with Schwarzschild like primaries, while those on the right are for the system with Kerr like primaries.

While for $\epsilon = 0.0$ both systems reduce to the Newtonian CRTBP, it is apparent that even for small perturbations to the Newtonian system, as represented by $\epsilon = 0.1$, the introduction of the spin destabilizes a number of initial conditions. For the biased-mass case, the Schwarzschild system shows all chosen initial conditions to be stable and quasi-periodic, with the Poincaré map showing KAM tori for all values of ϵ. The Poincaré map for the Kerr system differs radically from its Schwarzschild counterpart even for $\epsilon = 0.1$, showing a large sea of chaotic points surrounding an island of stable initial conditions. The island of stability grows smaller as ϵ is increased, as has already been discussed in Sect. 4. For the Copenhagen case, the Poincaré maps for both the Schwarzschild and Kerr systems feature an island of stability surrounded by a sea of chaos for all three values of ϵ. The Poincaré maps for both the systems look alike, implying a similar number of stable initial conditions. As ϵ is increased, the number of stable initial conditions for the Kerr system decreases rapidly, as evident from the smaller islands of stability on the Poincaré maps of the system for $\epsilon = 0.5$ and $\epsilon = 1.0$. However, for the same values of ϵ, the number of stable initial conditions for the Schwarzschild system remains approximately the same.

Thus, for both mass ratios, we observe that the introduction of spin in the CRTBP with Schwarzschild-like primaries destabilizes a number of initial conditions, with the amount of chaos in the system growing with increase in ϵ.

7 Conclusions

In the present paper, we present a beyond-Newtonian potential for the planar circular restricted three-body problem with Kerr like primaries. This is achieved by using the Fodor-Hoenselaers-Perjés procedure to expand the Kerr metric and by retaining corrections up to the first non-Newtonian term. The system is conservative, with the Hamiltonian being time independent. The parameter $\epsilon \in [0.0, 1.0]$ is introduced in order to facilitate the observation of the system as it transitions from the Newtonian to the beyond-Newtonian regime. The dynamics of a test particle in this potential for $\mu_1 = \mu_2 = 0.001$ (or the biased-mass system) and for $\mu_1 = \mu_2 = 0.5$ (or the Copenhagen system), are inspected for a Jacobi constant $C_j = 4.0$. For an initial investigation of the system, orbits for a few selected initial conditions are plotted. A short analysis of the fixed points of the systems and their stability is undertaken. A purely Newtonian CRTBP system is known to have five Lagrange points, as seen for $\epsilon = 0$ in our case. However, number of Lagrange points is not constant as the system transitions from the Newtonian to the beyond-Newtonian regime. It is observed that the number of fixed points strongly depends on the parameter ϵ as does their stability. Next, the stability of the orbits is also examined through the use of the Poincaré map of section for different values of ϵ. The Poincaré maps for all non-zero values of ϵ show islands of stability constructed of concentric Kolmogorov-Arnold-Moser (KAM) tori, embedded in a sea of chaos.

Thus we note that the introduction of the parameter ϵ helps us to conclude that even small perturbations to the Newtonian CRTBP destabilizes the system for both the cases. If we track the evolution of the system keeping the Jacobian constant fixed, a stable orbit in the Newtonian system is observed to become either chaotic or sometimes even remain regular in the beyond-Newtonian limit. In the limits $\epsilon = 0$ and $\epsilon = 1$, the phase space is seen to be filled mostly with periodic orbits, rarely interspersed with chaotic ones. However, as ϵ departs even slightly from zero, trajectories that were stable in the Newtonian system become unstable. It is seen that in most of the cases (for a given set of initial conditions) whose phase space is bounded in the classical regime, correspond to unbounded trajectories in the non-Newtonian regime. This implies that both systems become largely unstable for intermediate values of ϵ. The instability of the orbits can possibly be linked to the observed lack of stable fixed points in both the systems. This is also confirmed by the Lyapunov Characteristic Exponent, calculated for each value of ϵ, which is in accordance to the conclusions made by several authors earlier for different systems [33,53,84]. In conclusion, we may say that even the smallest corrections to the Newtonian circular restricted three-body problem could drastically change the stability and the dynamics of the system.

In addition, we would like to note that an in-depth study of the phase space using more rapid indicators of chaos, like Fast Lyapunov Indicators (FLI) [42,43,128], Small Alignment Index (SALI) [105] and General Alignment Index (GALI) [106] will facilitate a much more detailed analysis of the evolution of the Lagrange points of the proposed potential. Coupled with this, a detailed linear stability analysis of the Lagrange points as a function of the parameter ϵ and an analysis of the basins of convergence is expected to reveal more information about the attractors of the system. Further, we would also like to investigate the degree of equivalence of the potential constructed in our paper with the pseudo-Newtonian potential formulation of a binary with spinning primaries, for example that of a system modelled by the superposition of two Artemova potentials [4]. This would allow us to reproduce features like the Innermost Stable Circular Orbit (ISCO), maximally stable orbits, and the horizon radius, in our chosen scalings and units. This would in turn facilitate the calculation of physically relevant distances, for example, the coordinates of fixed points for different values of ϵ and primary masses in real physical units, thereby allowing us to predict real astrophysical scenarios using our present model (for a recent example refer to [136]). Thus, we would like to explore these issues in greater depth as part of our future work.

We also note that the current formalism is strictly valid for particles whose motion is restricted to the plane containing the primaries. However, a more general model for accreting particles should also include a study of the dynamics of such off-axis motion. Thus, we would like to direct our future studies to incorporate such effects for off-axis halo particles in a generalized beyond-Newtonian framework.

Acknowledgements We would like to acknowledge our collaborator, Mr. Shounak De, for his help and assistance for typesetting, proof reading and discussions. He was also an integral part of this work.

References

1. B.P. Abbott et al., Phys. Rev. Lett. **116**, 061102 (2016)
2. B.P. Abbott et al., Phys. Rev. X **6**, 041015 (2016)
3. P. Amaro-Seoane et al., Class. Quant. Grav. **29**, 124016 (2012)
4. V. Artemova, G. Bjoernsson, I.D. Novikov, ApJ **461**, 565 (1996)
5. A.N. Baltagiannis, K.E. Papadakis, Int. J. Bifurc. Chaos **21**, 2179 (2011)
6. R. Barrio, F. Blesa, S. Serrano, New J. Phys. **11**, 053004 (2009)
7. V. Beckmann, C.R. Shrader, *Active Galactic Nuclei* (Wiley-VCH Verlag GmbH, 2012)
8. M.C. Begelman, R.D. Blandford, M.J. Rees, Nature **287**, 307 (1980)

9. L. Benet, D. Trautman, T. Seligman, Celest. Mech. Dyn. Astron. **66**, 203 (1996)
10. L. Benet, T. Seligman, D. Trautman, Celest. Mech. Dyn. Astron. **71**, 167 (1998)
11. G. Benettin, L. Galgani, J.M. Strelcyn, Phys. Rev. A **14**, 2338 (1976)
12. S. Bleher, C. Grebogi, E. Ott, Physica D **46**, 87 (1990)
13. S. Bleher, E. Ott, C. Grebogi, Phys. Rev. Lett. **63**, 919 (1989)
14. S. Bleher, C. Grebogi, E. Ott, R. Brown, Phys. Rev. A **38**, 930 (1998)
15. H. Broer, F. Takens, *Dynamical Systems and Chaos* (Springer, New York, 2010)
16. R.C. Chen, X. Wu, Commun. Theor. Phys. **65**, 321 (2016)
17. R. Churchill, G. Pecelli, D.L. Rod, J. Differ. Equ. **17**, 329 (1975)
18. M. Connors, P. Wiegert, C. Veillet, Nature **475**, 481 (2011)
19. G. Contopoulos, in *In Memoriam D. Eginitis. D. Kotsakis* (Athens: Ed. Athens.) (1976)
20. G. Contopoulos, A&A **231**, 41 (1990)
21. G. Contopoulos, D. Kaufmann, A&A **253**, 379 (1992)
22. G. Contopoulos, M. Harsoula, G. Lukes-Gerakopoulos, Celest. Mech. Dyn. Astron. **113**, 255 (2012)
23. N.J. Cornish, J. Levin, Phys. Rev. Lett. **89**, 179001 (2002)
24. N.J. Cornish, J. Levin, Phys. Rev. D **68**, 024004 (2003)
25. T. Damour, P. Jaranowski, G. Schäfer, Phys. Rev. D **63**, 044021 (2001)
26. T. Damour, P. Jaranowski, G. Schäfer, Phys. Rev. D **66**, 029901 (2002)
27. S. De, S. Roychowdhury, R. Banerjee, Mon. Not. R. Astron. Soc. **501**, 713 (2021)
28. V.C. de Andrade, L. Blanchet, G. Faye, Class. Quant. Grav. **18**, 753 (2001)
29. S.C. de Assis, M.O. Terra, Celest. Mech. Dyn. Astron. **120**, 105 (2014)
30. G.M. de Castro, P.S. Letelier, Class. Quant. Grav. **28**, 225020 (2011)
31. C. Deng, X. Wu, E. Liang, MNRAS **496**, 2946 (2020)
32. C.N. Douskos, Ap&SS **326**, 263 (2010)
33. F.L. Dubeibe, F.D. Lora-Clavijo, G.A. González, Phys. Lett. A **381**, 563 (2016)
34. F.L. Dubeibe, F.D. Lora-Clavijo, G.A. González, Ap&SS **362**, 97 (2017)
35. R. Dvorak, Stickiness in Dynamical Systems, in *The Dynamics of Small Bodies in the Solar System*. NATO ASI Series (Series C: Mathematical and Physical Sciences), vol. 522, ed. by B.A. Steves, A.E. Roy (Springer, Dordrecht, 1999)
36. B. Eckhardt, Physica D **33**, 89 (1988)
37. A. Einstein, L. Infeld, B. Hoffmann, Ann. Math. **39**, 65 (1938)
38. A. Ernst, T. Peters, MNRAS **443**, 2579 (2014)
39. F.J. Ernst, Phys. Rev. **167**, 1175 (1968)
40. F.J. Ernst, Phys. Rev. **168**, 1415 (1968)
41. G. Fodor, C. Hoenselaers, Z. Perjés, J. Math. Phys. **30**, 2252 (1989)
42. C. Froeschlé, E. Lega, R. Gonczi, Celest. Mech. Dyn. Astr. **67**, 41 (1997)
43. C. Froeschlé, E. Lega, Celest. Mech. Dyn. Astr. **78**, 167 (2000)
44. R.J. Geroch, Math. Phys. **11**, 1955 (1970)
45. M. Gidea, F. Deppe, G. Anderson, AIP Conf. Proc. **886**, 139 (2007)
46. A. Gopakumar, C. Königsdörffer, Phys. Rev. D **72**, 121501 (2005)
47. E. Gueron, P.S. Letelier, A&A **368**, 716 (2001)
48. B. Gwak, Entropy **21**, 1017 (2019)
49. R.O. Hansen, J. Math. Phys. **15**, 46 (1974)
50. M.D. Hartl, A. Buonanno, Phys. Rev. D **71**, 024027 (2005)
51. M. Henon, *Generating Families in the Restricted Three-Body Problem* (Springer, Berlin, 1997)
52. G.W. Hill, Acta Math. **8**, 1 (1886)
53. G. Huang, X. Wu, Phys. Rev. D **89**, 124 (2014)
54. G. Huang, X. Wu, Gen. Relativ. Gravit. **46**, 1798 (2014)
55. G. Huang, X. Ni, X. Wu, Eur. Phys. J. C **74**, 3012 (2014)
56. L. Huang, X. Wu, D. Ma, Eur. Phys. J. C **76**, 488 (2016)
57. H.E. Kandrup, C. Siopis, G. Contopoulos, R. Dvorak, Chaos **9**, 381 (1999)
58. J. Klačka, M. Kocifaj, MNRAS **390**, 1491 (2008)
59. S. Komossa, V. Burwitz, G. Hasinger, P. Predehl, J.S. Kaastra, Y. Ikebe, ApJ **582**, L15 (2003)

60. O. Kopáček, V. Karas, ApJ **787**, 12 (2014)
61. O. Kopáček, V. Karas, J. Phys. Conf. Ser. 012070 (2015)
62. O. Kopáček et al., in *Mathematics and Astronomy: A Joint Long Journey: Proceedings of the International Conference*, vol. 1283, p. 278 (2010)
63. J. Kovář, Z. Stuchlík, V. Karas, Class. Quant. Grav. **25**, 095011 (2008)
64. J. Kovář, O. Kopáček, V. Karas, Z. Stuchlík, Class. Quant. Grav. **27**, 135006 (2010)
65. E. Krefetz, AJ **72**, 471 (1967)
66. R. Kumari, B.S. Kushvah, Ap&SS **349**, 693 (2014)
67. Y.C. Lai, T. Tél, *Transient Chaos* (Springer, New York, 2011)
68. P.S. Letelier, W.M. Vieira, Phys. Rev. D **56**, 8095 (1997)
69. M. Levi, J. Steinhoff, JCAP **12**, 003 (2014)
70. J. Levin, Phys. Rev. Lett. **84**, 3515 (2000)
71. C. Lhotka, A. Celletti, Icarus **250**, 249 (2015)
72. D. Li, X. Wu, E. Liang, Annalen der Physik **531**, 1900136 (2019)
73. D. Li, Y. Wang, C. Deng, X. Wu, Eur. Phys. J. Plus **135**, 390 (2020)
74. J. Luo, X. Wu, G. Huang, F. Liu, ApJ **834**, 64 (2017)
75. T.I. Maindl, *Astronomical Society of the Pacific Conference Proceedings*, vol. 107, p. 147 (1996)
76. D.Z. Ma, X. Wu, J.F. Zhu, New Astron. **13**, 216 (2008)
77. L. Mei, M. Ju, X. Wu, S. Liu, MNRAS **435**, 2246 (2013)
78. E. Meletlidou, S. Ichtiaroglou, F.J. Winterberg, Celestial Mech. Dynam Astron. **80**, 145 (2001)
79. A.E. Motter, Y.C. Lai, Phys. Rev. E **65**, 1 (2002)
80. B. Mukhopadhyay, ApJ **581**, 427 (2002)
81. F. Müller-Sánchez, J.M. Comerford, R. Nevin, R.S. Barrows, M.C. Cooper, J.E. Greene, ApJ **813**, 103 (2015)
82. C.D. Murray, S.F. Dermott, *Solar System Dynamics* (Cambridge University Press, Cambridge, 1999)
83. Z.E. Musielak, B. Quarles, Rep. Prog. Phys. **77**, 6 (2014)
84. S. Nag, S. Sinha, D.B. Ananda, Ap&SS **362**, 81 (1974)
85. J. Nagler, Phys. Rev. E **69**, 066218 (2004)
86. J. Nagler, Phys. Rev. E **71**, 026227 (2005)
87. J.F. Navarro, J. Henrard, A&A **369**, 1112 (2001)
88. E. Ott, T. Tel, Chaos **3**, 417 (1993)
89. B. Paczyński, P.J. Wiita, A&A **88**, 23 (1980)
90. T.S. Parker, L.O. Chua, in *Practical Numerical Algorithms for Chaotic Systems* (Springer, New York, NY, 1989)
91. J.M. Petit, M. Hénon, Icarus **66**, 536 (1986)
92. J.M. Petit, M. Hénon, A&A **173**, 389 (1987)
93. R. Ruffini et al., ApJ **831**, 174 (2016)
94. J.D. Schnittman, ApJ **724**, 39 (2010)
95. J.D. Schnittman, F.A. Rasio, Phys. Rev. Lett. **87**, 121101 (2001)
96. J.M. Seoane, J. Aguirre, M.A.F. Sanjuán, Y.C. Lai, Chaos **16**, 1 (2006)
97. O. Semerák, V. Karas, A&A **343**, 325 (1999)
98. O. Semerák, P. Suková, MNRAS **404**, 545 (2010)
99. O. Semerák, P. Suková, MNRAS **425**, 2455 (2012)
100. O. Semerák, P. Suková, MNRAS **436**, 978 (2013)
101. C. Simó, T.J. Stuchi, Physica D **140**, 1 (2000)
102. C.V. Siopis, G. Contopoulos, H.E. Kandrup, New York Acad. Sci. Ann. **751**, 205 (1995)
103. C.V. Siopis, H.E. Kandrup, G. Contopoulos, R. Dvorak, New York Acad. Sci. An **773**, 221 (1995)
104. C.V. Siopis, H.E. Kandrup, G. Contopoulos, R. Dvorak, Celest. Mech. Dyn. Astron. **65**, 57 (1996)
105. C. Skokos, J. Phys. A: Math. Gen. **34**, 10029 (2001)
106. C. Skokos, T.C. Bountis, C. Antonopoulos, Physica D: Nonlinear Phenomena **231**, 30 (2007)
107. T.P. Sotiriou, T.A. Apostolatos, Class. Quant. Grav. **21**, 5727 (2004)

108. A.F. Steklain, P.S. Letelier, Phys. Lett. A **352**, 398 (2006)
109. A.F. Steklain, P.S. Letelier, Phys. Lett. A **373**, 188 (2009)
110. S.H. Strogatz, *Nonlinear Dynamics and Chaos with Student Solutions Manual: With Applications to Physics, Biology, Chemistry, and Engineering* (CRC Press, 1994)
111. X. Su, X. Wu, F. Liu, Ap&SS **361**, 32 (2016)
112. V. Szebehely, AJ **68**, 147 (1963)
113. V. Szebehely, *Theory of Orbits* (Academic Press, NY, 1967)
114. M. Tabor, *Chaos and Integrability in Nonlinear Dynamics: An Introduction* (Wiley, New York, 1989)
115. M. Takahashi, H. Koyama, ApJ **693**, 472 (2009)
116. G. Tancredi, A. Sánchez, F. Roig, AJ **121**, 1171 (2001)
117. W.M. Vieira, P.S. Letelier, ApJ **513**, 383 (1999)
118. D. Vogt, P.S. Letelier, Phys. Rev. D **68**, 084010 (2003)
119. Y. Wang, X. Wu, Class. Quant. Grav. **28**, 025010 (2011)
120. H. Wang, G.Q. Huang, Commun. Theor. Phys. **64**, 159 (2015)
121. S. Wang, X. Wu, F. Liu, MNRAS **463**, 1352 (2016)
122. S. Wang, G. Huang, X. Wu, AJ **155**, 67 (2018)
123. O.C. Winter, C.D. Murray, QMW Maths Notes, No. 16, Queen Mary and Westfield College, London, UK (1994)
124. O.C. Winter, C.D. Murray, QMW Maths Notes, No. 17, Queen Mary and Westfield College, London, UK (1994)
125. V. Witzany, O. Semerák, P. Suková, MNRAS **451**, 1770 (2015)
126. X. Wu, T.Y. Huang, Phys. Lett. A **313**, 77 (2003)
127. X. Wu, H. Zhang, ApJ **652**, 1466 (2006)
128. X. Wu, T.Y. Huang, H. Zhang, Phys. Rev. D **74**, 083001 (2006)
129. X. Wu, Y. Xie, Phys. Rev. D **76**, 124004 (2007)
130. X. Wu, Y. Xie, Phys. Rev. D **77**, 103012 (2008)
131. X. Wu, Y. Xie, Phys. Rev. D **81**, 084045 (2010)
132. X. Wu, S.Y. Zhong, Gen. Rel. Grav. **43**, 2185 (2011)
133. X. Wu, L. Mei, G. Huang, S. Liu, Phys. Rev. D **91**, 024042 (2015)
134. X. Wu, G. Huang, MNRAS **452**, 3167 (2015)
135. K. Yamada, H. Asada, Phys. Rev. D **82**, 104019 (2010)
136. M. Yi, X. Wu, Phys. Scr. **95**, 8 (2020)
137. S.Y. Zhong, X. Wu, Phys. Rev. D **81**, 104037 (2010)
138. S.Y. Zhong, X. Wu, S.Q. Liu, X.F. Deng, Phys. Rev. D **82**, 124040 (2010)
139. E.E. Zotos, Nonlinear Dyn. **76**, 1301 (2014)
140. E.E. Zotos, Nonlinear Dyn. **78**, 1389 (2014)
141. E.E. Zotos, MNRAS **446**, 770 (2015)
142. E.E. Zotos, Nonlinear Dyn. **82**, 357 (2015)
143. E.E. Zotos, Nonlinear Dyn. **85**, 1613 (2016)
144. E.E. Zotos, Nonlinear Dyn. **89**, 2 (2017)
145. E.E. Zotos, Ap&SS **362**, 195 (2017)
146. E.E. Zotos, Ap&SS **362**, 2 (2017)
147. E.E. Zotos, Chaos. Solitons and Fractals **101**, 8 (2017)
148. E.E. Zotos, F.L. Dubeibe, G.A. González, MNRAS **477**, 5388 (2018)
149. E.E. Zotos, M.S. Suraj, M. Jain, R. Aggarwal, Int. J. Non-Linear Mech. **105**, 43 (2018)

Extended Gravity Models

Neutron Stars in General Relativity and in Alternative Gravity Theories: Quasinormal Modes and Universal Relations

Fech Scen Khoo

Abstract

We begin by briefly reviewing neutron stars in GR. Next we discuss some recent results of quasinormal modes of neutron stars in alternative gravity theories such as the scalar-tensor theories, in particular $\mathcal{R}^2$ gravity theory. We present the modes which are astrophysically relevant, and also new channel of modes arising from the additional scalar degree of freedom in the alternative theories. We devote our discussions mainly on polar perturbations. We propose several universal relations considering equations of state which cover nucleons, hyperons and quarks. Finally we compare the modes and the universal relations that we obtained between GR and the alternative theories, and discuss their possible signatures.

1 Introduction

Interests in alternative theories of gravity spark from our unsatisfactory understanding of predictions from General Relativity (GR) for dark matter and dark energy. On the other hand, we are still in the quest for a theory of quantum gravity which is able to fully describe both quantum mechanics and gravity. Evidently, GR might not be the ultimate gravitational theory. From an alternative gravity theory, the theory parameter can be studied and constrained according to the current observations. Predictions from the theory such as the speed of the gravitational waves is one of the fundamental properties of the theory. Such predictions when compared to observations can decide the fate of the respective alternative gravity theory proposal [7]. In 2017, the first direct detection of gravitational waves from a neutron star merger system (GW170817) was reported by LIGO/Virgo Collaboration [2]. GW170817 signal came from a low-mass system, and the gravitational waves were shortly followed by

F. S. Khoo (✉)
University of Oldenburg, D-26111 Oldenburg, Germany
e-mail: fech.scen.khoo@uni-oldenburg.de

B. Hartmann and J. Kunz (eds.), *Gravity, Cosmology, and Astrophysics*,
Lecture Notes in Physics 1022, https://doi.org/10.1007/978-3-031-42096-2_6

a γ-ray burst. This detection came about two years after the very first observation of gravitational waves emitted from a binary black hole system (GW150914) [1]. With the GW170817 detection, the era of multi-messenger gravitational wave astronomy was launched. Following the increasing number of detections we are getting, and many more to follow as more advanced detectors are under development (e.g. [23]), probing gravity in these strong gravity regimes as in neutron stars, and improving our understanding in the nature of cosmic acceleration have become possible than ever before.

We are particularly interested in testing GR and other alternative theories in the ringdown phase of a binary system, thus in the late part of the signal. This part of the signal critically carries the characteristics of the final object, and therefore evidences of the no-hair theorem. That is, we can extract the mass and spin parameters of the final object. These characteristics are contained in the quasinormal mode (QNM), which is a composition of a resonant frequency and an exponential damping time. To make such investigations, perturbation theory is required.

Here we work with a class of alternative gravity theories, a particular scalar-tensor theory (STT) which adds a dynamical scalar field to the gravity theory. We will consider cases when the scalar field is massive or massless, where in the massive case, it is known as $\mathcal{R}^2$ gravity. When there are additional fields coupled to gravity, the speed of the gravitational waves can differ from the standard GR's. From GW170817, an almost immediate observation of both gravitational wave and its electromagnetic counterpart (the burst that occurred about 2s later) places a stringent bound on the gravitational wave speed. This bound rules out STTs which predict an anomalous speed for the gravitational wave [17].

In this contribution, we will focus on a static and spherically symmetric neutron star in alternative gravity theories which belong to a certain class of viable STTs. Our discussions of the results on the QNMs and universal relations are mainly based on our previous work in [11–14].

2 Neutron Stars

Neutron star is a compact astrophysical object, and is considered highly compact when its radius R is less than or equal to three times of its mass M, i.e. $R \leq 3M$. A neutron star can be formed from a supernova, where the iron core of a massive star collapses and forms the neutron star. The matter density in the center of the star can exceed 10^{15}gcm^{-3}. At such high densities, neutron stars provide a natural example of cosmic laboratory to study extreme physics, for instance studies of matter under extreme density and pressure. The most internal structure of a neutron star is however still not fully understood. There have since been various nuclear physics models proposed to describe the matter content of the neutron star [18]. In general, the star has a layered structure from the surface to the center of it, with each layer of different densities, sizes and particle compositions. Starting from the outermost layer is the atmosphere dominated by magnetic field, followed by an outer crust, an inner crust where neutrons are in a superfluid phase, an outer core where many-body nucleon interaction occurs, and finally an inner core which is composed of unknown ultra-dense matter. Temperature has an effect on the behaviour of the external layers, while the behaviour of matter at high densities influences the stellar structure itself.

To build a neutron star model, an equation of state (EOS) that describes the matter of the star is required. Basically, the EOS provides the relation of energy density ρ and pressure p. How these quantities are related depends on the exact matter content and interactions which are present. The EOS becomes more model dependent at high densities. For simplicity, we will treat the matter inside the star as a continuous medium, and the fluid velocity u to be continuous, thus modeling the matter as a perfect fluid. The stress-energy tensor of a perfect fluid is given by

$$T^{\mu\nu} = (p + \rho)u^{\mu}u^{\nu} - pg^{\mu\nu}\,, \tag{1}$$

where

$$\nabla_{\nu}T^{\mu\nu} = 0\,, \tag{2}$$

with the background metric $g^{\mu\nu}$. A barotropic EOS is simply given by

$$\rho = \rho(p)\,. \tag{3}$$

At the surface of the star, the pressure p vanishes.

Below are some examples of EOS, categorized in different matter contents.

(a) Pure nuclear matter (i.e. neutrons, protons, electrons, and muons): SLy, APR4
(b) Mixed hyperon-nuclear matter: GNH3, H1, H4, BGN1H1, WCS1, WCS2, BHZBM
(c) Hybrid stars with hyperons and quark color-superconductivity: BS1, BS2, BS3, BS4
(d) Quark stars: WSPHS1, WSPHS2

Typically, stars composed of hyperon matter have a larger radius than the stars composed of nuclear matter, despite having the same mass [8]. For more discussions on neutron stars, see e.g. [5,16,20].

2.1 Neutron Stars in GR

From the well-known Einstein-Hilbert action ($G = c = 1$),

$$S[g] = \frac{1}{16\pi}\int d^4x\sqrt{-g}\mathcal{R} + S_{matter}\,, \tag{4}$$

where $\mathcal{R}$ is the curvature scalar and S_{matter} is the contribution from the matter in the star, the Einstein tensor is given by

$$\mathcal{R}_{\mu\nu} - \frac{1}{2}\mathcal{R}g_{\mu\nu} = 8\pi\big((p+\rho)u_{\mu}u_{\nu} - pg_{\mu\nu}\big)\,, \tag{5}$$

where a perfect fluid for the matter is assumed. Considering a spherically symmetric spacetime inside and around a static neutron star, the metric is given by

$$ds^2 = e^{2\nu(r)}dt^2 - e^{2\lambda(r)}dr^2 - r^2(d\theta^2 + sin^2\theta\, d\varphi^2)\,. \tag{6}$$

Inside the star, the metric functions $e^{2\nu(r)}$ and $e^{2\lambda(r)}$ are independent, while outside of the star, the spacetime is Schwarzschild with a gravitational mass M. The Tolman-Oppenheimer-Volkov (TOV) equations inside the star are

$$\frac{dm}{dr} = 4\pi r^2 \rho(r)\,, \tag{7}$$

$$\frac{dp}{dr} = -(p+\rho)\frac{m + 4\pi r^3 p}{(r-2m)r}\,, \tag{8}$$

$$\frac{d\nu}{dr} = -\frac{1}{p+\rho}\frac{dp}{dr}\,. \tag{9}$$

Integrating Eq. (7) from the center of the star to its radius $r = R$ gives us the total mass of the star M. The parameter m comes from the expression $e^{-2\lambda(r)} = 1 - 2m(r)/r$.

Assuming a small pulsation of the star, we can describe the oscillation using the first-order perturbation theory. The metric is perturbed by,

$$g_{\mu\nu} = g^{(0)}_{\mu\nu}(r) + h_{\mu\nu}(t, r, \theta, \varphi)\,. \tag{10}$$

$g^{(0)}$ is the static configuration, and h is the perturbation where $|h| << |g^{(0)}|$. The perturbation function depends on the radial, time, and angular coordinates . We can decompose these perturbations into a product of temporal-radial part and spherical harmonics Y_{lm},

$$h_{\mu\nu}(t, r, \theta, \varphi) = \sum_{l,m} h_{\mu\nu}(t, r)\, Y_{lm}(\theta, \varphi)\,. \tag{11}$$

The components of the stress-energy tensor are also perturbed with

$$p = p^{(0)} + \delta p\,, \quad \rho = \rho^{(0)} + \delta\rho\,, \quad u = u^{(0)} + \partial_t \xi\,, \tag{12}$$

where ξ is the Lagrangian displacement of a fluid element.

Based on the angular parts, under the parity transformation $\theta \to \pi - \theta$, $\varphi \to \pi + \varphi$, the perturbations can be grouped into two types:

(a) Axial perturbations, which transform with a factor of $(-1)^{l+1}$.
(b) Polar perturbations, which transform with a factor of $(-1)^l$.

These two types of perturbations are completely decoupled when the background is static. However, this is no longer true if the star is rotating. Axial perturbations are associated with the spacetime perturbations, while the pressure and density perturbations are polar perturbations.

The axial and polar perturbations of the metric are given by

$$h_{\mu\nu}^{\rm ax} = \begin{pmatrix} 0 & 0 & h_0^l S_\theta^l & h_0^l S_\varphi^l \\ 0 & 0 & h_1^l S_\theta^l & h_1^l S_\varphi^l \\ h_0^l S_\theta^l & h_1^l S_\theta^l & -h_2^l \frac{X^l}{\sin\theta} & h_2^l \sin\theta\, W^l \\ h_0^l S_\varphi^l & h_1^l S_\varphi^l & h_2^l \sin\theta\, W^l & h_2^l \sin\theta\, X^l \end{pmatrix}, \tag{13}$$

and

$$h_{\mu\nu}^{\rm pol} = \begin{pmatrix} H_0^l Y^l & H_1^l Y^l & L_0^l Y^l_{,\theta} & L_0^l Y^l_{,\varphi} \\ H_1^l Y^l & H_2^l Y^l & L_1^l Y^l_{,\theta} & L_1^l Y^l_{,\varphi} \\ L_0^l Y^l_{,\theta} & L_1^l Y^l_{,\theta} & r^2(K^l Y^l + G^l W^l) & r^2 G^l X^l \\ L_0^l Y^l_{,\varphi} & L_1^l Y^l_{,\varphi} & r^2 G^l X^l & r^2 \sin^2\theta A^l \end{pmatrix}, \tag{14}$$

where $A^l = K^l Y^l - G^l W^l$, $X^l = 2(Y^l_{,\theta\varphi} - Y^l_{,\varphi}\cot\theta)$, $W^l = Y^l_{,\theta\theta} - Y^l_{,\theta}\cot\theta - Y^l_{,\varphi\varphi}/\sin^2\theta$, $S^l_\theta = -Y^l_{,\varphi}/\sin\theta$, $S^l_\varphi = Y^l_{,\theta}\sin\theta$, and Y^l is the spherical harmonics. Due to the spherical symmetry, we could drop the angular number m. The functions h_0^l, h_1^l, h_2^l belong to the axial-parity metric perturbations, and H_0^l, H_1^l, H_2^l, K^l, G^l, L_0^l, L_1^l are the polar-parity metric perturbations. All these functions depend on t and r. This system of equations can be simplified further by applying the Regge-Wheeler gauge: $L_0^l = L_1^l = G^l = h_2^l = 0$ [24].

Outside the star, as it is vacuum, the solution is simply Schwarzschild. Using the harmonic time decomposition such that

$$h(t,r) = \int d\omega\, e^{-i\omega t} h(\omega, r)\,, \tag{15}$$

we obtain the following two coupled first-order axial perturbation equations from the standard Einstein equation,

$$\begin{aligned} \frac{d}{dr}h_0 = &-\frac{1}{\omega r^3}(r^3 h_1 \omega^2 + 2r^2 h_0 \omega + 2l^2 M h_1 - l^2 r h_1 + 2Mlh_1 - lrh_1 \\ &-4Mh_1 + 2rh_1)\,, \end{aligned} \tag{16}$$

$$\frac{d}{dr}h_1 = \frac{1}{r(2M-r)^2}(-r^3\omega h_0 + 4M^2 h_1 - 2Mrh_1)\,. \tag{17}$$

Formally, we can organize these equations into a single Schrödinger-like equation of second order. If we define $Z = \frac{h_1}{r}\left(1 - \frac{2M}{r}\right)$, we arrive at

$$\frac{d^2}{dr_*^2}Z + (\omega^2 - V_{RW}(r))Z = 0\,, \tag{18}$$

which is known as the Regge-Wheeler equation. The tortoise coordinate is given by $r_* = r + 2M \log\left(\frac{r}{2M} - 1\right)$, and the effective potential (Regge-Wheeler potential) is

$$V_{RW}(r) = \left(1 - \frac{2M}{r}\right)\left(\frac{l(l+1)}{r^2} - \frac{6M}{r^3}\right) . \tag{19}$$

The polar perturbation master equation also takes the same form as Eq. (18) but with the effective potential being the Zerilli potential [26],

$$V_{Ze}(r) = \left(1 - \frac{2M}{r}\right)\left(\frac{2n^2(n+1)r^3 + 6n^2Mr^2 + 18nM^2r + 18M^3}{r^3(nr+3M)^2}\right) ,$$

where $2n = (l-1)(l+2)$.

Inside the star there is pressure p, density ρ, and four-velocity u, they contribute to the perturbations as well. The pressure perturbation is

$$\delta p = \int d\omega\, e^{-i\omega t} \sum_l P_l(t,r) Y_l(\theta,\varphi) , \tag{20}$$

the energy density perturbation is

$$\delta \rho = \int d\omega\, e^{-i\omega t} \sum_l E_l(t,r) Y_l(\theta,\varphi) , \tag{21}$$

and the perturbation part of u is given by $\partial_t \xi$ where

$$\xi = \left(0\, , r^{l-1} e^{-(\nu+\lambda)} W_l Y_l\, , -r^{l-2} e^{-\nu} V_l \frac{\partial Y_l}{\partial \theta}\, , -\frac{r^l e^{-\nu} V_l}{r^2 \sin^2\theta} \frac{\partial Y_l}{\partial \varphi}\right) . \tag{22}$$

In total, the system of equations consists of four first-order differential equations for the functions H_1, K, W, V, three algebraic equations for H_0, H_2, P, and an EOS for E.

By now, we have transformed the time-dependent problem to a time-independent problem, through harmonic time decomposition, and the task becomes solving an eigenvalue equation (such as Eq. (18)). The QNM or eigenvalue ω is a complex number,

$$\omega = \omega_R + i\omega_I , \tag{23}$$

where the real part is the oscillation frequency and the imaginary part produces the damping time.

At infinity, the solution Z to the perturbation equation contains two components, incoming and outgoing waves which behave as

$$\lim_{r_* \to \infty} Z^{in} \sim e^{-i\omega r_*} , \quad \lim_{r_* \to \infty} Z^{out} \sim e^{i\omega r_*} . \tag{24}$$

Table 1 Frequency and damping time for some $l = 2$ QNMs in GR assuming APR1

Mode	Frequency ω_R	Damping time τ
f-mode	2047 Hz	144 ms
p_1-mode	6498 Hz	7.9 s
wI-mode	8303 Hz	40.5 μs

A physical solution should have the form of a purely outgoing wave at the infinity, without any contributions from the ingoing wave. By substituting the damping time,

$$\tau = -1/\omega_I \tag{25}$$

in Eq. (24), we get

$$\lim_{r_* \to \infty} Z^{in} \sim e^{-r_*/\tau} , \quad \lim_{r_* \to \infty} Z^{out} \sim e^{r_*/\tau} . \tag{26}$$

The incoming waves tend to zero exponentially as the the radial coordinate increases, while the outgoing waves are divergent at spatial infinity. The imaginary part ω_I of the QNM signifies an exponential damping when it is negative. This can be understood from the factor $e^{-i\omega t}$ in the perturbation functions (see e.g. Eq. (20)). The perturbations decay exponentially with time: $e^{-i(i|-\omega_I|)t} = e^{-\frac{t}{\tau}}$, energy is radiated away as gravitational radiation. In general, the QNM frequencies are labeled by an overtone number n. The fundamental mode is at $n = 0$, it is the least damped mode or in another word it is longer lived, and this mode dominates the ringdown waveform.

There are two families of modes in neutron stars. First are the spacetime modes, known as w-modes. They are found in both axial and polar perturbations. For instance, the curvature modes (wI-modes) which are the standard spacetime modes. These modes exist in every neutron star. Second are the fluid modes, which are associated with the energy and pressure oscillations (i.e. polar modes). An example of the fluid mode is the fundamental f-mode. It is a stable mode of non-radial oscillations, and is nearly independent of the details of the stellar structure. Another example is the pressure mode (p-mode) resulted from the pressure fluctuations. These modes on the other hand are sensitive to the EOS.

In Table 1 are some well-known results for $l = 2$ polar QNM in GR, for a neutron star with a mass of $2M_\odot$, assuming the EOS APR1.

Gravitational waves from the neutron star can be used to infer the mass and radius of the star, which are useful constraints on the EOS. Assuming we have extracted both the frequency and damping time from a detected f-mode or a p-mode, we could in principle deduce the stellar mass and radius [19]. To overcome the dependence of or the uncertainties in the EOS, we turn to constructing universal relations [3,4]. We can consider universal relations built out of rescaled frequencies and damping times. In [8], by scaling the real and imaginary part of the w-modes in terms of the central pressure of the star p_c (in cm^{-2}) such that

$$\bar{\omega}_R = 2\pi \frac{1}{\sqrt{p_c}} \frac{10^3}{c} \omega\,(\text{kHz})\,, \quad \bar{\omega}_I = \frac{1}{\sqrt{p_c}} \frac{10^6}{c} \frac{1}{\tau} (\mu\text{s})^{-1}\,, \tag{27}$$

where c is the speed of light in cm/s, the relation between the scaled quantities $\bar{\omega}_R$ and $\bar{\omega}_I$ becomes quite independent of the considered EOSs.

2.2 Neutron Stars in Alternative Gravity: $\mathcal{R}^2$ Gravity

$\mathcal{R}^2$ gravity stems from an extension of GR, namely as a special case of $f(\mathcal{R})$ gravity, $f(\mathcal{R}) = \mathcal{R} + a\mathcal{R}^2$ where a is the parameter of the alternative theory. An equivalence between $f(\mathcal{R})$ theories and a certain class of STTs was established in [25]. We consider the following form of the Einstein frame action in STT,

$$\begin{aligned} S[g,\phi] = &\frac{1}{16\pi} \int d^4x \sqrt{-g} \big(\mathcal{R} - 2g^{\mu\nu} \partial_\mu \phi \partial_\nu \phi - V(\phi)\big) \\ &+ S_{matter}(A^2(\phi) g_{\mu\nu}, \chi)\,, \end{aligned} \tag{28}$$

where the potential $V(\phi)$ for $\mathcal{R}^2$ gravity, and the standard Brans-Dicke coupling function $A(\phi)$ are

$$V(\phi) = \frac{1}{4a} \left(1 - e^{-\frac{2\phi}{\sqrt{3}}}\right)^2, \quad A(\phi) = e^{-\frac{1}{\sqrt{3}}\phi}\,. \tag{29}$$

The Einstein and Jordan frames are related by a conformal transformation of the metric. Both frames describe the same physics, and computations can be done in either frame [22]. The scalar field mass is related to the theory parameter by $m_\phi = 1/\sqrt{6a}$. There are two limits which can be taken in the $\mathcal{R}^2$ theory, one is approaching GR: $a \to 0$ (i.e. $m_\phi \to \infty$), and another is towards the massless Brans-Dicke theories: $a \to \infty$ (i.e. $m_\phi = 0$) which represents a maximal deviation from GR.

We consider a static and spherically symmetric background,

$$ds^2 = -e^{2\nu} dt^2 + e^{2\lambda} dr^2 + r^2 (d\theta^2 + \sin^2\theta d\varphi^2)\,. \tag{30}$$

To obtain the neutron star solutions, we need the field equations from the theory,

$$G_{\mu\nu} = T^{(\phi)}_{\mu\nu} + 8\pi T^{(matter)}_{\mu\nu} - \frac{1}{2} V(\phi) g_{\mu\nu}\,, \tag{31}$$

$$\nabla_\mu \nabla^\mu \phi = -4\pi \frac{1}{A} \frac{dA}{d\phi} T^{(matter)} + \frac{1}{4} \frac{dV}{d\phi}\,, \tag{32}$$

in addition to an EOS for the matter in Jordan frame $\bar{p}(\bar{\rho})$. Inside the star, the equations for the static functions are

$$\frac{1}{r^2}\frac{d}{dr}[r(1-e^{-2\lambda})] = 8\pi\rho + e^{-2\lambda}\left(\frac{d\phi}{dr}\right)^2 + \frac{1}{2}V\,, \tag{33}$$

$$\frac{2}{r}e^{-2\lambda}\frac{d\nu}{dr} - \frac{1}{r^2}(1-e^{-2\lambda}) = 8\pi p + e^{-2\lambda}\left(\frac{d\phi}{dr}\right)^2 - \frac{1}{2}V\,, \tag{34}$$

$$\frac{d^2\phi}{dr^2} + \left(\frac{d\nu}{dr} - \frac{d\lambda}{dr} + \frac{2}{r}\right)\frac{d\phi}{dr} = 4\pi(\rho - 3p)\frac{d\ln A}{d\phi}e^{2\lambda} + \frac{1}{4}\frac{dV}{d\phi}e^{2\lambda}\,, \tag{35}$$

$$\frac{dp}{dr} = -(\rho+p)\left(\frac{d\nu}{dr} + \frac{d\ln A}{d\phi}\frac{d\phi}{dr}\right)\,. \tag{36}$$

The density and pressure are given by $\rho = A^4\bar{\rho}$ and $p = A^4\bar{p}$, expressed in the physical Jordan frame (denoted by an overline).

For boundary conditions, we require regularity at the center of the star ($r = 0$), $\lambda(0) = 0$, $\frac{d\phi}{dr}(0) = 0$, and $\bar{\rho}(0) = \bar{\rho}_c$ where $\bar{\rho}_c$ is the central energy density. We also require asymptotic flatness at the infinity, $\lim_{r\to\infty}\nu(r) = 0$, and $\lim_{r\to\infty}\phi(r) = 0$.

Here we discuss briefly about the axial perturbations of a neutron star. The axial perturbation equations for $h_0(t,r)$ and $h_1(t,r)$ are

$$-e^{-2\nu}\partial_t^2 h_1 + e^{-2\nu}\left(\partial_r - \frac{2}{r}\right)\partial_t h_0 - \frac{(l-1)(l+2)}{r^2}h_1 = 0\,, \tag{37}$$

$$\partial_t h_0 - e^{\nu-\lambda}\partial_r(e^{\nu-\lambda}h_1) = 0\,, \tag{38}$$

which are derived from the field equations, using the metric perturbation Eq. (13) with the Regge-Wheeler gauge. We can substitute Eq. (38) into Eq. (37) and have a single equation for h_1. Furthermore, if we define $X = h_1\frac{e^{\nu-\lambda}}{r}$, the axial perturbations are described by the following time-dependent equation,

$$\begin{aligned}&\frac{\partial^2 X}{\partial t^2} - e^{\nu-\lambda}\frac{\partial}{\partial r}\left(e^{\nu-\lambda}\frac{\partial X}{\partial r}\right)\\ &+e^{2\nu}\left(\frac{l(l+1)}{r^2} - \frac{3}{r^2}(1-e^{-2\lambda}) + 4\pi A^4(\bar{\rho}-\bar{p}) + \frac{1}{2}V\right)X = 0\,.\end{aligned} \tag{39}$$

Next we let $X(r,t) = X(r)e^{i\omega t}$ to obtain a time-independent equation,

$$\begin{aligned}&e^{\nu-\lambda}\frac{d}{dr}\left(e^{\nu-\lambda}\frac{dX}{dr}\right) + \left(\omega^2 - e^{2\nu}\left(\frac{l(l+1)}{r^2} - \frac{3}{r^2}(1-e^{-2\lambda})\right.\right.\\ &\left.\left.+4\pi A^4(\bar{\rho}-\bar{p}) + \frac{1}{2}V\right)\right)X = 0\,.\end{aligned} \tag{40}$$

For detailed discussions and results on the axial perturbations, see [10]. In the next section, we will focus on the polar perturbations of the star and their corresponding results.

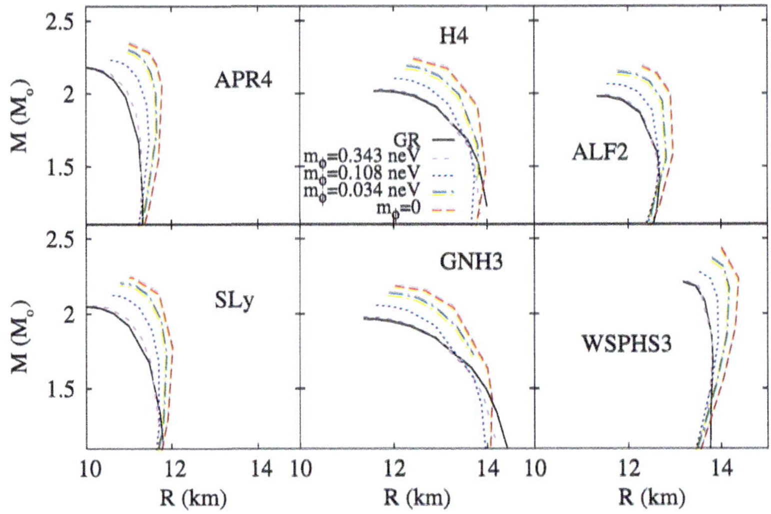

Fig. 1 Mass of neutron star M in solar masses versus its radius R in km for six EOSs under GR and several STTs

Finally we end this section by illustrating in Fig. 1 the profile of the star, given by its mass-radius relation which yields the compactness of the star M/R. In our work, we employ six contending EOSs of varying matter contents: SLy and APR4 of plain nuclear matter, GNH3 and H4 of a nucleon-hyperon mixture, ALF2 and WSPHS3 of a hybrid of nuclear-quark matter. In general, we consider the range of the scalar field mass m_ϕ from 0.011 to 0.343 neV that falls within the current observational constraints [15,21]. We can learn from Fig. 1 that a star composed of purely nuclear matter (SLy and APR4) tends to have a smaller radius than a star of the same mass with other compositions.

2.2.1 Polar Perturbations of a Neutron Star

We perturb all the dynamical fields in the theory, namely the metric, matter, and scalar field. The polar perturbations of the metric are,

$$h_{\mu\nu}^{\text{pol}} = \begin{pmatrix} 2Ne^{2\nu}Y_{lm} & -FY_{lm} & 0 & 0 \\ -FY_{lm} & -2e^{2\lambda}LY_{lm} & 0 & 0 \\ 0 & 0 & -2Tr^2Y_{lm} & 0 \\ 0 & 0 & 0 & -2Tr^2\sin^2\theta\, Y_{lm} \end{pmatrix}, \tag{41}$$

where Y_{lm} are the spherical harmonics, and F, L, N, T are functions of r, l, m, given by

$$F = i\omega r^{l+1}H_1\,,\; L = -\frac{1}{2}r^lH_2\,,\; N = \frac{1}{2}r^lH_0\,,\; T = -\frac{1}{2}r^lK\,. \tag{42}$$

The form of the matter perturbations is the same as in GR. In addition, we have perturbations of the scalar field from the theory,

$$\delta\phi = \int \sum_{l,m} r^l \phi_1 Y_{lm} e^{-i\omega t} d\omega \,, \tag{43}$$

where ϕ_1 is a function of r, l, m. Here let us define a perturbation function $\mathcal{X}$ in the following combinations of the functions V, W, H_0,

$$\mathcal{X} = \omega^2 \left(\bar{p}^{(0)} + \bar{\rho}^{(0)}\right) e^{-\nu} V - \frac{1}{r}\frac{d\bar{p}^{(0)}}{dr} e^{\nu-\lambda} W + \frac{1}{2}\left(\bar{p}^{(0)} + \bar{\rho}^{(0)}\right) e^{\nu} H_0 \,. \tag{44}$$

Then the system of first-order differential equations is given by

$$\frac{d}{dr}\Psi + \sigma\Psi = 0 \,, \tag{45}$$

for $\Psi = (K, H_1, W, \mathcal{X}, \phi_1, \frac{d\phi_1}{dr})$ where σ depends on the static functions ν, λ, $\phi^{(0)}$, $\bar{p}^{(0)}$, $\bar{\rho}^{(0)}$, the eigenvalue ω and angular number l.

Note that the asymptotic behaviour of the outgoing perturbation for the scalar field is $Z_s^{out} \sim e^{i\Omega r}$, where Ω satisfies the dispersion relation, $\Omega^2 = \omega^2 - m_\phi^2$. That is, the scalar field perturbations cannot propagate at the speed of light as the spacetime perturbations do.

We briefly describe here our numerical procedures in solving for the QNMs, and refer to [9] for more details. First we need to generate the static background solution. Next we generate an interior solution from the center up to a point outside the star, and an exterior solution from this point to infinity. QNMs are obtained when the interior and exterior solutions are continuous at this particular matching point, which can be changed as a test of the numerical stability. The results should nevertheless be independent of this auxiliary parameter.

In the next section, we present our main results for $l = 2$ quadrupole modes, the fundamental modes which dominate the ringdown spectrum after a merger, in particular the f-modes which are dominant in neutron stars. On top of that, we also present the results for $l = 2$ scalar-led ϕ-modes. These will be followed by the results from $l = 0$ radial modes which are propagating modes in this theory.

2.2.2 Quasinormal Modes and their Application in Universal Relations

In this section, we present some recent findings on the use of QNMs in the construction of universal relations. Earlier work can be found in [4]. We consider both scenarios when the scalar field carries a mass or when it is massless, and with a focus on the QNMs computed for two multipolar numbers, $l = 2, 0$. The quadrupole $l = 2$ modes are most astrophysically relevant and considered most dominant in the gravitational emissions, while the monopole $l = 0$ modes are normal modes in GR, their nature change in STTs as we will see in the following.

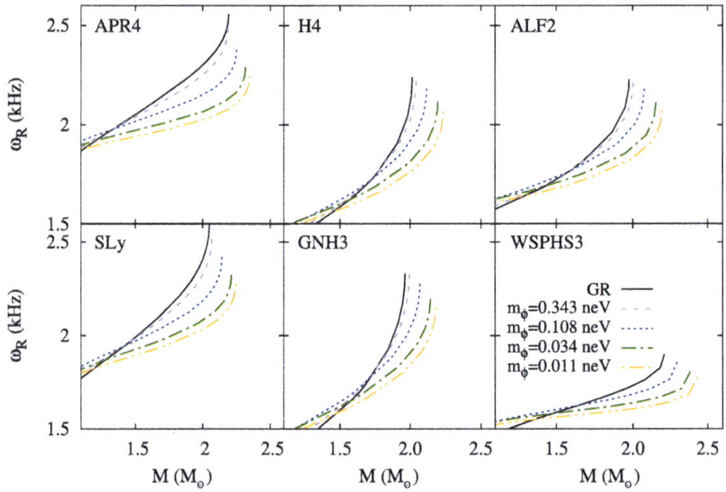

Fig. 2 Real frequency ω_R in kHz for $l = 2$ f-modes versus the mass of neutron star M in solar masses for six EOSs under GR and four STTs

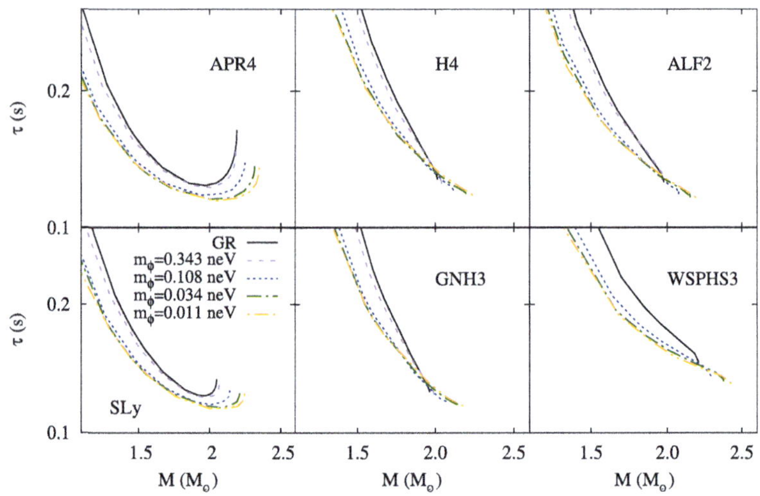

Fig. 3 Damping time τ in seconds for $l = 2$ f-modes versus the mass of neutron star M in solar masses for six EOSs under GR and four STTs

Figures 2 and 3 show the spectrum of the fundamental quadrupole $l = 2$ fluid f-mode. We show the comparison of the modes ranges from GR to $\mathcal{R}^2$ theories of different scalar field mass m_ϕ. Recall that m_ϕ is inversely proportional to the parameter a in the $\mathcal{R}^2$ theory. GR leads a higher frequency and a longer damping time for more compact neutron stars, while the modes produced by the theory with $m_\phi = 0.343$ neV closely follow GR. The deviation from GR grows as the mass of the scalar field decreases.

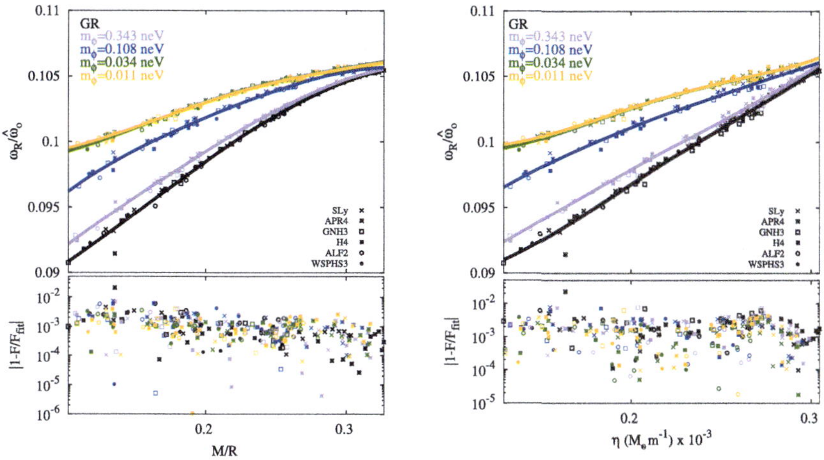

Fig. 4 Universal relations for dimensionless real frequency of $l = 2$ f-mode against compactness M/R (left) and generalized compactness η of the star (right)

Using these quadrupole modes, we obtain the following universal relations for neutron stars in the respective $\mathcal{R}^2$ theories, shown in Fig. 4. We observe an almost independence from the various EOSs, when we consider the real frequency part of the modes scaled with a reference frequency $\hat{\omega}_o$ against the (generalized) compactness of the star. The reference frequency $\hat{\omega}_o$ is given by

$$\hat{\omega}_o = \frac{c}{M}\sqrt{\frac{3}{4}\eta^3}\,, \tag{46}$$

where c is the speed of light, M is the mass of the star, and generalized compactness $\eta = \sqrt{M^3/I}$ where I is the moment of inertia. Under these relations, GR exhibits the lowest scaled frequency across the (generalized) compactness of the star, and as the scalar field mass decreases, the corresponding scaled frequency grows.

Next, we turn to another family of modes: the scalar ϕ-mode which is present due to the scalar degree of freedom from the alternative theory, coupled to the metric and matter perturbations. In GR limit this boils down to a minimally coupled scalar field in the background of the star. We present in Figs. 5 and 6 the massless ϕ-mode from a Brans-Dicke type STT, and its GR limit. The real frequencies of the modes show little dependence on the mass of the star, and the frequencies from GR remain larger than the massless alternative theory. The damping times for lighter neutron stars in the massless theory are close to GR but differ more significantly for heavier stars. These ϕ-modes possess shorter damping times (in ms) than the previously discussed f-modes, therefore they decay faster comparatively. Nonetheless, these scalar modes can provide a new channel of emission [6].

Similarly, with these computed QNMs, we obtain the universal relations shown in Fig. 7. We notice a good universality with respect to the six EOSs considered, by scaling the real frequency with the mass of the star against the star compactness, and

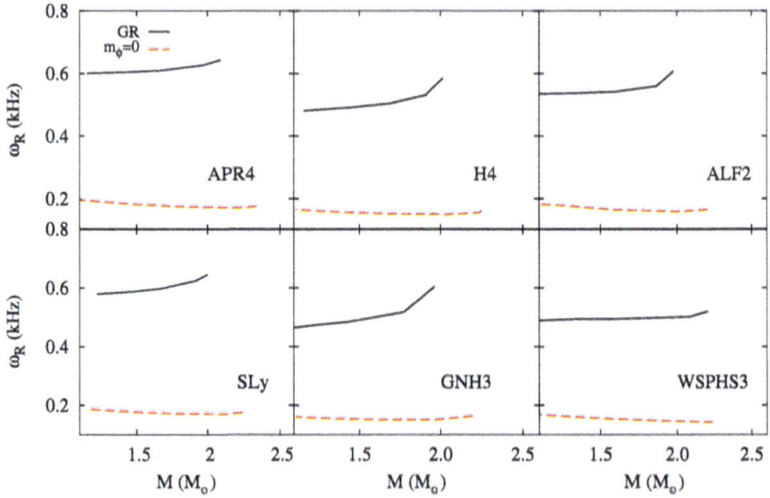

Fig. 5 Real frequency ω_R in kHz for $l = 2$ ϕ-modes versus the mass of neutron star M in solar masses for six EOSs under GR and a massless STT

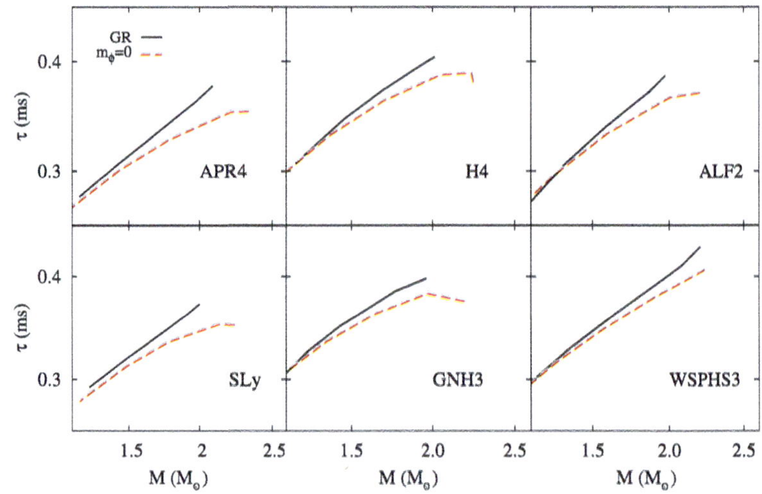

Fig. 6 Damping time τ in milliseconds for $l = 2$ ϕ-modes versus the mass of neutron star M in solar masses for six EOSs under GR and a massless STT

another by scaling with the physical radius of the star against the generalized compactness. In general, in both relations, GR always exhibits a larger scaled frequency than the massless theory.

Finally let us discuss the $l = 0$ perturbations. In GR these modes are normal modes, confined to the interior of the star. However, in STTs, they propagate outside the star and are damped, which we can show by computing their QNMs. Therefore in another word, $l = 0$ gravitational radiation is permitted in STTs. In Figs. 8 and 9 we present the pressure-led $l = 0$ F-mode for SLy EOS, in GR, several $\mathcal{R}^2$ theories, and

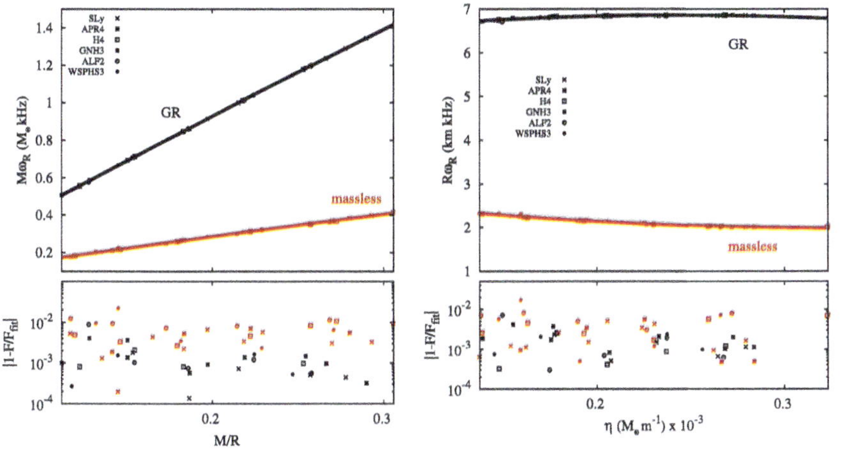

Fig. 7 Universal relations for the scaled real frequency of $l = 2$ ϕ-mode against compactness M/R (left) and generalized compactness η of the star (right)

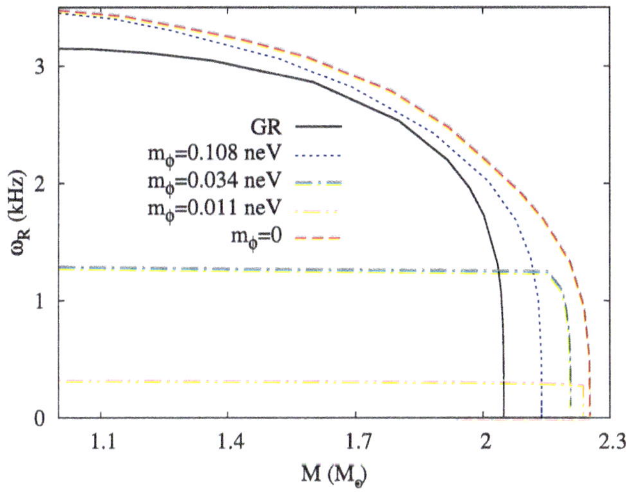

Fig. 8 Real frequency ω_R in kHz for $l = 0$ F-modes versus the mass of neutron star M in solar masses for SLy EOS under GR and several STTs

a massless Brans-Dicke theory. The real frequencies become zero at the maximal mass of the star, and the damping times indicate an order of 10^5 years or more for stars below the maximal mass. Thus these stable F-modes are ultra long-lived. Beyond the maximal mass of the star, the modes become purely imaginary and the star is unstable. In the limit of $m_\phi \to \infty$, i.e. reaching the GR limit, the ultra long-lived modes become normal modes.

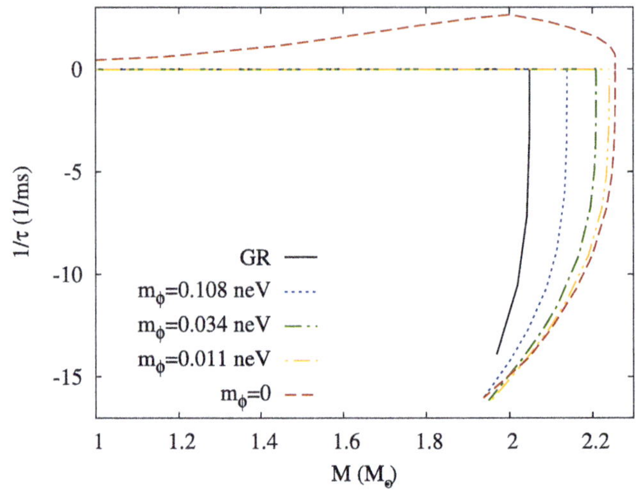

Fig. 9 Inverse damping time τ in milliseconds for $l = 0$ F-modes versus the mass of neutron star M in solar masses for SLy EOS under GR and several STTs

3 Conclusions

Neutron stars are compact objects where we can put our understanding in especially nuclear physics and astrophysics to test. With the unknown matter composition of the star at the core, universal relations become important tools to study the global properties of the star, and hence constraints on alternative theories. In regard to the gravitational waves from the star, its quasinormal mode spectrum is rich with the coupled perturbations of the spacetime and matter of the star. Here, contrasting GR with several viable STTs as alternative gravity theories, we present our results for $l = 2$ fluid f-mode, $l = 2$ scalar ϕ-mode, and $l = 0$ pressure F-mode, discussing the effects of the scalar field mass from the STTs on the behaviour of the modes, and presenting several universal relations for the star in the theories by using the properties and observables of the star. We find that there exist gravitational radiations corresponding to $l < 2$ modes in STTs, unlike GR. For instance, the $l = 0$ pressure normal modes of GR turn into propagating and ultra long-lived QNMs in STTs. The presence of the scalar field in STTs also leads to additional ϕ-mode emissions. The scalar field mass gives rise to a dispersion relation between the frequency of the spacetime and scalar field oscillations at infinity, indicating that the scalar field perturbations do not propagate at the speed of light.

Acknowledgements I would like to thank the organizers for the invitation to give a talk in London-Oldenburg Relativity Seminar from which this contribution is partially based on. I gratefully acknowledge support by the DFG Research Training Group 1620 *Models of Gravity*, DFG project Ku612/18-1, and MICINN project PID2021-125617NB-I00 "QuasiMode". I thank Jose Luis Blázquez-Salcedo for comments on the manuscript. Finally I thank all my collaborators.

References

1. B.P. Abbott et al., LIGO Scientific and Virgo. Phys. Rev. Lett. **116**(6), 061102 (2016)
2. B.P. Abbott et al., LIGO Scientific and Virgo. Phys. Rev. Lett. **119**(16), 161101 (2017)
3. N. Andersson, K.D. Kokkotas, Phys. Rev. Lett. **77**, 4134–4137 (1996)
4. N. Andersson, K.D. Kokkotas, Mon. Not. R. Astron. Soc. **299**, 1059–1068 (1998)
5. N. Andersson, V. Ferrari, D.I. Jones, K.D. Kokkotas, B. Krishnan, J.S. Read, L. Rezzolla, B. Zink, Gen. Relativ. Gravit. **43**, 409–436 (2011)
6. S. Barsanti, A. Maselli, T.P. Sotiriou, L. Gualtieri, Phys. Rev. Lett. **131**(5), 051401 (2023)
7. D. Bettoni, J.M. Ezquiaga, K. Hinterbichler, M. Zumalacárregui, Phys. Rev. D. **95**(8), 084029 (2017)
8. J.L. Blazquez-Salcedo, L.M. Gonzalez-Romero, F. Navarro-Lerida, Phys. Rev. D. **87**(10), 104042 (2013)
9. J.L. Blázquez-Salcedo, Z. Altaha Motahar, D.D. Doneva, F.S. Khoo, J. Kunz, S. Mojica, K.V. Staykov, S.S. Yazadjiev, Eur. Phys. J. Plus. **134**(1), 46 (2019)
10. J.L. Blázquez-Salcedo, D.D. Doneva, J. Kunz, K.V. Staykov, S.S. Yazadjiev, Phys. Rev. D. **98**(10), 104047 (2018)
11. J.L. Blázquez-Salcedo, F.S. Khoo, J. Kunz, EPL. **130**(5), 50002 (2020)
12. J.L. Blázquez-Salcedo, F.S. Khoo, J. Kunz, V. Preut, Front. Phys. **9**, 741427 (2021)
13. J.L. Blázquez-Salcedo, L.M. González-Romero, F.S. Khoo, J. Kunz, V. Preut, Front. Astron. Space Sci. **9**, 1005108 (2022)
14. J.L. Blázquez-Salcedo, L.M. González-Romero, F.S. Khoo, J. Kunz, V. Preut, Phys. Rev. D. **106**(4), 044007 (2022)
15. R. Brito, S. Ghosh, E. Barausse, E. Berti, V. Cardoso, I. Dvorkin, A. Klein, P. Pani, Phys. Rev. D. **96**(6), 064050 (2017)
16. G.F. Burgio, H.J. Schulze, I. Vidana, J.B. Wei, Prog. Part. Nucl. Phys. **120**, 103879 (2021)
17. J.M. Ezquiaga, M. Zumalacárregui, Phys. Rev. Lett. **119**(25), 251304 (2017)
18. P. Haensel, A.Y. Potekhin, D.G. Yakovlev, Astrophys. Space Sci. Libr. **326**, 1–619 (2007), (Springer, 2007)
19. K.D. Kokkotas, T.A. Apostolatos, N. Andersson, Mon. Not. R. Astron. Soc. **320**, 307–315 (2001)
20. J. Kunz, arXiv:2204.12520 [gr-qc]
21. J. Naf, P. Jetzer, Phys. Rev. D. **81**, 104003 (2010)
22. M. Postma, M. Volponi, Phys. Rev. D. **90**(10), 103516 (2014)
23. M. Punturo, M. Abernathy, F. Acernese, B. Allen, N. Andersson, K. Arun, F. Barone, B. Barr, M. Barsuglia, M. Beker et al., Class. Quantum Gravity. **27**, 194002 (2010)
24. T. Regge, J.A. Wheeler, Phys. Rev. **108**, 1063 (1957)
25. S.S. Yazadjiev, D.D. Doneva, K.D. Kokkotas, K.V. Staykov, JCAP. **06**, 003 (2014)
26. F.J. Zerilli, Phys. Rev. Lett. **24**, 737 (1970)

Introduction to Stellar and Substellar Physics in Modified Gravity

Aneta Wojnar

Abstract

We discuss the standard Lane-Emden formalism as well as the one related to the slowly rotating objects. It is preceded by a brief introduction of different forms of the polytropic equation of state. This allows to study a wide class of astrophysical objects in the framework of a given theory of gravity, as demonstrated in a few examples. We will discuss light elements burning processes and cooling models in stars and substellar objects with the use of the Lane-Emden formalism.

1 Introduction

In the spherical symmetric spacetime (hence all physical quantities are dependent on the radial coordinate r only), the hydrostatic equilibrium equations are given by

$$\frac{dP}{dr} = \rho \frac{d\Phi}{dr} \tag{1}$$

$$\frac{dm}{dr} = 4\pi r^2 \rho(r), \tag{2}$$

where $\Phi = \Phi(r)$ is the gravitational potential, $P = P(r)$ and $\rho = \rho(r)$ are pressure and energy density, respectively, while $m = m(r)$ is mass enclosed in a spherical-symmetric ball.

In order to solve the above hydrostatic equations, we need to have boundary conditions which are given by: $\rho(0) = \rho_c$, with ρ_c being the central density, $m(0) = 0$, and $P(R) = 0$, where R is the radius of the object. Apart from them, we also need a relation between the pressure and energy density which is given by an equation of state (EoS). In the next section, we will discuss some particular forms of EoS's.

A. Wojnar (✉)
Departamento de Física Teórica, Universidad Complutense de Madrid, E-28040 Madrid, Spain
e-mail: awojnar@ucm.es

B. Hartmann and J. Kunz (eds.), *Gravity, Cosmology, and Astrophysics*,
Lecture Notes in Physics 1022, https://doi.org/10.1007/978-3-031-42096-2_7

2 Matter Description in Stellar Interiors

2.1 Equation of State

As briefly discussed before, an equation of state is the crucial ingredient to solve the set of Eqs. (1) and (2). It cannot be *any* equation relating thermodynamics quantities such as pressure P and energy density ρ–it must satisfy a few particular conditions[1] such that it can indeed describe matter inside a stellar object. One of them is the the weak energy condition $\rho > 0$ and $\rho + P > 0$. Apart from it, matter must not spontaneous collapse–it means, it must be microscopically stable (Le Chatelier's principle)–which provides the conditions $P \geq 0$ and $dP/d\rho > 0$. Moreover, the speed of perturbations cannot exceed the speed of light $c_s \equiv (dP/d\rho)^{1/2} \leq 1$.

There are many EoS's which fulfill those conditions and are used to study various astrophysical objects. However, we will focus now on two specific ones, given in analytical forms: ideal gas and polytropic EoS. The ideal gas is often used as an approximated description of the atmosphere, but also as one of many components of a mixture of non-interacting gases, described by the Dalton law $p_{\text{total}} = \sum_{i=1}^{m} p_i$. The ideal gas has a well known form

$$\rho = \frac{\mu p}{N_A k_B T}. \tag{3}$$

However, we will mainly focus our attention on the polytropic form of EoS's:

$$P = K\rho^{1+\frac{1}{n}}, \tag{4}$$

where K and n are polytropic parameters or functions. Their exact forms depend on an astrophysical object we are interested in, and, as it will turn out, they also depend on a lot of physical processes and information which can be incorporated into their expressions. This EoS is widely used to approximate the matter description in substellar and stellar objects and also in compact stars, such white dwarfs or neutron ones. Often, it is an equation of state which one uses to analyze an astrophysical object in modified theories of gravity [12,44,45,65], to trace eventual problems as well as to understand the structure of equations.

Before discussing more interesting features of the polytropic EoS, let us just briefly present the simplest forms, related to the non-relativistic degenerate electron gas for which $n = 3/2$, and to relativistic one, given by $n = 3$. Then, the polytropic parameters K's are constants, given as [34]

$$K_{n=3/2} = \frac{1}{20}\left(\frac{3}{\pi}\right)^{\frac{2}{3}} \frac{h^2}{m_e} \frac{1}{(\mu_e m_u)^{\frac{5}{3}}}, \quad K_{n=3} = \frac{hc}{8}\left(\frac{3}{\pi}\right)^{\frac{1}{3}} \frac{1}{m_H(\mu_e)^{\frac{4}{3}}}, \tag{5}$$

[1] We will focus on the barotropic EoS only, that is, when $P = P(\rho)$.

where

$$\mu_e^{-1} = X + \frac{Y}{2} + (1 + X + Y)\left\langle \frac{Z}{A} \right\rangle \tag{6}$$

with $\langle Z/A \rangle$ being the average number of electrons per nucleons in metals, X and Y the mass fractions of hydrogen and helium, respectively, while the other symbols have their standard meaning. Usually, one assumes $\langle Z/A \rangle = 1/2$ and that the star or substellar object consists of 70% of hydrogen (that is, $X = 0.7$).

However, the above forms do not catch many interesting effects, such as for example more realistic description of the electron degeneracy and its time evolution, strongly coupled plasma [66], finite gas temperatures with phase transition points between metallic hydrogen and molecular state. It turns out that those effects can be mimicked by the polytropic EoS (or its slight modifications), which significantly simplify the calculations, especially if we do them in a framework of MG. Let us now briefly discuss some of them. In most of the cases, if not indicated, we will deal with the non-relativistic degenerate electron gas, that is, $n = 3/2$.

The degeneracy parameter Ψ is defined as follows

$$\Psi = \frac{\mu_F}{k_B T} = \frac{\left(3\pi^2\hbar^3\right)^{2/3}}{2m_e k_B T}\left[\frac{\rho N_A}{\mu_e}\right]^{2/3}, \tag{7}$$

where N_A is the Avogadro number while the other constants have the standard meaning. Since it is dependent on the density and temperature, which change during an object's evolution, for instance, when it gravitationally contracts, its time evolution can have a non-trivial effect on the stellar and substellar properties. To take it into account, there have been a few improvements proposed such that the simple forms (5) acquired additional terms. For example, to cover high densities with low temperatures (degenerated gas, $\Psi >> 1$) and ideal gas (3) ($\Psi << 1$), the polytropic constant (5) is [11]

$$K = \frac{(3\pi^2)^{2/3}\hbar}{5m_e m_H^{5/3}\mu_e^{5/3}}\left(1 + \frac{\alpha_d}{\Psi}\right), \tag{8}$$

with $\alpha_d \equiv 5\mu_e/2\mu \approx 4.82$, where μ is the mean molecular weight of ionized hydrogen/helium mixtures, and the provided value is for $X = 0.75$ and $Y = 0.25$. However, this EoS cannot be used for a partially degenerate gas, since Ψ is considered here as a constant. Nevertheless, it has been used to get the minimum Main Sequence masses via obtaining the luminosity produced by the hydrogen burning [11,22,27,46,55,56] or maximal mass of a fully convective star on the Main Sequence [1,75] in various theories of gravity. Similar modifications to the polytropic EoS are also used to study light elements' burning in fully convective stars [9,70,76].

To consider a mixture of degenerate and ideal gas states at finite temperature as well as to take into account ionization and phase transition points, the polytropic parameter K becomes a function, with the following form [3]

$$K = C\mu_e^{-5/3}(1 + b + a\eta), \tag{9}$$

where we have redefined the electron degeneracy as $\eta = \Psi^{-1}$. The constant $C = 10^{13}\text{cm}^4\text{g}^{-2/3}\text{s}^{-2}$ while $a = \frac{5}{2}\mu_e\mu_1^{-1}$, where μ_1 is defined as

$$\frac{1}{\mu_1} = (1 + x_{H^+})X + \frac{Y}{4}. \tag{10}$$

It takes into account ionization of hydrogen X, represented by the ionization fraction x_{H^+} and depends on the phase transition points [16]. On the other hand, the quantity b reads

$$b = -\frac{5}{16}\eta\ln(1 + e^{-1/\eta}) + \frac{15}{8}\eta^2\left(\frac{\pi^2}{3} + \text{Li}_2[-e^{-1/\eta}]\right), \tag{11}$$

where Li_2 denotes the second order polylogarithm function. Such an EoS is used to described matter properties, in very low-mass stars, brown dwarfs and giant exoplanets [3,8,38,51].

? Exercises

1. Show that a combination of pressures $p = p_1 + p_2$, where p_1 is the simple polytropic EoS with $n = 3/2$ (5) and p_2 is the ideal gas one (3), can be written as a polytropic EoS with the polytropic index $n = 3/2$.
2. Show that the polytropic equation of state with (9) in the approximation of negligible degeneracy reduces to the ideal gas.

Various forms of the polytropic EoS are also used to study terrestrial (exo-)planets. To describe a complex planets' interiors, one needs to consider a merger of the third-order finite strain Birch-Murgnagham equation of state [10] with Thomas-Fermi-Dirac one [23–25,57,68]. Such a merger is well approximated by the polytropic EoS up to the pressure range $p < 10^7$ GPa [64]

$$\rho(p) = \rho_0 + cp^n, \tag{12}$$

whose best-fit parameters ρ_0, c, and n depend on a material the layer is composed of. The presence of ρ_0 in the above EoS allows to include the incompressibility of solids and liquids at low pressures. One may also make the polytropic index n a variable, since it is a derivative of the inverse of the compressibility, which is a property of the layer's material, to describe rocky and gaseous (exo-)planets [72].

2.2 Other Properties

Although the equation of state is enough to solve the system of the differential Eqs. (1) and (2) to get the basic properties of an astrophysical object such as mass and radius, as well as its core properties if we deal with a polytrope (see the next section), it is not sufficient to study processes happening in its interior, such us thermonuclear reactions, cooling and crystallization processes, or its evolution. Let us then discuss additional equations which are crucial for those physical problems.

If a low-mass star or a brown dwarf is massive enough[2] to burn light elements in its core, the outcome of such an energy generation process is, roughly speaking, luminosity. It is obtained by the integration of the below expression:

$$\frac{dL_{burning}}{dr} = 4\pi r^2 \dot{\epsilon} \rho, \tag{13}$$

where the energy generation rate $\dot{\epsilon}$ is a function of energy density, temperature, and stellar composition. It is often approximated as a power-low function of the two first [26]. Its (analytical) form depends on a given reaction and here we will not focus on any particular form (see Sect. 4.1 for some examples). Nevertheless of the type of the reaction, such an energy is transported from the deep interior to the surface of the object. Depending on the type of the object (which can be translated as a mass criterion), one deals with the heat transport through object's interior and its atmosphere. The most common criterion determining the class of the energy transport is provided by the Schwarzschild one [62,62]:

$$\nabla_{rad} \leq \nabla_{ad} \quad \text{pure diffusive radiative or conductive transport} \tag{14}$$

$$\nabla_{rad} > \nabla_{ad} \quad \text{adiabatic convection is present locally,} \tag{15}$$

where the gradient denotes the temperature T variation with depth

$$\nabla_{rad} := \left(\frac{d\ln T}{d\ln p}\right)_{rad}, \tag{16}$$

and ∇_{ad} is the adiabatic temperature gradient, which in case of perfect, monatomic gas has a constant value $\nabla_{ad} = 0.4$. In the case of Newtonian gravity, the criterion is given by

$$\nabla_{rad} = \frac{3\kappa_{rc} l p}{16\pi \bar{a} c G m T^4}, \tag{17}$$

where l is the local luminosity, the constant $\bar{a} = 7.57 \times 10^{-15} \frac{erg}{cm^3 K^4}$ is the radiation density and κ_{rc} is the radiative and/or conductive opacity which we will discuss later. However it was demonstrated that for MG the Schwarzschild criterion becomes

[2] It will turn out that you will find a lot of terms as "critical mass" for something when you go deeper into some of the branches of the astrophysics.

modified [75] by additional terms which have a stabilizing or destabilizing effects (therefore, for instance, one can deal with more massive fully convective stars on the Main Sequence in comparison to Newtonian gravity.)

When the produced energy reaches the object's surface, it is radiated away through it. This process is well described by the Stefan-Boltzmann law (L stands for luminosity)

$$L = 4\pi f \sigma T_{eff}^4 R^2, \tag{18}$$

where σ is the Stefan-Boltzmann constant and the factor $f \leq 1$ which allows to include planets since they radiate less than the black-body with the same effective temperature T_{eff}.

Unfortunately, the effective temperature as well as other parts of the atmospheric modelling are not only complicated to determine, but they often carry significant uncertainties. Moreover, they properties and forms also depend on the interior properties and matching conditions. However, we can keep them as parameters to be determined by some particular methods, or we can derive them from scratch. To do so, we can use for example the optical depth τ, averaged over the object's atmosphere[3] [28,34]:

$$\tau(r) = \bar{\kappa} \int_r^\infty \rho dr, \tag{19}$$

where $\bar{\kappa}$ is a mean opacity. For objects possessing atmospheres with low temperatures, the most common mean opacity is the Rosseland one. It is given by the Kramers' law

$$\bar{\kappa} = \kappa_0 p^u T^w, \tag{20}$$

where κ_0, u and w are values depending on different opacity regimes [2,43]. Another useful approximation is the assumption is the atmospheric particles satisfy the ideal gas relation (3).

Another quantity which is crucial for studying processes in the stellar and substellar interiors is the thermal energy

$$U = \bar{c}_v \frac{\mathcal{M}}{A m_{\rm p}} T, \tag{21}$$

where T is the temperature of the isothermal core and $\mathcal{M}/(A m_{\rm p})$ is the number of ions with A being the mean atomic weight. The mean specific heat is taken for the whole stellar configuration since it depends on the density

$$\bar{c}_v = \frac{1}{\mathcal{M}} \int_0^{\mathcal{M}} c_v(T, \rho) \frac{dm}{dr} dr, \tag{22}$$

[3] However, it is useful for analytical studies only when we can assume that the surface gravity $g = \frac{Gm(r)}{r^2}$ is constant.

where $c_v = c_v^{\rm el} + c_v^{\rm ion}$, with $c_v^{\rm ion}$ being the specific heats of ions and $c_v^{\rm el}$ the electrons per ions depend on density. If the main contribution comes from the ions, for which the specific heat is $c_v = (3/2)k_{\rm B}$, the thermal energy is reduced to the well-known form

$$U = \frac{3}{2}k_{\rm B}T\frac{\mathcal{M}}{Am_{\rm p}}. \tag{23}$$

However, if one wants to consider a more realistic case, the dependency on the Debye temperature $\Theta_{\rm D}$ and the ratio of Coulomb to thermal energy Γ

$$\Gamma = 2.28 \times 10^5 \frac{Z^2}{A^{1/3}}\frac{\rho_s^{1/3}}{T}, \tag{24}$$

with $\rho_s(T)$ being the density of the crystallized mass at a temperature T, must be taken into account. They can be incorporated already in the expression for the specific heat of ions. The exact forms of the specific heats are as follows

$$c_v^{\rm el} = \frac{3}{2}\frac{k_{\rm B}\pi^2}{3}Z\frac{k_{\rm B}T}{\epsilon_{\rm F}}, \quad c_v^{\rm ion} = 9k_{\rm B}\left(\frac{T}{\Theta_{\rm D}}\right)^3\int_0^{\Theta_{\rm D}/T}\frac{x^4e^x}{(e^x-1)^2}dx, \tag{25}$$

where Z is the charge and $\epsilon_{\rm F}$ is the Fermi energy ($p_{\rm F}$ is the Fermi momentum)

$$\epsilon_{\rm F}^2 = p_{\rm F}^2c^2 + m_{\rm e}^2c^4, \quad p_{\rm F}^3 = \frac{3h^3}{8\pi}\frac{\rho}{\mu_{\rm e}m_{\rm p}}. \tag{26}$$

As evident from the form of the specific heat of ions $c_v^{\rm ion}$, it depends on the crystallization properties of matter. More specifically, it depends on the critical value of the ratio of Coulomb to thermal energy Γ, denoted by Γ_m. For $\Gamma < \Gamma_m$, $c_v^{\rm ion}$ reduces to $(3/2)k_{\rm B}$ while above this value, the specific heat of ions is given by the expression (25), in which the Debye temperature is given by

$$\Theta_{\rm D} = 0.174 \times 10^4\frac{2Z}{A}\sqrt{\rho}. \tag{27}$$

We see that all those quantities depend on density, which is obtained from the hydrostatic equilibrium equation. Because of that fact, depending on the gravity framework one works, those matter properties do depend on it. As we will see in the Sect. 4.2.2, it has an important consequence on the cooling and crystallization processes in white dwarfs.

3 Lane-Emden Formalism

Although the Lane-Emden formalism is well-known and it can be find in almost every standard textbook, let us briefly recall it before moving to its MG counterpart. Roughly speaking, it is a dimensionless Poisson equation with the hydrostatic equilibrium equation and polytropic form of the equation of state in the spherical-symmetric spacetime. As we will see in the further part, it is widely used in many astrophysical problems.

3.1 Newtonian Case

Let us consider the Poisson equation

$$\nabla^2\Phi = -4\pi G\rho, \tag{28}$$

where Φ is the gravitational potential, ρ is energy density while G the Newton's gravitational constant. Considering spherical-symmetric spacetime such that all variables are the functions of the radial coordinates, that is, $\Phi \equiv \Phi(r)$ and $\rho \equiv \rho(r)$, the Poisson Eq. (28) and corresponding hydrostatic equilibrium equation take the following form

$$\frac{1}{r^2}\frac{d}{dr}\left(r^2\frac{d\Phi}{dr}\right) = -4\pi G\rho, \qquad \frac{dP}{dr} = \rho\frac{d\Phi}{dr}, \tag{29}$$

where $P \equiv P(r)$. Applying to both the polytropic equation of state (4), these equations can be written as one

$$\frac{1}{\xi^2}\frac{d}{d\xi}\left(\xi^2\frac{d\theta}{d\xi}\right) = -\theta^n, \tag{30}$$

where θ is a function of ξ. It satisfies the boundary conditions $\theta(0) = 1, \theta'(0) = 0$, where the prime $' \equiv \frac{d}{d\xi}$. The above equation is the Lane-Emden equation (LEE). The relations between the dimensionless variables θ and ξ and energy density ρ and radial coordinate r are like follows:

$$\rho = \rho_c\theta^n, \qquad r = r_c\xi \quad \text{with} \quad r_c^2 = \frac{K(n+1)\rho_c^{(\frac{1}{n}-1)}}{4\pi G}. \tag{31}$$

where ρ_c denotes the central density. Notice that from (4) we have

$$P = P_c\theta^{n+1}, \tag{32}$$

where $P_c := K\rho_c^{\frac{n+1}{n}}$.

? Exercises

1. Derive the Lane-Emden Eq. (30).
2. It turns out that the LEE possesses three exact solutions for some specific values of the polytropic index n. Find the exact solutions of the LEE (30) for $n = 0$, $n = 1$, and $n = 5$.
3. In some particular cases, such as for instance studying processes happening in the stellar core (see the Sect. 4.1), the near-center solution is needed and quite easy to find with respect to the the general one. Therefore, find the approximated solution of (30) around $\xi \approx 0$, that is, an analytic form of $\theta(\xi \approx 0)$.

Knowing the solution θ of the LEE for the chosen polytropic index n and polytropic (constant or variable) K, one can immediately get the characteristics of a given astrophysical object. Therefore, the total stellar mass M and its radius R are given as

$$M = 4\pi r_c^3 \rho_c \omega_n, \quad R = \gamma_n \left(\frac{K}{G}\right)^{\frac{n}{3-n}} M^{\frac{1-n}{n-3}}, \tag{33}$$

where

$$\omega_n = -\xi_1^2 \frac{d\theta}{d\xi}\bigg|_{\xi=\xi_1}, \quad \gamma_n = (4\pi)^{\frac{1}{n-3}} (n+1)^{\frac{n}{3-n}} \omega_n^{\frac{n-1}{3-n}} \xi_1 \tag{34}$$

and ξ_R is a value for which $\theta(\xi_R) = 0$. Thus, the first zero of θ indicates the radius of the object.

? Exercise

Derive the mass and radius given by (33).

The density and pressure profiles are provided by (31) and (4), respectively, while the temperature profile yields

$$T = K \frac{m_H \mu}{k_B} \rho_c^{\frac{1}{n}} \theta, \tag{35}$$

with k_B being Boltzmann's constant, m_H the mass of Hydrogen atom, and μ the mean molecular weigh.

? Exercise

Derive the temperature profile (35).

Moreover, one can also derive the core quantities, such as the core temperature T_c, core density ρ_c, and core pressure P_c:

$$\rho_c = \delta_n \left(\frac{3M}{4\pi R^3}\right), \quad T_c = K\frac{m_H \mu}{k_B}\rho_c^{\frac{1}{n}}, \quad P_c := K\rho_c^{\frac{n+1}{n}} \tag{36}$$

where $\delta_n = -\frac{\xi_1}{3\frac{d\theta}{d\xi}|_{\xi=\xi_1}}$.

? Exercise

Find mass, radius, central temperature and density for a polytropic star with respect to the solutions of the Lane-Emden equation for $n = 1.5$ and $n = 3$.

To know more about the LEE, its generalization and further applications, see [29]. Now on, we will focus on its MG form.

3.2 Modified Gravity

If you are interested in modified theories of gravity, you already now that some of those proposal can also modify the Newtonian limit equations. Therefore, the Poisson and hydrostatic equilibrium equations acquire additional terms whose forms depend on a particular theory [5,37,47,48,69]. We can then write the first equation in a generic way for the spherical-symmetric spacetime as

$$\frac{1}{r^2}\frac{d}{dr}\left(r^2\frac{d\Phi}{dr}\right) = -4\pi G\rho + \text{mgt}(r), \tag{37}$$

where the modified gravity term mgt(r) is a general function, characteristic for a given theory of gravity. Here we also assume that all additional elements of that function are $r-$coordinate dependent.

Analogously, employing the polytropic EoS (4) and the hydrostatic equilibrium Eq. (1), we can rewrite it as the modified Lane-Emden equation (MLEE)

$$\frac{1}{\xi^2}\frac{d}{d\xi}\left(\xi^2\frac{d\theta}{d\xi}\right) = -\theta^n + g_{mod0}(\xi) \tag{38}$$

where

$$g_{mod0} = \frac{\text{mgt}(r)}{4\pi G\rho_c} \tag{39}$$

is a dimensionless term induced by a given theory. Notice its dependence on ρ_c. As you will see in the following examples, usually the core density can be hidden into a re-scaled theory parameter.

? Exercise

Derive the MLEE (38).

Examples: Modified Lane-Emden equations

- Scalar-tensor theories (Horndeski and beyond,...) [36,53,54]

$$\frac{1}{\xi^2}\frac{\mathrm{d}}{\mathrm{d}\xi}\left[\left(1+\frac{n}{4}\Upsilon\xi^2\theta^{n-1}\right)\xi^2\frac{\mathrm{d}\theta}{\mathrm{d}\xi}+\frac{\Upsilon}{2}\xi^3\theta^n\right]=-\theta^n, \tag{40}$$

  where Υ in the theory parameter.
- Eddington-inspired Born Infeld gravity (EiBI) [49]

$$\frac{d}{d\xi}\left(\xi^2\frac{d\theta}{d\xi}\left[1+\alpha\theta^{n-1}\right]\right)=-\xi^2\theta^n\,, \tag{41}$$

  where the EiBI corrections are hidden in the dimensionless parameter $\alpha=\frac{\epsilon n}{2r_c^2}$. Notice that it depends on the polytropic parameter and also on the star's central energy density ρ_c.
- Palatini $f(R)$ gravity [74]

$$\frac{1}{\xi}\frac{d^2}{d\xi^2}\left[\sqrt{\phi}\xi\left(\theta-\frac{2\alpha}{n+1}\theta^{n+1}\right)\right]=-\frac{(\phi+\frac{1}{2}\xi\frac{d\phi}{d\xi})^2}{\sqrt{\phi}}\theta^n, \tag{42}$$

  where $\phi=1+2\alpha\theta^n$ while $\alpha=\kappa c^2\beta\rho_c$ is the rescaled modifed gravity parameter.
- Metric $f(R)$ gravity has a much more complex form, see [12].

3.3 Slowly Rotating Astrophysical Objects

The formalism presented above turn out to be very useful (see the Sect. 4) in studying non-relativistic objects in modified gravity. However, everything in the Universe rotates, therefore if one day you would like to consider a more realistic models to, for example, test your model of gravity (or other fundamental interactions) against observational data, the rotation[4] should be included.

Can we however have a similar formalism in modified gravity but for the rotating polytropes? The answer is yes–we can adopt the derivation presented in [17] and write down the MLEE for a rotating object, as it was done in [19]. Moreover, it turns out that one can even provide a generic solution of such an equation. Let us see how it can be done.

Let us consider then a slowly rotating object along the z-axis with the uniform angular speed ω. Adopting the polar coordinates $\{r, \mu(= \cos\vartheta), \phi\}$, the equations of hydrostatic equilibrium in that case are written as

$$\frac{\partial P}{\partial r} = \rho\frac{\partial \Phi}{\partial r} + \rho\omega^2 r(1-\mu^2)\,, \qquad \frac{\partial P}{\partial \mu} = \rho\frac{\partial \Phi}{\partial \mu} - \rho\omega^2 r^2 \mu \tag{43}$$

in which we have neglected ϕ by assuming the axial symmetry. The gravitational potential Φ, as previously, carries additional terms introduced by the given theory of gravity. Moreover, the extra term in the Poisson Eq. (37) also gains the $\mu-$dependence due to the rotation, that is:

$$\nabla^2\Phi = -4\pi G\rho + \text{mgt}(r,\mu), \tag{44}$$

where, analogously to the non-rotating case, the term $\text{mgt}(r, \mu)$ is a general function. Apart from it, the energy density also depends now on the angular coordinate μ, that is, $\rho = \rho(r, \mu)$ as well as pressure, such that the polytropic EoS is now

$$P(r,\mu) = K\rho^{1+\frac{1}{n}}. \tag{45}$$

Introducing the dimensionless variables, we also need to remember about the $\mu-$dependence, thus

$$\rho = \rho_c\Theta^n, \quad r = r_c\xi \ \text{ with } \ r_c^2 = \frac{K(n+1)\rho_c^{(\frac{1}{n}-1)}}{4\pi G} \tag{46}$$

where $\Theta \equiv (\xi, \mu)$ is a function of both ξ and μ. Then, we can derive the MLEE for the rotating polytrope of the form:

$$\frac{1}{\xi^2}\frac{\partial}{\partial\xi}\left(\xi^2\frac{\partial\Theta}{\partial\xi}\right) + \frac{1}{\xi^2}\frac{\partial}{\partial\mu}\left((1-\mu^2)\frac{\partial\Theta}{\partial\mu}\right) = \upsilon + g_{mod}(\xi,\mu) - \Theta^n \tag{47}$$

where we have introduced a dimensional parameter $\upsilon = \omega^2/2\pi G\rho_c$. It can be interpreted as a measure of the outward centrifugal force compared to the self-gravity of the rotating polytrope. On the other hand, $g_{mod}(\xi, \mu) = \text{mgt}(r, \mu)/4\pi G\rho_c$ is the dimensionless modification term whose form is explicit when a gravitational proposal is chosen.

[4] As well as magnetic field, energy transport properties, time dependency and many others...

? Exercise

Derive the MLEE for the rotating polytrope (47).

As already mentioned, one can find an exact solution $\Theta(\xi, \mu)$ of (47) in the form of the solutions of the non-rotating MLEE which is usually much easier to solve than its rotating counterpart. It turns out that such a solution for a slow rotation has the following form [19]:

$$\Theta(\xi, \mu) = \theta(\xi) + v\Big[\psi_0(\xi) + A_2\psi_2(\xi)P_2(\mu)\Big], \tag{48}$$

where $P_2(\mu)$ is the Legendre function while the quantity A_2 is given by

$$A_2 = -\frac{5}{6}\frac{\xi_R^2}{[3\psi_2(\xi_R) + \xi_R\psi_2'(\xi_R)]}. \tag{49}$$

The dimensionless radius ξ_1 is the first zero of $\theta(\xi)$ and $'$ denotes derivative with respect to ξ. The functions ψ_0 and ψ_2 satisfy particular differential equations which depend on modified gravity terms. You may check how they look like in [19]. The derivation of (48) is quite tedious and technical, and we will skip it here.

Since the differential equations the function ψ_0 and ψ_2 need to satisfy depend on a model of gravity, it can happen that one cannot solve those equations to get their exact form. However, it was shown [19] that for some models of gravity such as

Examples: Modified Poisson equations

- Scalar-tensor theories (Horndeski and beyond,...)

$$\nabla^2\Phi \sim -\frac{\kappa}{2}\Big(\rho + \frac{\Upsilon}{4}\nabla^2(r^2\rho)\Big) \tag{50}$$

- Eddington-inspired Born Infeld gravity

$$\nabla^2\Phi \sim -\frac{\kappa}{2}\Big(\rho + \frac{\epsilon}{2}\nabla^2\rho\Big) \tag{51}$$

- Palatini $f(R)$ gravity [69]

$$\nabla^2 V \sim -\frac{\kappa}{2}\Big(\rho + 2\beta\nabla^2\rho\Big) \tag{52}$$

one can indeed solve them. It seems that the formalism presented in that section can be applied to any theory of gravity of which the additional term appearing in the Poisson equation is a function of density, its higher order radial-derivatives, or its Laplacian [19].

4 Applications

Since an application of the introduced formalism (and its extension to, for instant, even more general EoS) is plentiful, we will briefly review the existing literature and have a closer look on three astrophysical processes in which the LE formalism has a direct use. Therefore, we will focus only on some parts of the stellar and substellar evolution, with the main focus on the objects with masses not exceeding $0.5M_\odot$, where $M_\odot$ is the solar mass. The reason for such a limitation is that the interiors of such objects, even if they are true stars on the Main Sequence, are fully convective. This means that their interior properties are well described by the polytropic EoS which we have discussed at the beginning of that chapter, while the LE formalism presented afterwards allows us to obtain necessary ingredients to study non-trivial problems in astrophysics.

Roughly speaking, before reaching the Main Sequence, a stellar object, called pre-Main Sequence (PMS) star, is still contracting. On the HR diagram it follows the so-called Hayashi track and it decreases its luminosity but does not change too much its surface temperature. Depending on the core conditions (mainly on its temperature, but also, as it turns out, a model of gravity), the PMS star can already burn light elements such as deuterium and lithium. The temperature of the hydrogen ignition, on the other hand, is much higher than that of the other light elements. Moreover, when a PMS star starts burning this element stably, it becomes a true star and enters the Main Sequence phase of the stellar evolution. The energy generated in the core induces pressure which balances the gravitational pull such that the star stops contracting. However, if the central temperature and other core conditions are not enough to enable the stable hydrogen burning, the object further contracts till the electron degeneracy pressure is high enough to counterbalance the gravitational attraction. Such an object, called a brown dwarf, radiates its energy away, and since it does not possess any other energy source apart from the one gained during its contraction phase, it cools down with time. As we will see, all those process are gravity model dependent. Before however doing it, let us briefly mention other works in which polytropes and the LE formalism were or can be used.

Since in modified gravity the hydrostatic equilibrium equations [40,48,52] (for review, see [47]) and matter description [73] are modified, the changes in the astrophysical object's internal properties and in its evolution are also expected to happen. Because of that fact, a few tests with the use of stars, brown dwarfs, (exo-)planets and white dwarfs have been proposed. The most common feature of those objects which is affected by these modifications are limited masses, such as the Chandrasekhar mass-limit of WDs [6,7,18,30–32,60,61,78], the minimum Main Sequence mass [22,46,55,56], minimum mass for deuterium burning [51], Jeans [13] and opacity mass [77]. Another tool which is used to constrain different theories of gravity are seismic data obtained from the helioseismology [58,59] or seismic analysis from the terrestrial planets [39,41,42]. It turns out that also the light elements' abundances in stellar atmospheres [76] are affected when the gravitational framework is different from the Newtonian one. The evolutionary phases of such objects as non-relativistic stars [1,20,21,27,67,75], brown dwarfs [8,38], WDs [33,61], and giant planets [77,79] considered in modified gravity also differ with respect to their Newtonian description. Moreover, those objects can be used to constrain different theories when more accurate data provided by GAIA, James Webb Space Telescope, or Nancy Grace Roman Space Telescope are available.

4.1 Light Elements' Burning

As an example of the use of the introduced formalism, let us discuss two important thermonuclear processes which happen in the PMS and stellar interiors. To examine them, we will need solutions of the LE equation. Since they depend on a model of gravity, the results one will obtain will differ with respect to Newtonian ones. As we will also discuss later on, this conclusion carries important consequences and possibilities.

4.1.1 Lithium

Since we will deal with a low-mass PMS star with mass M as already mentioned, it is fully convective (apart from the radiative atmosphere). Because of that fact we can safely assume that it is well-mixed, so we do not need to worry about the non-homogeneous distribution of various elements. It also means that the matter properties are well-described by the polytropic EoS and the LE formalism. Denoting the hydrogen fraction as X, the depletion rate of ^{7}Li is expressed by

$$M\frac{df}{dt} = -\frac{Xf}{m_H}\int_0^M \rho\langle\sigma v\rangle dM, \tag{53}$$

where f is the lithium-to-hydrogen ratio. Since we deal with a non-resonant reaction rate in the temperature range $T < 6\times 10^6$ K, it is given by

$$N_A\langle\sigma v\rangle = Sf_{scr}T_{c6}^{-2/3}\exp\left[-aT_{c6}^{-\frac{1}{3}}\right]\frac{\text{cm}^3}{\text{s g}}, \tag{54}$$

where $T_{c6}\equiv T_c/10^6$ K is the core temperature and f_{scr} is the screening correction factor. The dimensionless parameters $S = 7.2\times 10^{10}$ and $a = 84.72$ were fitted to the reaction rate $^7\text{Li}(p,\alpha)^4\text{He}$ [15,50,70].

Immediately we see from the above formulas that in order to calculate the lithium depletion rate, we need to find out the central temperature T_c and density ρ. Fortunately, from the Sect. 3.1 and for a given theory of gravity we can write that the core characteristic are given by

$$T_c = 1.15\times 10^6\left(\frac{\mu_{\text{eff}}}{0.6}\right)\left(\frac{M}{0.1M_\odot}\right)\left(\frac{R_\odot}{R}\right)\frac{\delta^{\frac{2}{3}}}{\xi_R^{\frac{5}{3}}(-\theta'(\xi_R))^{\frac{1}{3}}}\text{K} \tag{55}$$

$$\rho_c = 0.141\left(\frac{M}{0.1M_\odot}\right)\left(\frac{R_\odot}{R}\right)^3\delta\,\frac{\text{g}}{\text{cm}^3}, \tag{56}$$

where the mean molecular weight μ_{eff} with the electron degeneracy taken into account is given by the expression

$$\frac{1}{\mu_{\text{eff}}}=\frac{1}{\mu_i}+\frac{1}{\mu_e}\frac{2F_{3/2}(\eta)}{3F_{1/2}(\eta)} \tag{57}$$

in which $\mu_i=\rho N_A/n_i$ is the mean molecular weight of the gas, η the electron degeneracy and $F_n(\eta)$ the nth order Fermi-Dirac function. We see that the values of the core quantities are MG dependent via the solutions of the modified LE, represented here by δ, ξ_R, and θ'.

For such a case, the radius of the PMS star is given by [70]

$$\frac{R}{R_\odot}\approx\frac{7.1\times10^{-2}\gamma}{\mu_{\text{eff}}\mu_e^{\frac{2}{3}}F_{1/2}^{\frac{2}{3}}(\eta)}\left(\frac{0.1M_\odot}{M}\right)^{\frac{1}{3}}. \tag{58}$$

? Exercises

1. Find the polytropic function K.
2. Change the integration variable to the spatial ones in (53).
3. Use the LE formalism to rewrite (53) in the form of the LE variables.

Following the steps given in the above exercises, the depletion rate of lithium will have the following form ($u\equiv aT_{c6}^{-1/3}$)

$$\frac{\mathrm{d}}{\mathrm{d}t}\ln f\sim\left(\frac{u}{a}\right)^2\int_0^{\xi_R}f_{\text{scr}}h(\theta,u)d\xi\ \frac{1}{\mathrm{s}}, \tag{59}$$

where $h(\theta,u)$ is a function of u and the solution of the LE equation, to be determined in the above exercise. It is very improbable that we will be able to solve the modified LE equation with $n=3/2$ analytically, and to be able to go forward with our calculations without the need of numerical methods. However, we should notice that the burning processes in low-mass stars happen in the star's core, therefore the near-center approximation $\theta(\xi\approx0)$, as obtained in one of the exercise in the Sect. 3.1 is sufficient for our purposes. Depending on the theory of gravity, its form can also depend on the theory parameter.

Integration in the Eq. (59) can be easily performed with the assumption that the time evolution of the degeneracy parameter η is not significant with respect to the changes in the star's size (that is, the radius). Therefore, for the PMS stars with masses $M>0.2M_\odot$ the lithium depletion $\mathcal{F}$ is

$$\mathcal{F} = \ln \frac{f_0}{f} = 5.6 \times 10^{14} T_{eff}^{-4} \left(\frac{X}{0.7}\right) \left(\frac{0.1 M_\odot}{M}\right)^3 \left(\frac{0.6}{\mu_{eff}}\right)^6 \times S f_{scr} a^{16} g(u) \Big(1 + \text{eventual MG term}\Big) \frac{\xi_R^7 (-\theta'(\xi_R))^2}{\delta^2}, \tag{60}$$

where f_0 is the initial abundance, T_{eff} the effective temperature, $g(u) = u^{-37/2} e^{-u} - 29\Gamma(-37/2, u)$, and $\Gamma(-37/2, u)$ is the upper incomplete gamma function. The effect of MG is clearly visible, even without the eventual MG modification term, since the last fraction includes the modified LE solutions.

? Exercises

1. Find the radius and luminosity as functions of time using the Stefan-Boltzmann equation and the virial theorem.
2. Write down the contraction time:

$$t_{cont} = -\frac{R}{dR/dt} \tag{61}$$

in terms of the central temperature T_c.

3. Assuming that the PMS star depletes its lithium when it reaches the Main Sequence, such that the contraction time t_{cont} is comparable to the destruction time t_{dest}, find the central temperature, age, radius, and luminosity of a $0.5 M_\odot$ star. The destruction time is given by

$$t_{dest} = \frac{m_p}{X\rho < \sigma v >}. \tag{62}$$

4.1.2 Hydrogen

When the young star reaches the Main Sequence, it means that in its core the temperature is high enough to start the hydrogen ignition. The energy generated by this process is radiated away through the star's surface. It turns out that this stable process happens for sufficiently massive objects. This critical mass, called minimum Main Sequence mass, we are going to calculate now.

The energy generation rate per unit mass for the hydrogen ignition process has the power law form [11,26]

$$\dot{\epsilon}_{pp} = \dot{\epsilon}_c \left(\frac{T}{T_c}\right)^s \left(\frac{\rho}{\rho_c}\right)^{u-1}, \quad \dot{\epsilon}_c = \epsilon_0 T_c^s \rho_c^{u-1}, \tag{63}$$

where the two exponents can be approximated as $s \approx 6.31$ and $u \approx 2.28$, while $\dot{\epsilon}_0 \approx 3.4 \times 10^{-9}$ ergs $\text{g}^{-1}\text{s}^{-1}$. Considering again a PMS star with the hydrogen fraction

$X = 0.75$, the number of baryons per electron in low-mass stars is $\mu_e \approx 1.143$. The luminosity being an effect of the hydrogen burning reads

$$L_{HB} = \int \dot{\epsilon}_{pp}\, dM, \tag{64}$$

in which we can again use the LE formalism. Considering the polytrope with the polytopic function K given by (8) and the near-zero solution $\theta(\xi \approx 0)$, we find that

$$\frac{L_{HB}}{L_\odot} = \frac{1.53 \times 10^7 \Psi^{10.15}}{(\Psi + \alpha_d)^{16.46}} \frac{\delta_{3/2}^{5.487} M_{-1}^{11.977}}{\omega_{3/2} \gamma_{3/2}^{16.46}}, \tag{65}$$

where $L_\odot$ is the solar luminosity while $M_{-1} = M/(0.1 M_\odot)$. As previously, the MG gravity effect is present via the solutions of the modified LE equation.

? Exercise

Following the similar steps as in the Sect. 4.1.1, derive (65).

To find the minimum Main Sequence mass, we need to write down the Stefan-Boltzmann Eq. (18) (with $f = 1$ for the black body approximation) as a function of mass M_{-1}. Equaling this two luminosities, that is, $L_{HB} = L$ will provide the critical mass M_{HB} which a PMS star needs to have in order to ignite hydrogen and burn it in a stable way.

From the LE formalism we have already the radius written as a function of mass (33) but the effective temperature, which we assume to be the temperature of the photosphere from which the energy is radiated away, is much more complicated to be obtained. Usually, one uses the matching procedure of the specific entropies calculated in the stellar interior and in the photosphere [11]

$$T_{ph} = \frac{1.8 \times 10^6 \rho_{ph}^{0.42}}{\Psi^{1.545}}. \tag{66}$$

? Exercises

Using the definition of the photopshere (for which the optical depth (19) is equaled to 2/3) and the assumption on the constant surface gravity

1. Find the photosheric pressure p_{eff}.
2. Assuming that the photosphere matter can be approximated by the ideal gas, use (66) and obtained before p_{eff} to find the photospheric density ρ_{eff}.
3. Write down the Stefan-Boltzmann law as a function of mass M_{-1}.

4. Show that the minimum Main Sequence mass is given by

$$M_{-1}^{MMSM} = 0.290\frac{\gamma_{3/2}^{1.32}\omega_{3/2}^{0.09}}{\delta_{3/2}^{0.51}}\frac{(\alpha_d + \Psi)^{1.509}}{\Psi^{1.325}}(1 + \text{eventual MG term}) \quad (67)$$

4.2 Cooling Processes

4.2.1 Brown Dwarfs

Since the conditions occurring in the core are not enough to ignite hydrogen and to produce radiation pressure to balance the gravitational pull, PMS objects (from now on called brown dwarfs) with masses lower than M_{-1}^{MMSM} continue contracting. There is no stable energy production, thus a brown dwarf radiates the stored energy away and cools down with time. Because of the ongoing contraction, the degenerate gas' contribution starts being relevant and moreover, it changes with time as the brown dwarf is still contracting before reaching the equilibrium, that is, when the electron degeneracy pressure balances the gravitational one. Therefore, we can still deal with the polytrope, however with the function K adjusted to the mixture of degenerate and ideal gas states at finite temperature (and other improvements, see the Sect. 2.1), given by (9). Similarly as in the case of low-mass stars, the biggest challenge is to derive the effective temperature which appears in the Stefan-Boltzmann Eq. (18). Again using the entropy matching procedure, the effective temperature T_{eff} is obtained in terms of the degeneracy parameter $\eta(=\Psi^{-1})$ and the photospheric density ρ_{ph} as [3]

$$T_{\text{eff}} = b_1 \times 10^6 \rho_{\text{ph}}^{0.4}\eta^{\nu} \text{ K}, \quad (68)$$

where the parameters b_1 and ν takes different values since they depend on the specific model adopted for describing the phase transition between a metallic hydrogen and helium state (the brown dwarf's interior) and the photosphere (composed of molecular hydrogen and helium).

? Exercise

Following the analogous steps from the Sect. 4.1.2, derived the effective temperature and show that the luminosity is given by

$$L = \frac{0.0721 L_{\odot}}{\kappa_R^{1.1424}\gamma^{0.286}}\left(\frac{M}{M_{\odot}}\right)^{1.239}\frac{\eta^{2.856\nu}b_1^{2.856}}{(1 + b + a\eta)^{0.2848}}(1 + \text{eventual MG term}). \quad (69)$$

However, the electron degeneracy η is a function of time, hence we need to find that dependence. This can be done again with the interior entropy (which is a function of η) and the thermodynamics laws. The result of this procedure yields

$$\frac{\mathrm{d}\eta}{\mathrm{dt}} = -\frac{1.1634 \times 10^{-18} b_1^{2.856} \mu_{1mod}}{\kappa_R^{1.1424} \mu_e^{8/3}} \left(\frac{M_\odot}{M}\right)^{1.094} \times \eta^{2.856\nu} (1 + b + a\eta)^{1.715} \gamma^{0.7143} (1 + \text{eventual MG term}), \tag{70}$$

where $\frac{1}{\mu_{1mod}} = \frac{1}{\mu_1} + \frac{3}{2} \frac{x_{H^+}(1-x_{H^+})}{2-x_{H^+}}$ and μ_1 is given by (10). The Eqs. (70) and (69) with the initial conditions $\eta = 1$ at $t = 0$, provides the model of cooling process for a brown dwarf star in a given theory of gravity.

4.2.2 White Dwarfs

The final state of a Main Sequence star with mass $\lesssim (10 \pm 2)\, M_\odot$ is a white dwarf–a very compact star which mainly consists of a core of the progenitor star. The white dwarf's mass ranges from a bit above of the solar mass, but with a much smaller radius. Since in that stage of the stellar evolution the main energy source is the energy stored during the previous active phases (that is, when the thermonuclear reactions were taking place in the stellar interior), such a stellar remnant also cools down with time. Here, we will focus on the process of cooling in which we take into account crystallization. When the white dwarf's core starts crystallizing, latent heat is released, contributing the thermal energy which is radiated away. This delays the cooling process. As soon as the star solidifies, the specific heat follows the Debye law (known as Debye cooling). The simplest model of cooling in which one takes into account only the radiation of the total thermal energy of the star (23) in modified gravity was given in [33]. In what follows, we will demonstrate how to include dependence on the Debye temperature and the ratio of Coulomb to thermal energy to it, and we will also insert the crystallization process.

Therefore, the luminosity resulting from a decrease in thermal energy (21) of ions and electrons in time t is given by

$$L_{\text{thermal}} = -\frac{dU}{dt} = -\frac{\mathcal{M}}{Am_\text{p}} \bar{c}_v \frac{dT}{dt}, \tag{71}$$

where the mean specific heat is given by (22) with (25). However, we need to also take into account the latent heat which is released during the crystallization process. Assuming that it is given $\sim k_\text{B}T$ (we will take the numerical coefficient equaled 1 later on), the contribution to the luminosity resulting from this energy release is given by [71]

$$L_{\text{latent}} = k_\text{B}T \frac{d(m_s/Am_\text{p})}{dt}, \tag{72}$$

where m_s is the amount of mass that is already crystallized. Let us write it as

$$L_{\text{latent}} = k_B T \frac{\mathcal{M}}{Am_p} \frac{1}{\mathcal{M}} \frac{dm}{dr} \frac{dr}{d\rho} \frac{d\rho_s(T)}{dT} \frac{dT}{dt}, \tag{73}$$

where we have introduced $\rho_s(T)$ as the density of the crystallized mass at a temperature T.

? Exercise

Derive the Eq. (73).

The density of the crystallized mass $\rho_s(T)$ is related to the ratio of Coulomb to thermal energy Γ by [35]

$$\Gamma = 2.28 \times 10^5 \frac{Z^2}{A^{1/3}} \frac{\rho_s^{1/3}}{T}. \tag{74}$$

The crystallization process starts when the above ratio reaches the critical value, it means when $\Gamma = \Gamma_m$.

? Exercise

Show that when the crystallization process is taken into account, one has

$$\frac{d\rho_s}{dT} = \frac{3\rho_s}{T}. \tag{75}$$

Using the above relation, the luminosity (73) takes the following form

$$L_{\text{latent}} = 3\rho_s k_B \frac{\mathcal{M}}{Am_p} \frac{1}{\mathcal{M}} \frac{dm}{dr} \frac{dr}{d\rho} \frac{dT}{dt}. \tag{76}$$

Let us note that $\frac{dm}{dr}$ and $\frac{d\rho}{dr}$ are taken at radius r_* for which $\rho(r_*) = \rho_s(T)$ is satisfied.
Therefore, we have finally derived the cooling equation

$$L = \frac{3k_B\mathcal{M}}{Am_p}\left(-\frac{\bar{c}_v}{3k_B} + \rho_s \frac{1}{\mathcal{M}} \frac{dm}{dr} \frac{dr}{d\rho}\right)\frac{dT}{dt}, \tag{77}$$

in which the effect of modified gravity are given by the density profile $\rho(r)$ which is a solution of the modified equilibrium equation (or the modified LE equation). Similarly as in the case of brown dwarfs, such an equation allows to obtain the age of the object. It can be shown that white dwarfs in modified gravity one deals with younger objects than in Newtonian framework (that is, the stars cool down faster).

5 Conclusions

We have introduced a standard formalism which is used to provide an hydrostatical equilibrium of stellar and substellar objects in a framework of modified gravity. Moreover, we have also briefly discussed other equations, mainly related to the matter properties, which allow to describe internal processes happening in low-mass stars, brown dwarfs, giant planets as well as white dwarfs. As examples, we discussed light elements' ignition in the stellar cores, such as hydrogen and lithium burning. Regarding evolutionary phases, we demonstrated how to use the modified Lane-Emden formalism to examine cooling processes of brown and white dwarf stars.

There is however still a lot to do. We have just read about a few particular processes whose description was undertaken in the framework of modified gravity. As shown in [4], we do not have too many tests of our models in the curvature regime of stars, which lies between the problematic one, that is, cosmological scales, and the well tested, that is, Solar System's and compact objects' one. This gap could potentially hide the onset of corrections to General Relativity, urgely searched by the modified gravity community.

Acknowledgements This work is supported by MICINN (Spain) *Ayuda Juan de la Cierva–incorporacíon* 2020 No. IJC2020-044751-I.

References

1. D. Aguiar Gomes, A. Wojnar, Eur. Phys. J. C **83**(6), 492 (2023)
2. P.J. Armitage, *Astrophysics of Planet Formation* (Cambridge University Press, 2010)
3. S. Auddy, S. Basu, S.R. Valluri, Adv. Astron. **2016** (2016)
4. T. Baker et al., Astrophys. J. **802**, 63 (2015)
5. M. Banados, P.G. Ferreira, Phys. Rev. Lett. **105**, 011101 (2010)
6. S. Banerjee et al., J. Cosmol. Astropart. Phys. **2017**(10), 004 (2017)
7. I.S. Belfaqih, H. Maulana, A. Sulaksono, Int. J. Mod. Phys. D **30**, 2150064 (2021)
8. M. Benito, A. Wojnar, Phys. Rev. D **103**, 064032 (2021)
9. L. Bildsten et al., Astrophys. J. **482**, 442 (1997)
10. F. Birch, Phys. Rev. **71**, 809 (1947)
11. A. Burrows, Adam. J. Liebert, Rev. Mod. Phys. **65**, 301 (1993)
12. S. Capozziello et al., Phys. Rev. D **83**, 064004 (2011)
13. S. Capozziello et al., Phys. Rev. D **85**, 044022 (2012)
14. S. Capozziello, M. De Laurentis, Phys. Rep. **509**, 167–321 (2011)
15. G.R. Caughlan, W.A. Fowler, At. Data Nucl. Data Tables **40**, 283 (1998)
16. G. Chabrier, D. Saumon, W.B. Hubbard, J.I. Lunine, Astrophys. J. **391**, 817 (1992)
17. S. Chandrasekhar, Mon. Not. R. Astron. Soc. **93**, 390–406 (1933)
18. S. Chandrasekhar, Mon. Not. R. Astron. Soc. **95**, 207–225 (1935)
19. S. Chowdhury, P. Banerjee, A. Wojnar, arXiv:2212.11620 (2022)
20. S. Chowshury, T. Sarkar, Mon. Not. R. Astron. Soc. **523**(1), 518–526
21. S. Chowdhury, T. Sarkar, J. Cosmol. Astropart. Phys. **2021**(05), 040 (2021)
22. M. Crisostomi, M. Lewandowski, F. Vernizzi, Phys. Rev. D **100**, 024025 (2019)
23. P.A.M. Dirac, *Mathematical Proceedings of the Cambridge Philosophical Society*, vol. 26, 3rd edn. (Cambridge University Press, 1930)
24. E. Fermi, Zeitschrift für Physik **48**, 73–79 (1928)

25. R.P. Feynman, N., E. Teller, Phys. Rev. **75**, 1561 (1949)
26. W.A. Fowler, G.R. Caughlan, B.A. Zimmerman, Annu. Rev. Astron. Astrophys. **13**, 69 (1975)
27. M. Guerrero, D. Rubiera-Garcia, A. Wojnar, Eur. Phys. J. C **82**, 707 (2022)
28. C.J. Hansen, S.D. Kawaler, V. Trimble, *Stellar Interiors: Physical Principles, Structure, and Evolution* (Springer Science & Business Media, 2012)
29. G.P. Horedt, *Polytropes: Applications in Astrophysics and Related Fields*, vol. 306. (Springer Science & Business Media, 2004)
30. R.K. Jain, C. Kouvaris, N. Nielsen, Phys. Rev. Lett. **116**, 151103 (2016)
31. S. Kalita, B. Mukhopadhyay, Astrophys. J. **909**, 65 (2021)
32. S. Kalita, L. Sarmah, Phys. Lett. B **827**, 136942 (2022)
33. S. Kalita, L. Sarmah, A. Wojnar, Universe **8**, 647 (2022)
34. R. Kippenhahn, A. Weigert, A. Weiss, *Stellar Structure and Evolution*, vol. 192, 2nd edn. (Springer-Verlag, Berlin. 1990)
35. D. Koester, Astron. Astrophys. **16**, 459 (1972)
36. K. Koyama, J. Sakstein, Phys. Rev. D **91**, 124066 (2015)
37. K. Koyama, J. Sakstein, Phys. Rev. D **108**(2), 024016 (2023)
38. A. Kozak, K. Soieva, A. Wojnar, arXiv:2205.12812 (2022)
39. A. Kozak, A. Wojnar, Phys. Rev. D **104**, 084097 (2021)
40. A. Kozak, A. Wojnar, Eur. Phys. J. C **81**, 492 (2021)
41. A. Kozak, A. Wojnar, Universe **8**, 3 (2022)
42. A. Kozak, A. Wojnar, Int. J. Geom. Methods Mod. Phys. **19**(Supp01), 2250157 (2022)
43. T.W.A. Müller, W. Kley, Astron. Astrophys. **539**, A18 (2012)
44. Sh. Nojiri, S.D. Odintsov, Phys. Rep. **505**, 59–144 (2011)
45. Sh. Nojiri, S.D. Odintsov, V.K. Oikonomou, Phys. Rep. **692**, 1–104 (2017)
46. G.J. Olmo, D. Rubiera-Garcia, A. Wojnar, Phys. Rev. D **100**, 044020 (2019)
47. G.J. Olmo, D. Rubiera-Garcia, A. Wojnar, Phys. Rep. **876**, 1–75 (2020)
48. G.J. Olmo, D. Rubiera-Garcia, A. Wojnar, Phys. Rev. D **104**, 024045 (2021)
49. P. Pani, V. Cardoso, T. Delsate, Phys. Rev. Lett. **107**, 031101 (2011)
50. G. Raimann, Z. Phys, Hadrons Nucl. **347**, 73–74 (1993)
51. A.S. Rosyadi et al., Eur. Phys. J. C **79**, 1030 (2019)
52. R. Saito et al., J. Cosmol. Astropart. Phys. **2015**(06), 008 (2015)
53. J. Sakstein, Phys. Rev. D **92**, 124045 (2015)
54. J. Sakstein, Phys. Rev. Lett. **115**, 201101 (2015)
55. J. Sakstein, Phys. Rev. Lett. **115**, 201101 (2015)
56. J. Sakstein, Phys. Rev. D **92**, 124045 (2015)
57. E.E. Salpeter, H.S. Zapolsky, Phys. Rev. **158**, 876 (1967)
58. I.D. Saltas, J. Christensen-Dalsgaard, Astron. Astrophys. **667**, A115 (2022)
59. I.D. Saltas, I. Lopes, Phys. Rev. Lett. **123**, 091103 (2019)
60. I.D. Saltas, I. Sawicki, I. Lopes, J. Cosmol. Astropart. Phys. **2018**(05), 028 (2018)
61. L. Sarmah, S. Kalita, A. Wojnar, Phys. Rev. D **105**, 024028 (2022)
62. M. Schwarzschild, *Structure and Evolution of Stars* (Princeton University Press, 2015)
63. K. Schwarzschild, Nachrichten göttingen. Math. Phys. Klasse **195**, 41–53 (1906)
64. S. Seager et al., Astrophys. J. **669**, 1279 (2007)
65. T.P. Sotiriou, V. Faraoni, Rev. Mod. Phys. **82**, 451 (2010)
66. D.J. Stevenson, Ann. Rev. Astron. Astrophys. **29**, 163–193 (1991)
67. M.C. Straight, J. Sakstein, E.J. Baxter, Phys. Rev. D **102**, 124018 (2020)
68. L.H. Thomas, *Mathematical Proceedings of the Cambridge Philosophical Society*, vol. 23, 5th edn. (Cambridge University Press, 1927)
69. J.D. Toniato, D.C. Rodrigues, A. Wojnar, Phys. Rev. D **101**, 064050 (2020)
70. G. Ushomirsky et al., Astrophys. J. **497**, 253 (1998)
71. H.M. Van Horn, Astrophys. J. **151**, 227 (1968)
72. S.P. Weppner et al., Mon. Not. R. Astron. Soc. **452**, 1375–1393 (2015)
73. A. Wojnar, Phys. Rev. D **107**(4), 044025 (2023)
74. A. Wojnar, Eur. Phys. J. C **79**, 51 (2019)

75. A. Wojnar, Phys. Rev. D **102**, 124045 (2020)
76. A. Wojnar, Phys. Rev. D **103**, 044037 (2021)
77. A. Wojnar, Phys. Rev. D **104**, 104058 (2021)
78. A. Wojnar, Int. J. Geom. Methods Mod. Phys. **18**(Supp01), 2140006 (2021)
79. A. Wojnar, Phys. Rev. D **105**, 124053 (2022)

Cosmology

Chameleon Perfect Scalar Field as a Geometric Correction in $f(R)$ Gravity

Laura L. Parrilla and Celia Escamilla-Rivera

Abstract

In this work, we derive the analytical form for a $f(R)$ model that describes a perfect scalar field ϕ by assuming the existence of a chameleon mechanism. Based on four statements, at the background and perturbative level, it is possible to relate the extra terms from this theory as a geometrical perfect fluid term, whose has been expressed as possible candidates to explain the nature of the dark sector, and possibly, in the case of a perfect scalar chameleon during inflation, satisfy the big bang nucleosynthesis (BBN) constraints until late times.

1 Introduction

Since the discovery of the current accelerated cosmic expansion of the Universe using observational data from astrophysical objects like Supernovae type Ia independently by the high-redshift Supernovae Search Team [29] and the Supernovae Cosmology Project Team [27], has been attributed the source of this phenomena to the so-called dark energy. However, even with the many efforts to pursue the understanding of the nature of such a dark component, this has not been observationally identified yet. One of the main characteristics of dark energy is attributed to a negative pressure, which leads to an accelerated expansion phenomenon by counteracting the force of gravity. The fact that the negative pressure leads to the cosmic acceleration may look

L. L. Parrilla · C. Escamilla-Rivera (✉)
Instituto de Ciencias Nucleares, Universidad Nacional Autónoma de México, Circuito Exterior C.U., A.P. 70-543, México D.F. 04510, México
e-mail: celia.escamilla@nucleares.unam.mx

L. L. Parrilla
e-mail: laura.parrilla@correo.nucleares.unam.mx

B. Hartmann and J. Kunz (eds.), *Gravity, Cosmology, and Astrophysics*, Lecture Notes in Physics 1022, https://doi.org/10.1007/978-3-031-42096-2_8

counter-intuitive.[1] The time-dependent pressure $p(t)$ in homogeneous and isotropic spacetimes appears in general relativity (GR) and mechanisms that generate this negative pressure and cosmic acceleration are one of the main research topics in Cosmology.

The straightforward candidate proposed for dark energy is the cosmological constant Λ, whose energy density remains constant and allows Einstein field equations to preserve the conservation of energy. The Λ term can be interpreted as a *perfect* fluid by shifting to the right-hand side (*r.h.s*),[2] and if dark energy modelled as Λ, for which equation-of-state $w_{\text{fld}} = -1$, is interpreted as a perfect fluid with $p_\Lambda \propto -\rho_\Lambda$. From the point of view of particle physics, the Cosmological Constant can be related to a vacuum energy density, where if we sum up zero-point energies of all normal modes of some field and consider the cut-off of the momentum at Planck scale, the vacuum energy density is around 10^{121} times larger than the observed Λ density, the so-called vacuum catastrophe. On the one hand, if the Cosmological Constant is truly the consequence of the present current cosmic acceleration, we need to find a mechanism to obtain the tiny value of Λ which could be consistent with observations.

On the other hand, if the origin of dark energy is not the cosmological constant, we need to search for some alternative (or extended) gravity models to explain the current cosmic acceleration. There are two approaches to building dark energy models other than using Λ. The first approach is to modify the *r.h.s* of Einstein field equations, $(G_{\mu\nu} = 8\pi G T_{\mu\nu})$, by considering specific forms of the energy-momentum tensor $T_{\mu\nu}$ that includes the possibility of a fluid with negative pressure. The most popular models that belong to this class are the so-called quintessence [7], k-essence [28] and *perfect* fluid models [19]. On this path, the methodology to follow is to consider the existence of scalar fields[3] with slowly/smooth varying potentials, whereas in k-essence it is the scalar field kinetic energy that drives the acceleration. The latter class is based on a perfect fluid of a specific equation-of-state.

The second approach to constructing dark energy models is to modify the *l.h.s* of Einstein field equations. We denote the representative models that belong to this class as *modified gravity* [14]. While scalar field models correspond to a modification of the energy-momentum tensor, the approach in this path corresponds to the modified gravity in which the gravitational theory is modified compared to general relativity. For example, the Lagrangian density for general relativity is given by $f(R) = R - 2\Lambda$, where R is the Ricci scalar and the constant allows us to have the acceleration phenomena required. A possible modification of this scenario can be described by a non-linear arbitrary function f in terms of R, which is called $f(R)$ theories of gravity.

[1] In Newtonian gravity the pressure is related to a force associated with a local potential that depends on the position in space.

[2] There are several alternative models of dark energy also on the *r.h.s* of the field equations, Λ being the simplest of all of them.

[3] In quantum field theory, several species of elementary particles corresponding to a field are produced. The fields are classified as boson or fermionic depending on the spin of the particle. In the case of scalar fields, this spin has a value of zero. At the quantum level, each field corresponds to an operator, however, bosonic fields can be considered classical in a suitable regime.

Dark energy models based on these kinds of theories have been studied extensively including metric formalism [8,9], observational test [1,3,20] and modifications to the spectra of galaxy clustering [10,16].

Historically, Brans-Dicke theory, an important class of scalar-tensor theories, gives rise to a constant coupling Q between the scalar field ϕ and the matter component. In this sense, modified gravity theories can be regarded as a coupled quintessence scenario in Einstein's frame. In the absence of a scalar field potential $V(\phi)$, the Solar system tests constraint the strength of this coupling Q to be smaller than the order of $\approx 10^{-3}$. In this case, is not possible to satisfy the local gravity constraints unless we have a $V(\phi)$ with a large mass that can be capable to suppress the Q coupling in the regions of high density. Furthermore, if the same field ϕ is responsible for the current cosmic acceleration phenomena, the potential $V(\phi)$ needs to be sufficiently flat in the regions of low density. On this line of thought, these requirements are possible to fulfill for large coupling models that satisfy the local gravity constraints through the chameleon mechanism [22]. The existence of a matter coupling gives rise to an extremum of the scalar field potential where the field can be stabilised. In high-density regions, such as the interiors of astrophysical objects, the field mass would be sufficiently large to avoid the propagation of the fifth force. Meanwhile, the field would have a much lighter mass in low-density environments, far away from compact objects, so it could be responsible for the present cosmic acceleration. In this work, we are going to consider a particular class of the chameleon field with an inverse power-law potential of the form $V(\phi) = M^{4+n}\phi^{-n}$, with $n \geq 1$, so local gravity constraints can be satisfied for M $\sim 10^{-2}$eV. Interestingly, these constrictions correspond to the energy scale required for the current cosmic acceleration observed.

Recently, the correspondence between the modified gravity geometrical terms inspired by perfect fluid components has been pointed out as a natural way to understand the nature of dark energy, and by extension, the full dark sector [12,33]. In resume, beyond the Ricci curvature scalar R or any other geometric invariant in a gravitational action, we can modelled these extra terms as perfect fluids. As an extension of these ideas, in this work, we propose to extend the *conditions* that allow us to model the scalar field as perfect fluids using a chameleon scalar field. Perfect fluids play a crucial role in general relativity being the natural sources of Einstein field equations compatible with Bianchi identities. This characteristic allows any source of field equations to be reformulated in adequate perfect fluid form, in principle, to solve the dynamics related to the Cauchy problem. In Cosmology, perfect fluids can represent, at least in a coarse-grained image Hubble's effective flow behaviour ranging from inflation to dark energy epochs. For these reasons, the compatibility of perfect fluid solutions with modified (or extended) theories of gravity is a crucial topic to be investigated.

This work is divided as follows: In Sect. 2 we discuss the perfect fluid form scheme for a FLRW spacetime and the three conditions to fulfill in order to have the perfect fluid description relationship. In Sect. 3 we present a fourth condition at a perturbative level to describe a perfect scalar field. In Sect. 4 it is described the $f(R)$ theory at the background and perturbative level. In Sect. 5 we introduce the chameleon scalar field in $f(R)$ theories and the analytical form of $f(R)$ to obtain a perfect scalar field. Finally, our discussions are given in Sect. 6.

2 Background Conditions: Perfect Scalar Field in FLRW Spacetimes

We can define a FLRW spacetime by considering a zero Weyl tensor, $C_{jklm} = 0$, with a time-like unit vector field in a covariant configuration as $u^k u_k = -1$ [24], as

$$ds^2 = -dt^2 + a^2(t)\left[\frac{dr^2}{1-kr^2} + r^2(d\theta^2 + \sin^2\theta d\phi^2)\right], \tag{1}$$

in spherical coordinates and $k = 0, -1, +1$. Under such a description, we can compute the following covariant derivative

$$\nabla_j u_k = H(g_{ij} + u_j u_k), \quad \nabla_j H = -u_j u^k \nabla_k H, \tag{2}$$

where $H \equiv \dot{a}/a$. Here we are considering the notation $\mu, \nu = 1, 2, 3$ and $i, j, k, l, m = 0, 1, 2, 3$. In a previous analysis [12] was considered the approach given by $h_{jk} = g_{jk} + u_j u_k$, as the projection on the vector space locally orthogonal to u_k, therefore, we can compute the Riemann tensor from $R_{jklm}u^m = (\nabla_j \nabla_k - \nabla k \nabla_j)u_l$, and its contraction, the Ricci tensor as

$$R_{kl} = \frac{1}{3}(R - 4\zeta)\, u_k u_l + \frac{1}{3}(R - 3\zeta)\, g_{kl}, \tag{3}$$

where $R = R^k R_k$ is the curvature scalar and $\zeta = 3\ddot{a}/a$. Notice that this form of the Ricci tensor has a perfect fluid form since the terms related to the time-like vector and the shear/vorticity/acceleration-free term are separated.

In order to reach a perfect fluid form like in Eq. (3), we need to consider the following three geometrical background tests on each of the parameters involved [12] :

1. *If ϕ is perfect, therefore a function $f(\phi)$ is perfect as long as f is smooth. This characteristic applies also to products and time-derivatives of such functions, e.g. if H is perfect, H^2 and $\dot{H}$ are perfect.*
2. *Considering (1), the derivatives of a perfect scalar field ϕ, $\nabla_i \nabla_j \phi$, also has a perfect fluid form.*
3. *Under (1) and (2), we conclude from Eq. (3) that R is a perfect scalar.*

3 Scalar Perturbation Conditions: Non-adiabatic Pressure

While the three conditions described in the latter section restrict us to work with perfect scalar forms, we need to be sure that such statements are also conserved at the perturbative level. In this way, we need to consider the FLRW (1) as a perturbed metric:

$$ds^2 = (1+2A)dt^2 - 2a(t)(\partial_i B)dtdx^i - a^2t\left[(1-2\psi)\delta_{ij} + 2(\partial_i\partial_j E)dx^i dx^j\right] \quad (4)$$

where A, B, ψ, and E will be related to the Bardeen functions in the standard scalar perturbations [4]. From this point forward we are going to consider linear order perturbations, therefore, we can derive the standard conformal time coordinate potential equation as:

$$\Psi'' - 3\mathcal{H}(1+c_a^2)\Psi' - c_a^2\nabla^2\Psi + [2\mathcal{H}' + (1+3c_a^2)\mathcal{H}^2]\Psi = (4\pi G a^2)\delta p, \quad (5)$$

where the prime denotes conformal time derivatives, $\mathcal{H} = aH$ and $c_a^2 = p'/\rho'$ denotes the adiabatic speed,[4] where p and ρ are the pressure and density, respectively. δp are the perturbations related to the pressure of the matter field. If we consider Eq. (5) in the standard perturbed Einstein field equations at first order we can recover

$$\Psi'' - 3\mathcal{H}(1+c_a^2)\Psi' - c_a^2\nabla^2\Psi + [2\mathcal{H}' + (1+3c_a^2)\mathcal{H}^2]\Psi = \left(c_s^2 - c_a^2\right)\nabla^2\Psi, \quad (6)$$

where c_s is referred to the speed of the perturbations. If we compare Eqs. (5) and (6) we can derive the expression that relates both speed of the perturbations as:

$$\delta p = \left(\frac{c_s^2 - c_a^2}{4\pi G a^2}\right)\nabla^2\Psi. \quad (7)$$

By using the conservation equation $\rho' + 3\mathcal{H}(\rho + p) = 0$, derived from the energy momentum tensor $T^\mu_\nu = (\partial\mathcal{L}/\partial X)(\partial^\mu\phi\partial_\nu\phi) - \delta^\mu_\nu\mathcal{L}$, with $X = \frac{1}{2}\partial_\mu\phi\partial^\mu\phi$, we can obtain

$$c_a^2 = \left(\frac{p'}{\rho'}\right) = \left(\frac{\dot{p}}{3H(\rho + p)}\right) = -\left(\frac{\mathcal{L}_\phi + \ddot{\phi}\mathcal{L}_X}{3H\dot{\phi}\mathcal{L}_X}\right). \quad (8)$$

We performed a change rule to transform to physical time, and the subindex denotes derivatives with respect to X and ϕ. And

$$c_s^2 = \frac{\left(\frac{\partial\mathcal{L}}{\partial X}\right)}{\frac{\partial\mathcal{L}}{\partial X} + 2X\left(\frac{\partial^2\mathcal{L}}{\partial X^2}\right)}. \quad (9)$$

[4] This definition is also preserved in physical time units.

In order to have a perfect fluid described by an equation-of-state in the background, we need to have a vanishing δp in the perturbative scheme, therefore, to achieve such a scenario we notice from Eq. (7) that $c_s^2 = c_a^2$, this is the so-called non-adiabatic condition.

The latter condition implies that [33]

$$\frac{\partial}{\partial X}\left[\frac{\mathcal{L}_\phi}{X\mathcal{L}_X}\right] = 0, \tag{10}$$

therefore, to satisfy the condition for a perfect fluid form like we need to consider, at the perturbative level, the following geometrical perturbed test:

4. *If Eq. (10) is satisfied, therefore ϕ is a perfect scalar field.*

4 Geometric Perfect Fluid from $f(R)$

Our next step is to follow the analogies from the statements *1–4*, in the context of $f(R)$ gravity. First, we need to verify if the high derivatives of R in this scheme are fulfilled according to *1–3*, this will denote that the extra terms rising from the geometrical part can be associated with a perfect fluid form by comparing them with the standard matter terms. Second, at the perturbative level, we need to derive the equations related to the non-adiabatic condition in $f(R)$ and compute the possible constrictions on a $f(R)$ smooth function in order to satisfy a modified version of Eq. (10), and therefore fulfill the condition *4*.

4.1 Background Conditions: Perfect Fluid in $f(R)$ Gravity

We need to consider an effective fluid approach to add fluids in extended theories from Einstein's gravity, as $f(R)$ landscape. Many works have been done in this direction [2, 18, 26, 32], are references cited in there. However, it is standard in all of them, to begin with, a fluid of the form $\rho = \rho_{\text{fld}} + \rho_m$, where ρ_{fld} is usually associated with dark energy and m already includes the dark matter rate in the baryonic component. In this context, we specify the modified Einstein-Hilbert action as:

$$S = \int d^4x\sqrt{-g}\left[\frac{1}{2\kappa}f(R) + \mathcal{L}_{\text{m}}\right], \tag{11}$$

where $\mathcal{L}_{\text{m}}$ is the matter contribution Lagrangian and $\kappa \equiv 8\pi G$, where G is the Newton's constant. Varying Eq. (11) with respect to the metric $g_{\mu\nu}$, we can derive the following field equations:

$$f_R G_{\mu\nu} - \frac{1}{2}\left[f - Rf_R\right]g_{\mu\nu} + \left(g_{\mu\nu}\Box - \nabla_\mu\nabla_\nu\right)f_R = \kappa T_{\mu\nu}^{(m)}, \tag{12}$$

where $G_{\mu\nu}$ is the standard Einstein tensor, $f_R = \partial f/\partial R$, and $T^{(m)}_{\mu\nu}$ is the energy-momentum tensor of the matter fields, which is one contribution of a general $T_{\mu\nu} = T^{(m)}_{\mu\nu} + T^{(\text{fld})}_{\mu\nu}$. According to this latter, we can consider that $T^{(\text{fld})}_{\mu\nu}$ can be a contribution related to the geometrical part of Eq. (12) as

$$\kappa T^{(\text{fld})}_{\mu\nu} \equiv [1 - f_R]G_{\mu\nu} + \frac{1}{2}[f - Rf_R]g_{\mu\nu} - \left(g_{\mu\nu}\Box - \nabla_\mu\nabla_\nu\right) f_R, \tag{13}$$

which satisfy the conservation equation $\nabla^\mu T^{(\text{fld})}_{\mu\nu} = 0$. The evolution equations associated to Eq. (12) considering Eq. (13) are given by

$$3\mathcal{H}^2 = \kappa a^2 \left(\rho_{\text{m}} + \rho_{\text{fld}}\right), \tag{14}$$

$$6\dot{\mathcal{H}} = -\kappa a^2\left[\left(\rho_{\text{m}} + 3p_{\text{m}}\right) + \left(\rho_{\text{fld}} + 3p_{\text{fld}}\right)\right], \tag{15}$$

where

$$\kappa\rho_{\text{fld}} = -\frac{f}{2} + 3\frac{\mathcal{H}^2}{a^2} - 3\frac{\mathcal{H}\dot{f}_R}{a^2} + 3\frac{f_R\dot{\mathcal{H}}}{a^2}, \tag{16}$$

$$\kappa p_{\text{fld}} = \frac{f}{2} - \frac{\mathcal{H}^2}{a^2} - \frac{2f_R\mathcal{H}^2}{a^2} + \frac{\mathcal{H}\dot{f}_R}{a^2} - \frac{\dot{\mathcal{H}}}{a^2} - \frac{f_R\dot{\mathcal{H}}}{a^2} + \frac{\ddot{f}_R}{a^2}. \tag{17}$$

Using Eqs. (16) and (17) we can derive the equation-of-state (EoS) for the fluid as:

$$\begin{aligned} w_{\text{fld}} &= \frac{p_{\text{fld}}}{\rho_{\text{fld}}} \\ &= \frac{2\left[(1+2f_R)\mathcal{H}^2 - \mathcal{H}\dot{f}_R + (2+f_R)\dot{\mathcal{H}} - \ddot{f}_R\right] - a^2 f}{a^2 f - 6(\mathcal{H}^2 - \mathcal{H}\dot{f}_R + f_R\dot{\mathcal{H}})}, \end{aligned} \tag{18}$$

which for $f(R) = R$, we obtain the standard EoS from Einstein's gravity [15]. Furthermore, if we consider $\dot{R} = 0$ and a flat space $k = 0$, we obtain the Eos for a perfect fluid $p = w\rho$. Following this idea and the conditions from *3–1*, Eq. (12) describe perfect fluids through the contributions from the geometric part in $f(R)$.

4.2 Scalar Perturbation Conditions: Non-adiabatic Pressure in $f(R)$

Since the contribution due to $T^{(m)}_{\mu\nu}$ is effectively associated with a standard perfect fluid, in such case with the standard matter, and we proved that $T^{(\text{fld})}_{\mu\nu}$ can describe a perfect fluid with respect to the extra geometric contributions, then we are ready to demonstrate is this condition is fulfilled at the perturbative level. Full analyses on $f(R)$ scalar perturbations have been presented in [14, 17]. In this work, we redo these calculations in order to obtain the quantities to satisfy the condition *4*.

We start with the gravitational field constraint equations given by

$$\Psi' + \mathcal{H}\Phi = -\frac{1}{2}\kappa^2 a^2 \sum (\rho + p)\, V, \tag{19}$$

$$\nabla^2\Psi - 3\mathcal{H}\left(\mathcal{H}\Phi + \Psi'\right) = \frac{1}{2}\kappa^2 a^2 \sum \delta\rho, \tag{20}$$

where ρ and p denote the density and pressure of all the fluid/matter components, in our case, matter (m) and the fluid (fld), and V is the effective velocity potential. Since we are associated our perfect fluid with the geometrical contribution from a $f(R)$ gravity, we define in conformal time

$$\kappa^2(\rho_{\text{fld}} + p_{\text{fld}})V_{\text{fld}} \equiv a^{-2}\left(\Phi f_R' + \mathcal{H}\delta f_R - \delta f_R'\right) + 2a^{-2}\left(\Psi' + \mathcal{H}\Phi\right) f_R, \tag{21}$$

where ρ_{fld} and p_{fld} are given by Eqs. (16) and (17). The fluid density perturbation can be derived as

$$\begin{aligned}\kappa^2\delta\rho_{\text{fld}} \equiv & 2a^{-2}\left[3\mathcal{H}(\Psi' + \mathcal{H}\Phi) - \nabla^2\Psi\right] f_R + a^{-2}\left(\nabla^2 + 3\mathcal{H}'\right)\delta f_R - 3a^{-2}\mathcal{H}\delta f_R' \\ & + 3a^{-2}\left(\Psi' + 2\mathcal{H}\Phi\right) f_R'.\end{aligned} \tag{22}$$

In order to arrive to an analogous expression as Eq. (7), we can combine Eqs. (19) and (20) to obtain

$$\nabla^2\Psi = \frac{1}{2}\kappa^2 a^2 \sum \rho\xi, \tag{23}$$

where $\xi = \Omega_m\xi_m + \Omega_{\text{fld}}\xi_{\text{fld}}$, denotes the comoving overdensity with

$$\kappa^2 a^2 \rho_{\text{fld}}\xi_{\text{fld}} = 3\left(\Psi' + \mathcal{H}\Phi\right) f_R' - 2 f_R \nabla^2\Psi + \left[\nabla^2 + 3\left(\mathcal{H}' - \mathcal{H}^2\right)\right]\delta f_R, \tag{24}$$

by using Eqs. (21) and (22). As in (5), we can establish the relation between Bardeen potentials through $\Psi - \Phi = \kappa^2 a^2 \sum (\rho + p)\,\Pi$, where Π denotes the anisotropic stress or anisotropic pressure dimensionless and define the evolution equation for the fluid as

$$\begin{aligned}\kappa^2\left[\delta p_{\text{fld}} + \frac{2}{3}(\rho_{\text{fld}} + p_{\text{fld}})\nabla^2\Pi_{\text{fld}}\right] \equiv a^{-2}\Big\{ & \delta f_R'' + \mathcal{H}\delta f_R' - 2\Phi f_R'' \\ & - \left(4\nabla^2 + \mathcal{H}^2 + a^2 R\right)\frac{\delta f_R}{6} \\ & - 2 f_R\Big[\Psi'' + 2\mathcal{H}\Psi' + \mathcal{H}\Phi' \\ & + \left(\mathcal{H}^2 + 2\mathcal{H}'\right)\Phi + \frac{1}{3}\nabla^2(\Phi - \Psi)\Big] \\ & - \left[\Phi' + 2(\Psi' + \mathcal{H}\Phi)\right] f_R'\Big\},\end{aligned} \tag{25}$$

and using (5) we obtain

$$\Psi'' + \mathcal{H}\left(2+3c_a^2\right)\Psi' + \mathcal{H}\Phi' + \left[2\mathcal{H}' + \left(1+3c_a^2\right)\mathcal{H}^2\right]\Phi = \frac{1}{3}\nabla^2\left(\Psi - \Phi\right) + \frac{3}{2}\mathcal{H}^2 c_s^2 \xi, \tag{26}$$

where

$$\delta p = c_a^2 \delta\rho + \left(c_s^2 - c_a^2\right)\rho\xi. \tag{27}$$

Under the non-adiabatic condition we recover from the latter that $c_s^2 = c_a^2$, henceforth, by assuming pressureless matter ($p_m = \delta p_m = 0$), the fluid *fld* comes from a geometrical $f(R)$ contribution. Furthermore, such scenarios have been directly related to barotropic perfect fluids, i.e with vanishing non-adiabatic pressure perturbations. Interesting scenarios on this matter have been discussed in [33]. However, this assumption is restricted to standard gravity models. In this work, we extended this assumption by coupling minimally a scalar field to the action (11). We will adopt this analysis from a chameleon scalar field scheme in what follows.

5 Chameleon Perfect Scalar Field in $f(R)$

In the context of scalar field-driven expansion dynamics, e.g. inflation epoch, a possible candidate so-called chameleon scalar field [21–23,30,34] has been suggested as to drive an inflationary expansion. One particular characteristic of this scalar field is that its mass depends on the matter density effects, i.e a scalar field with a varying mass in a dense scenario, where the scalar field can acquire a large mass in a short range. Some observational tests have been performed at CMB scales [31] in order to associate this scalar field to an early accelerated expansion phase. In the context of $f(R)$, the chameleon scalar field has been studied in [5,6]. However, if the chameleon scalar field can have the property a perfect scalar needs to be proved.

To develop the latter, we are going to consider the following action:

$$S = \int d^4x\sqrt{-g}\frac{1}{2}\left[M_{\rm pl}^2 f(R) - \partial_\mu\phi\partial^\mu\phi - 2V(\phi) + \mathcal{L}_{\rm m}\right]. \tag{28}$$

In this landscape, we are going to consider a conformal transformation to relate this action with a standard one in a scalar-tensor theory through

$$\exp\left(-\frac{2\beta\phi}{M_{\rm pl}}\right) = f'(R). \tag{29}$$

where $\beta = \sqrt{\frac{1}{6}}$ [5]. In Einstein frame we can write the metric as $\bar{g}_{\mu\nu}$ by a conformal transformation defined as:

$$\bar{g}_{\mu\nu} = e^{-\frac{2\beta\phi}{M_{\rm pl}}} g_{\mu\nu}, \tag{30}$$

therefore, we can rewrite (28) as

$$S = \int d^4x\sqrt{-g}\left(\frac{R}{2\kappa} - \frac{1}{2}\partial_\mu\phi\partial_\nu\phi - V(\phi)\right) + S_{\rm m}[e^{\frac{2\beta\phi}{M_{\rm pl}}}], \tag{31}$$

where

$$V(\phi) = \frac{M_{\rm pl}^2\left[Rf'(R) - f(R)\right]}{2f'(R)^2}. \tag{32}$$

Notice that $f(R)$ theories are equivalent to scalar-tensor theories in this chameleon scheme. Now, from the conditions described in Sect. 4, we can compute the explicit form for the chameleon geometric scheme in $f(R)$. First, we need to consider a Lagrangian density of the form

$$\mathcal{L}(X, \phi) = f(X) - V(\phi). \tag{33}$$

While a form of $f(X) \propto \log(X)$ satisfies directly the condition *4*, in this scheme we need to verify this condition on the chameleon potential. In a $f(R)$ theory to have a chameleon mechanism it is required that the derivatives of the potential behave as: $V'(\phi) < 0,\ V''(\phi) > 0,$ and, $V'''(\phi) < 0$. From (32) we can derive

$$V'(\phi) = \frac{\beta M_{\rm pl}}{f'^2}\left[Rf' - 2f\right], \tag{34}$$

$$V''(\phi) = \frac{1}{3}\left[\frac{R}{f'} + \frac{1}{f''} - \frac{4f}{f'^2}\right], \tag{35}$$

$$V'''(\phi) = \frac{2\beta}{3M_{\rm pl}}\left[\frac{3}{f''} + \frac{f'f'''}{f''^3} + \frac{R}{f'} - \frac{8f}{f'^2}\right]. \tag{36}$$

Generally, these functions give tight constraints on the form selected for $f(R)$, therefore for a specific form of $V(\phi)$ we can find a specific $f(R)$. To perform this calculation, a potential that follows the chameleon mechanism given by Eqs. (34), (35) and (36) is a power law potential of the form:

$$V(\phi) = M^{4+n}\phi^{-n}, \tag{37}$$

and introduce this expression in Eq. (32) and solve the differential equation to obtain an exact form for $f(R)$:

$$f(R) = \frac{A + 2^{2/3}B - 32M^{n+4}M_{\rm pl}^4 - 32M_{\rm pl}^4\phi^n + 4R^2M_{\rm pl}^8}{16M_{\rm pl}^6}, \tag{38}$$

where

$$A = \frac{2\sqrt[3]{2} R M_{\text{pl}}^{10}\left(R^3 M_{\text{pl}}^6 + 2\right)}{B}, \tag{39}$$

$$B = \left(C - 2R^6 M_{\text{pl}}^{24} + 10 R^3 M_{\text{pl}}^{18} + M_{\text{pl}}^{12}\right)^{1/3}, \tag{40}$$

$$C = \sqrt{-M_{\text{pl}}^{24}\left(4R^3 M_{\text{pl}}^6 - 1\right)^3}. \tag{41}$$

Equation (38) is the specific $f(R)$ form for a perfect chameleon scalar field. Also, we can perform the derivation of this latter expression to obtain:

$$f_R = \frac{1}{8} M_{\text{pl}}^2 (f_1 + f_2 - f_3 + f_4 + 4R), \tag{42}$$

where

$$f_1 = \frac{3\sqrt[3]{2} R^3 M_{\text{pl}}^8}{\sqrt[3]{-2R^6 M_{\text{pl}}^{24} + 10R^3 M_{\text{pl}}^{18} + \sqrt{-M_{\text{pl}}^{24}\left(4R^3 M_{\text{pl}}^6 - 1\right)^3} + M_{\text{pl}}^{12}}}, \tag{43}$$

$$f_2 = \frac{\sqrt[3]{2} M_{\text{pl}}^2 \left(R^3 M_{\text{pl}}^6 + 2\right)}{\sqrt[3]{-2R^6 M_{\text{pl}}^{24} + 10R^3 M_{\text{pl}}^{18} + \sqrt{-M_{\text{pl}}^{24}\left(4R^3 M_{\text{pl}}^6 - 1\right)^3} + M_{\text{pl}}^{12}}}, \tag{44}$$

$$f_3 = \frac{2\sqrt[3]{2} R^3 M_{\text{pl}}^{20}\left(R^3 M_{\text{pl}}^6 + 2\right)\left(-2R^3 M_{\text{pl}}^6 - \frac{3\sqrt{-M_{\text{pl}}^{24}\left(4R^3 M_{\text{pl}}^6 - 1\right)^3}}{M_{\text{pl}}^{12} - 4R^3 M_{\text{pl}}^{18}} + 5\right)}{\left(-2R^6 M_{\text{pl}}^{24} + 10R^3 M_{\text{pl}}^{18} + \sqrt{-M_{\text{pl}}^{24}\left(4R^3 M_{\text{pl}}^6 - 1\right)^3} + M_{\text{pl}}^{12}\right)^{4/3}}, \tag{45}$$

$$f_4 = \frac{2^{2/3} R^2 M_{\text{pl}}^{10}\left(-2R^3 M_{\text{pl}}^6 - \frac{3\sqrt{-M_{\text{pl}}^{24}\left(4R^3 M_{\text{pl}}^6 - 1\right)^3}}{M_{\text{pl}}^{12} - 4R^3 M_{\text{pl}}^{18}} + 5\right)}{\left(-2R^6 M_{\text{pl}}^{24} + 10R^3 M_{\text{pl}}^{18} + \sqrt{-M_{\text{pl}}^{24}\left(4R^3 M_{\text{pl}}^6 - 1\right)^3} + M_{\text{pl}}^{12}\right)^{2/3}}. \tag{46}$$

Introducing Eqs. (38)–(42) in Eq. (13), we obtain for the time-time component:

$$\begin{aligned} &-6\kappa\,(g_3 + g_4 - g_5 + g_6 + 4R)\, M_{\text{pl}}^2\,(3p_{\text{fld}} + \rho_{\text{fld}}) + 32\kappa\rho_{\text{fld}} \\ &= \frac{3\left(g_1 + 2^{2/3} g_2 - 32M^{n+4} M_{\text{pl}}^4 - 32M_{\text{pl}}^4 \phi^n + 4R^2 M_{\text{pl}}^8\right)}{M_{\text{pl}}^6}, \end{aligned} \tag{47}$$

where

$$g_1 = \frac{2\sqrt[3]{2}RM_{\rm pl}^{10}\left(R^3M_{\rm pl}^6+2\right)}{\sqrt[3]{-2R^6M_{\rm pl}^{24}+10R^3M_{\rm pl}^{18}+\sqrt{-M_{\rm pl}^{24}\left(4R^3M_{\rm pl}^6-1\right)^3}+M_{\rm pl}^{12}}}, \tag{48}$$

$$g_2 = \sqrt[3]{-2R^6M_{\rm pl}^{24}+10R^3M_{\rm pl}^{18}+\sqrt{-M_{\rm pl}^{24}\left(4R^3M_{\rm pl}^6-1\right)^3}+M_{\rm pl}^{12}}, \tag{49}$$

$$g_3 = \frac{3\sqrt[3]{2}R^3M_{\rm pl}^{8}}{\sqrt[3]{-2R^6M_{\rm pl}^{24}+10R^3M_{\rm pl}^{18}+\sqrt{-M_{\rm pl}^{24}\left(4R^3M_{\rm pl}^6-1\right)^3}+M_{\rm pl}^{12}}}, \tag{50}$$

$$g_4 = \frac{\sqrt[3]{2}M_{\rm pl}^{2}\left(R^3M_{\rm pl}^6+2\right)}{\sqrt[3]{-2R^6M_{\rm pl}^{24}+10R^3M_{\rm pl}^{18}+\sqrt{-M_{\rm pl}^{24}\left(4R^3M_{\rm pl}^6-1\right)^3}+M_{\rm pl}^{12}}}, \tag{51}$$

$$g_5 = \frac{2\sqrt[3]{2}R^3M_{\rm pl}^{20}\left(R^3M_{\rm pl}^6+2\right)\left(-2R^3M_{\rm pl}^6-\frac{3\sqrt{-M_{\rm pl}^{24}\left(4R^3M_{\rm pl}^6-1\right)^3}}{M_{\rm pl}^{12}-4R^3M_{\rm pl}^{18}}+5\right)}{\left(-2R^6M_{\rm pl}^{24}+10R^3M_{\rm pl}^{18}+\sqrt{-M_{\rm pl}^{24}\left(4R^3M_{\rm pl}^6-1\right)^3}+M_{\rm pl}^{12}\right)^{4/3}}, \tag{52}$$

$$g_6 = \frac{2^{2/3}R^2M_{\rm pl}^{10}\left(-2R^3M_{\rm pl}^6-\frac{3\sqrt{-M_{\rm pl}^{24}\left(4R^3M_{\rm pl}^6-1\right)^3}}{M_{\rm pl}^{12}-4R^3M_{\rm pl}^{18}}+5\right)}{\left(-2R^6M_{\rm pl}^{24}+10R^3M_{\rm pl}^{18}+\sqrt{-M_{\rm pl}^{24}\left(4R^3M_{\rm pl}^6-1\right)^3}+M_{\rm pl}^{12}\right)^{2/3}}. \tag{53}$$

Notice that the expression (47) and its second order derivative satisfy the conditions from *1–3*, therefore the chameleon field ϕ with potential Eq. (32) describe a perfect fluid through the contributions from the geometric part in $f(R)$.

6 Discussion

In this work, we considered a particular class of chameleon field ϕ with a power-law potential of the form $V(\phi) = M^{4+n}\phi^{-n}$. Under this assumption, we found that the chameleon can be associated with a perfect scalar field since its behaviour on the evolution and conservative background equations satisfy the conditions described by *1–3*. Furthermore, at the perturbative level, ϕ fulfill directly the condition *4* if we consider a Lagrangian of the form (33). For $n \geq 1$, we notice from Eq. (47) that local gravity constraints are satisfied for M $\sim 10^{-2}$eV. As we mentioned, this is a restricted condition on the energy scale required for the current cosmic acceleration observed. While the conditions at the background evolution seem a *natural* candidate for dark energy for this kind of potential, it is interesting to notice that at a perturbative level, a chameleon perfect scalar ϕ could emerge at first during a phase transition during the inflation era and, finally end with a random position within its potential. Following this evolution, there has been a study on attractors conditions in which the chameleon ϕ could satisfy the big bang nucleosynthesis (BBN) constraints until today [25], which behind the idea of a perfect scalar field at perturbative level can help to set up an ensemble of initial conditions to study furthermore its behaviour at the early universe. This study will be reported elsewhere.

Acknowledgements CE-R acknowledges the Royal Astronomical Society as FRAS 10147 and is supported by DGAPA-PAPIIT-UNAM Project TA100122. This work is based upon work from COST Action CA21136 Addressing observational tensions in cosmology with systematics and fundamental physics (CosmoVerse) supported by COST (European Cooperation in Science and Technology).

References

1. L. Amendola, R. Gannouji, D. Polarski, S. Tsujikawa, Phys. Rev. D. **75**, 083504 (2007)
2. S. Bahamonde, C. Böhmer, S. Carloni, E. Copeland, W. Fang, N. Tamanini, Phys. Rep. **775–777**, 1–122 (2018)
3. F. Bajardi, R. D'Agostino, M. Benetti, V. De Falco, S. Capozziello, Eur. Phys. J. Plus. **137**, 1239 (2022)
4. J. Bardeen, Phys. Rev. D. **22**, 1882–1905 (1980)
5. P. Brax, C. Bruck, A. Davis, D. Shaw, Phys. Rev. D. **78**, 104021 (2008)
6. C. Burrage, J. Sakstein, Living Rev. Relativ. **21**, 1 (2018)
7. R. Caldwell, Braz. J. Phys. **30**, 215–229 (2000)
8. S. Capozziello, Int. J. Mod. Phys. D. **11**, 483–492 (2002)
9. S. Capozziello, V. Cardone, S. Carloni, A. Troisi, Int. J. Mod. Phys. D. **12**, 1969–1982 (2003)
10. S. Capozziello, M. Faizal, M. Hameeda, B. Pourhassan, V. Salzano, S. Upadhyay, Mon. Not. R. Astron. Soc. **474**, 2430–2443 (2018)
11. S. Capozziello, C. Mantica, L. Molinari, Gen. Relativ. Gravit. **52**, 36 (2020)
12. S. Capozziello, C. Mantica, L. Molinari, EPL. **137**, 19001 (2022)
13. S. Chervon, A. Nikolaev, T. Mayorova, S. Odintsov, V. Oikonomou, Nucl. Phys. B. **936**, 597–614 (2018)
14. T. Clifton, P. Ferreira, A. Padilla, C. Skordis, Phys. Rep. **513**, 1–189 (2012)
15. E. Copeland, M. Sami, S. Tsujikawa, Int. J. Mod. Phys. D. **15**, 1753–1936 (2006)
16. I. De Martino, M. De Laurentis, F. Atrio-Barandela, S. Capozziello, Mon. Not. R. Astron. Soc. **442**, 921–928 (2014)
17. D. Duniya, A. Abebe, A. Cruz-Dombriz, P. Dunsby, Mon. Not. R. Astron. Soc. **518**, 6102–6113 (2022)
18. C. Escamilla-Rivera, L. Casarini, J. Fabris, J. Alcaniz, JCAP. **11**, 010 (2016)
19. V. Faraoni, S. Giardino, A. Giusti, R. Vanderwee, Eur. Phys. J. C. **83**, 24 (2023)
20. L. Jaime, M. Jaber, C. Escamilla-Rivera, Phys. Rev. D. **98**, 083530 (2018)
21. J. Khoury, A. Weltman, Phys. Rev. Lett. **93**, 171104 (2004)
22. J. Khoury, A. Weltman, Phys. Rev. D. **69**, 044026 (2004)
23. J. Khoury, Class. Quant. Grav. **30**, 214004 (2013)
24. C. Mantica, L. Molinari, J. Math. Phys. **57**, 102502 (2016)
25. D. Mota, C. Schelpe, Phys. Rev. D. **86**, 123002 (2012)
26. S. Nesseris, Universe. **9**, 2022 (2023)
27. S. Perlmutter et al., Astrophys. J. **517**, 565–586 (1999)
28. A. Rendall, Lect. Notes Phys. **692**, 141–155 (2006)
29. A. Riess et al., Astron. J. **116**, 1009–1038 (1998)
30. N. Saba, M. Farhoudi, Int. J. Mod. Phys. D. **27**, 1850041 (2017)

31. H. Sheikhahmadi, A. Mohammadi, A. Aghamohammadi, T. Harko, R. Herrera, C. Corda, A. Abebe, K. Saaidi, Eur. Phys. J. C. **79**, 1038 (2019)
32. S. Tsujikawa, Phys. Rev. D. **76**, 023514 (2007)
33. S. Unnikrishnan, L. Sriramkumar, Phys. Rev. D. **81**, 103511 (2010)
34. T. Waterhouse, An introduction to chameleon gravity (2006)

Probing the Nature of Cosmic Strings with Gravitational Waves

Lara Sousa

There is now compelling evidence that the fundamental interactions were once unified, but that, as the universe expanded and cooled, a series of symmetry-breaking phase transitions lead to their separation. Many Grand Unified Theories, that attempt to unify the description of the Standard Model interactions, predict the formation of line-like topological defects known as *cosmic strings* as remnants of (some) these phase transitions [81]—as do several beyond-the-standard-model scenarios. These cosmic strings are vortex-like configurations of the fields that experience the symmetry breaking and correspond to regions in which the fields cannot relax into a vacuum state. They thus still exhibit the unbroken symmetry at their cores.

The study of cosmic strings may allow us to use the universe as a particle physics laboratory. Their mass per unit length μ (which in the simplest models is similar to their tension) is determined by the energy scale η of the phase transition that originated them, with $\mu \sim \eta^2$. A detection would then help us reconstruct the series of phase transitions that happened in the distant past. However, since cosmic strings are predicted in a large variety of particle physics scenarios—supersymmetric grand unified theories [49], brane-inflation [50,52], axion models [38] and many others—there may be more information to be gained in this endeavour, as their properties and phenomenology should depend on the nature of the fields that constitute them.

Cosmic strings are generally stable and expected to survive throughout cosmic history and they may leave behind imprints on different observational probes (see

L. Sousa (✉)
Centro de Astrofísica da Universidade do Porto, Rua das Estrelas, 4150-762 Porto, Portugal
e-mail: lara.sousa@astro.up.pt

Instituto de Astrofísica e Ciências do Espaço, Rua das Estrelas, 4150-762 Porto, Portugal

Departamento de Física e Astronomia, Faculdade de Ciências, Universidade do Porto, Rua do Campo Alegre 687, PT4169-007 Porto, Portugal

B. Hartmann and J. Kunz (eds.), *Gravity, Cosmology, and Astrophysics*,
Lecture Notes in Physics 1022, https://doi.org/10.1007/978-3-031-42096-2_9

e.g. [81] for a comprehensive review). Despite this, they have evaded detection as of yet. The advent of Gravitational Wave Astronomy, however, opens the possibility of probing cosmic strings with unprecedented detail through their stochastic gravitational wave background [1,79].

Most studies of the gravitational wave signatures of cosmic string networks [1, 9,21,27,69,74,75,79] have focused on the simplest models, in which strings can be assumed to be infinitely thin, with no internal degrees of freedom. However several key aspects affecting their stochastic gravitational background—their large-scale dynamics, their interactions and the main energy loss mechanism—may strongly depend on the nature of cosmic strings and of the fields that constitute them. The shape and amplitude of the stochastic gravitational wave background generated by string networks may then also depend on the underlying particle physics model and there may then be distinct observational signatures that may allow us to gain information about the nature of cosmic strings. Here, by considering the illustrative example of cosmic superstrings—fundamental strings and 1-dimensional D-branes from string theory that may be stretched to macroscopic sizes and play a cosmological role akin to that of cosmic strings—I show that this may indeed be the case.

1 Cosmic Strings and Gravitational Waves

Although cosmic strings are created in the early universe, they are in general expected to survive throughout cosmic history until the present time. As strings evolve, they move under the effect of their tension and are stretched and damped by cosmological expansion. Cosmic strings often collide and interact too and, when this happens, strings break at the point of collision, exchange partners and reconnect (see Fig. 1). This process—known as intercommutation—leads to the copious production of closed cosmic strings or *loops* that detach from the Hubble flow and start to oscillate with ultra relativistic velocities. These loops are expected to emit bursts of gravitational radiation until they eventually completely evaporate. There are many such cosmic string loops at any moment in cosmic history, emitting gravitational wave bursts in different directions, thus giving rise to a stochastic gravitational wave background [1,79].

The stochastic gravitational wave background generated by loops provides us with a window to probe cosmic strings and to unveil through them the physics of the early universe. However, in order to maximize this potential, accurately characterizing this background is crucial. Usually, its amplitude is quantified in terms of the spectral density of gravitational waves,

$$\Omega_{\rm gw}(f) = \frac{1}{\rho_c}\frac{d\rho_{\rm gw}}{d\log f}\,, \tag{1}$$

in units of the critical density of the universe $\rho_c = 3H_0^2/8\pi G$, where G is the gravitational constant, $H = (da/dt)/a$ is the Hubble parameter, a is the cosmological scale factor, t is the physical time and the subscript '0' is used to label the values

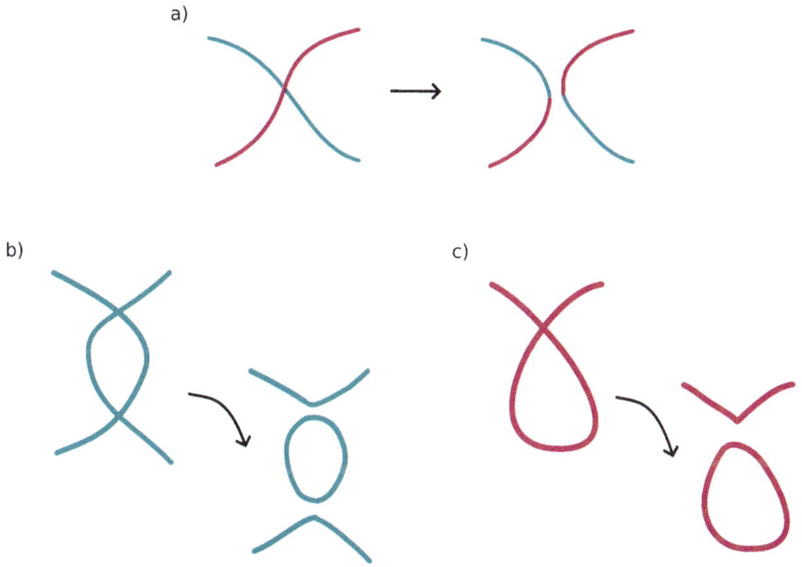

Fig. 1 Intercommutation **a** and loop production in collisions between curved strings **b** and in string self-intersections **c**. Here different colours were used merely for illustration purposes

of the corresponding quantities at the present time. Since, in the case of cosmic strings, this background has contributions from all loops that have emitted, during the history of the universe, gravitational waves that reach us at t_0 with a frequency f, characterizing it involves not only understanding the spectrum of emission of cosmic string loops, but also knowing how many loops, and with what length, exist and emit gravitational waves throughout cosmic history.

1.1 Emission of Gravitational Waves by Cosmic String Loops

The emission spectrum of cosmic string loops was thoroughly studied in the literature [3,20,25,36,37,43,79]. The frequency of the gravitational waves emitted by a cosmic string loop is determined by harmonics of its length:

$$f_j = \frac{2j}{\ell}, \tag{2}$$

where f_j is the frequency corresponding to the jth harmonic mode of emission and ℓ is the length of the loop. The power emitted in each mode, on average, is roughly given by

$$\frac{dE_j}{dt} = \Gamma_j G\mu^2, \quad \text{with } \Gamma_j = \frac{\Gamma}{\zeta(q)} j^{-q}, \tag{3}$$

where ζ is the Riemann zeta-function, $E = \mu\ell$ is the energy of the cosmic string loop and the subscript 'j' is used to refer to the contribution of the jth harmonic mode. Here, $\Gamma = \sum_{j=1}^{+\infty} \Gamma_j$ is the total power emitted in gravitational waves in units of $G\mu^2$, or gravitational radiation emission efficiency, and, for standard cosmic strings, we expect $\Gamma \approx 50$ [3–5,20,25,28,65,73]. The spectral index q depends on the shape of loops and what type of small-scale structure dominates their emission of gravitational radiation. The intercommutation process introduces discontinuities in the string tangent known as *kinks* that travel along the string at the speed of light.[1] Loops are also generally expected to have *cusps*, or points in which the string reaches the speed of light, which should contribute to their gravitational wave emission as well.[2] For loops with cusps, one should have $q = 4/3$, while if the loop has kinks, $q = 5/3$. Finally, collisions between kinks give rise to a power-law spectrum with $q = 2$. In [20], the average power spectrum was computed using approximately 1000 loops obtained from numerical simulation. They found that, for large enough j, it is well described by $q = 4/3$ and thus dominated by the emission of cusps.

1.2 Loop Number Density

Although, in principle, the loop number density $n(\ell, t)d\ell$—describing the number of loops with lengths between ℓ and $\ell + d\ell$ per unit volume—may be measured or inferred from numerical simulations, this was only possible for the simplest string models. Nambu-Goto simulations—which, despite approximating cosmic strings as infinitely-thin 1-dimensional objects, were generally expected to provide a realistic description of the evolution of these networks—systematically predict a significant production of large loops [19,23,56]. There is, however, a significant discrepancy in the number of small loops predicted by the models inferred from simulations performed by different groups (see [22,23]). Numerical simulations of Abelian-Higgs string networks (in which the equations of motion of the fields that constitute the strings are evolved on a lattice), however, paint a significantly different picture: they show no evidence of a stable population of loops [46]. The loops formed in these simulations, in fact, decay as fast as causality allows by emitting classical scalar and gauge radiation (a decay mechanism that is artificially turned off when one uses the Nambu-Goto approximation). More recently, however, it was shown that, for special initial conditions, loops can, in fact, be long lived in Abelian-Higgs simulations and thus decay by emitting gravitational waves [47]. Such loops could only constitute a fraction of the loops produced by the network, which they estimate to be smaller than 10% at a 95% confidence level.

Given these disparities in the results of simulations, an alternative approach is to resort to analytical modeling to describe the loop number density. This was actually the first approach used to study the stochastic gravitational wave background

---

[1] This only applies to cosmic strings with no internal degrees of freedom. For current-carrying strings, for instance, the velocity of kinks is expected to be subluminal (see e.g. [66]).

[2] It has been argued in [43], however, that kinks may inhibit the formation of cusps. Moreover, if strings have internal degrees of freedom, cusps may also be weaker and have velocities smaller than the speed of light.

generated by cosmic strings [1,27,53,80] and its basic premise is very simple: if one knows how the energy density of the network evolves, one may infer (or at least estimate) how much energy is lost as a result of loop production.[3] Originally, these models relied on the simplifying assumption that cosmic string networks maintain a linear scaling regime throughout their evolution and thus that their energy density remains a fixed fraction of the background energy density. This analytical approach, however, was extended in [74,75] to relax this assumption—which actually made it a semi-analytical approach instead—in order to improve the accuracy of computations of the stochastic gravitational wave background. The existence of the linear scaling regime is well established, after being observed in radiation and matter era simulations and shown to be an attractor solution in analytical models for a power-law evolution of the scale factor $a \propto t^{\beta}$, where $0 \leq \beta < 1$ is constant. But, in a realistic cosmological background, cosmic strings are expected to evolve somewhat differently. Deep in the radiation era, during which we effectively have that $a \propto t^{1/2}$, the cosmic string network may indeed be assumed to be in a linear scaling regime.[4] But, as the radiation-matter transition is triggered, the network is expected to enter a long-lasting transitional phase during which it is not scaling. Instead the network has to adapt to the change of the evolution of the cosmological background and starts to evolve towards the scaling regime of the matter era (which is characterized by a different energy density) [10]. This transition between the radiation and matter era scaling regimes is expected to be rather slow and, in fact, dark energy is expected to become relevant to the dynamics of the universe before the cosmic string network is able to re-establish scaling. When this happens, the network starts to be diluted by the accelerated expansion of the universe and experiences a stretching regime instead. The cosmic string network is then unable to maintain a linear scaling evolution after the onset of the radiation-matter transition. The assumption that the network maintains linear scaling (which is usually coupled to the assumption that there is a sudden shift between the radiation- and matter-era scaling regimes at the time of the radiation-matter equality) although convenient—since, for instance, it allows for a complete analytical description of the stochastic gravitational wave background spectrum [77]—does not paint an accurate picture of the later stages of the evolution of a cosmic string network in a realistic cosmological background.

To describe the large-scale dynamics of a cosmic string network, the approach proposed in [74,75] resorts to the Velocity-dependent One-Scale (VOS) model [58, 59], which describes the cosmological evolution of the characteristic length of the network L and its root-mean-squared (RMS) velocity $\bar{v}$:

[3] This, off course, has the underlying assumption that loop production, and not the direct emission of radiation (gravitational or otherwise), is the main energy loss mechanism by long strings; this assumption is relaxed later, but one generally expects this to be the case for standard strings.

[4] But, even in this case, it should be temporarily "knocked out" of scaling whenever there is a decrease of the effective number of relativistic degrees of freedom as the temperature of the universe decreases.

$$\frac{d\bar{v}}{dt} = \left(1-\bar{v}^2\right)\left[\frac{k(\bar{v})}{L} - 2H\bar{v}\right], \tag{4}$$

$$\frac{dL}{dt} = \left(1+\bar{v}^2\right)HL + \frac{\tilde{c}}{2}\bar{v}, \tag{5}$$

where $k(\bar{v})$ is a momentum parameter describing the average curvature of the strings and that accounts, to some extent, for the impact of small-scale structure (here, we shall use the *ansatz* proposed in [59]). The characteristic length L in these equations is a measure of the energy density of the cosmic string network, with $\rho \equiv \mu/L^2$, and, for standard cosmic strings, it is also a measure of the average distance between strings. The last term in (5)—originally proposed in [53]—is a phenomenological term describing the energy loss caused by loop production and $\tilde{c}$ is a parameter describing the efficiency of this energy loss mechanism. It was shown in [59] that $\tilde{c} = 0.23$ provides a good fit to radiation- and matter-era simulations of standard cosmic strings. It is through this term that we may describe the amount of energy that goes into the production of loops throughout cosmic history (which requires solving (4) and (5) coupled to the Friedmann equation).

Once this is known, the only other ingredient needed to compute the number of loops produced is the length of the loops that are created. Given the uncertainties about the length of loops described previously, this approach is often used to study different loop production scenarios. As a matter of fact, it is usually assumed that all loops are created equal[5] and that their length is a fixed fraction of the characteristic length of the network at the time of creation:

$$\ell_b = \alpha L(t_b), \quad \text{with } \alpha < 1, \tag{6}$$

where the subscript 'b' is used to indicate that we are considering the value of these quantities at the time of birth of loops t_b. Note that originally, in the first analytical approaches developed, it was assumed that ℓ_b scaled with physical time t instead of L—which in a linear scaling regime, during which $L \propto t$, is equivalent—since this behaviour was observed in simulations. However, this was merely a consequence of the fact that the long string network was in a linear scaling regime in these simulations and, since L is a measure of the physical length of strings and of the average interstring distance, it should be the relevant physical quantity determining the length of loops. A scaling of loop length with L is expected to provide a better description of how the length of the loops at the time of creation evolves when the network is not in a linear scaling regime.[6]

[5] Note that this assumption is not as strong as it seems: the results obtained under this assumption may be used to obtain the spectrum for any distribution of loop lengths at the time of creation [69,74].

[6] The results of Nambu-Goto numerical simulations actually support this assumption: the length of the large loops created both in radiation and matter eras are well described by a single value of α, while ℓ_b/t is different almost by a factor of 2. Thus, the scaling of the length of loops with L instead of t seems to naturally describe the impact of the radiation-matter transition in the length of loops.

Under these assumptions, the number of loops produced per unit volume is given by [74,77]

$$\frac{dn_\ell}{dt} = \frac{1}{\ell_b}\frac{d\rho_\ell}{dt} = \mathcal{F}\frac{\tilde{c}}{\sqrt{2}\alpha}\frac{\tilde{v}}{L^4}, \tag{7}$$

where the last term in (5) was used to compute the energy density lost due to loop chopping $d\rho_\ell/dt$. Moreover, the factor of $\sqrt{2}$ was introduced in order to account for the fact that, as a result of the redshifting of their peculiar velocities by cosmological expansion, part of the energy of loops is lost [81]. Moreover, we have introduced an additional parameter, $\mathcal{F}$, as a way to relax some of the assumptions of the model, as several phenomena may be parameterized this way. For instance, if the length of loops follows a peaked distribution at the moment of creation, this factor may be used to, to some extent, accommodate for this effect [19,77]. Moreover, if not all of the energy of the loops goes into gravitational radiation and/or only a fraction of the loops decay by emitting gravitational waves (as is the case of Abelian-Higgs strings [46,47]), this may be parameterized through this factor as well. In this approach, then, we have essentially two free parameters, α and $\mathcal{F}$, that may be calibrated to describe the results of simulations—for instance, the simulations of [19,23] are well described by $\alpha \approx 0.34$ and $\mathcal{F} = 0.1$; the simulations in [56] require an additional contribution from very small loops [8]—or they may be left as as free to study a variety of scenarios with minimal assumptions.

Once one knows the number of loops that are created throughout cosmic history, the loop number density is simply given by [74]:

$$n(\ell, t) = \left(\alpha\left.\frac{dL}{dt}\right|_{t=t_b} + \Gamma G\mu\right)^{-1} \left.\frac{dn_\ell}{dt}\right|_{t=t_b}\left(\frac{a(t_b)}{a(t)}\right)^3, \tag{8}$$

which takes into account not only loop production, but also the dilution of loops caused by expansion of the background. So, the problem of finding the number density of loops with a length ℓ at a time t actually reduces to finding the times of births t_b of these loops. Taking into account that we are considering subhorizon loops (as larger loops behave effectively as long strings), for which the impact of expansion is expected to be negligible, we then simply need to find the $t_b < t$ that satisfies

$$\ell(t) = \alpha L(t_b) + \Gamma G\mu(t_b - t), \tag{9}$$

where the last term takes into account the decrease in the length of loops as it radiates gravitational waves. If the network is, at the time of creation of the loops, in a linear scaling regime, this may be obtained analytically, but, in general, this has to be solved numerically.

1.3 The Stochastic Gravitational Wave Background

With all the necessary ingredients set up, the amplitude of the stochastic gravitational wave background generated by the cosmic string loops created throughout the realistic evolution of the universe can now be obtained (see e.g. [9,69,74]):

$$\Omega_{\rm gw}(f) = \frac{16\pi}{3}\left(\frac{G\mu_0}{H_0}\right)^2 \frac{1}{f}\int_{t_i}^{t_0} dt' \sum_{j=1}^{+\infty} j\Gamma_j n(\ell_j(t'),t')\left(\frac{a(t')}{a(t_0)}\right)^5 , \quad (10)$$

where the integration is carried out from the time in which significant loop production starts t_i until t_0 and $\ell_j \equiv (2ja(t')/(fa(t_0))$ is the length that a loop should have at a time t' to emit, in the jth harmonic mode, gravitational waves that reach us with a frequency f.

Figures 2 and 3 show examples of the stochastic gravitational wave background generated by the cosmic string loops produced throughout the evolution of a string network for different values of the loop-size parameter α and of cosmic string tension $G\mu$ (respectively) in a realistic cosmological background. As these figures show, these spectra share roughly the same shape. They all have a flat region in the high-frequency range generated by the loops that decay in the radiation era. This plateau, however, is perturbed by the decrease of the number of relativistic degrees of freedom as the universe expands, which causes a sudden change in the expansion rate that knocks strings out of scaling. When this happens, there is a sudden increase in the amplitude of the plateau, which leads to the appearance of "steps" in the spectrum (these are more evident for smaller values of α in Fig. 2). On the low-frequency range, there is a peak generated by the loops that decay in the matter era, be it loops that were created in radiation and survive into the matter era or those produced in the matter era. Despite these similarities, the overall amplitude of this spectra, as well as the location, shape and broadness of the peak are highly dependent on the value of α and $G\mu$. Larger loops survive for longer and emit their energy in the form of gravitational radiation closer to the present time. These gravitational waves then reach us less diluted by expansion, resulting in a higher amplitude of the background. As a matter of fact, the amplitude of the background scales roughly as $\alpha^{1/2}$ for large enough α, but this decrease slows down as one reaches values closer to the gravitational backreaction scale $\Gamma G\mu$.[7] A decrease of the tension of strings, however, not only leads to a decrease of the amplitude of the spectrum, but also to a shift of the peak of the spectrum towards higher frequencies.

This strong dependence on α means that the observational constraints on cosmic string tension derived using gravitational wave detectors are—unlike those resulting from the cosmic microwave background radiation (see however [68])—are also highly dependent on the size of loops created. As a matter of fact, while the loop production scenarios inferred from Nambu-Goto simulations (corresponding to 10% of the energy lost by the network going into loops with a size $\alpha = 0.34$) are already strongly constrained by pulsar timing data[8]—which exclude Nambu-Goto

[7] For $\alpha < \Gamma G\mu$, in fact, the shape and amplitude of the background is independent of α and a decrease of this parameter merely causes a shift of the spectrum towards higher frequencies (see e.g. [75,77].

[8] The major pulsar timing arrays have all recently reported a common-spectrum stochastic process that could potentially correspond to a stochastic gravitational wave background [7,30,44]. A Nambu-

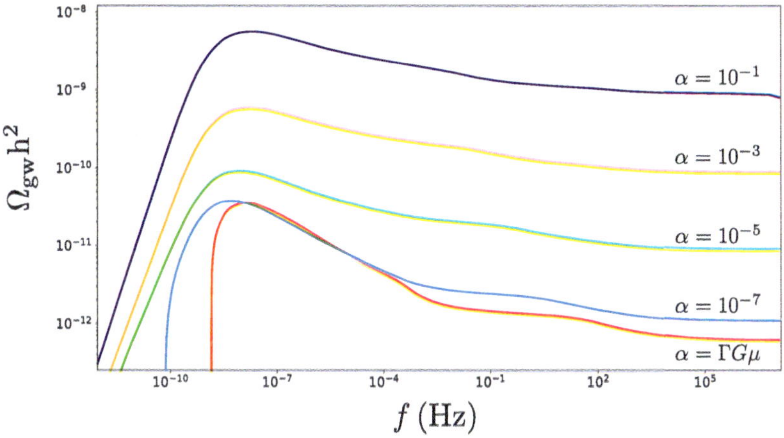

Fig. 2 Stochastic gravitational wave background generated by cosmic string networks for different values of loop-size parameter α. Here, the emission of loops is assumed to be dominated by cusps and $G\mu = 10^{-10}$, $\Gamma = 50$ and $\mathcal{F} = 1$

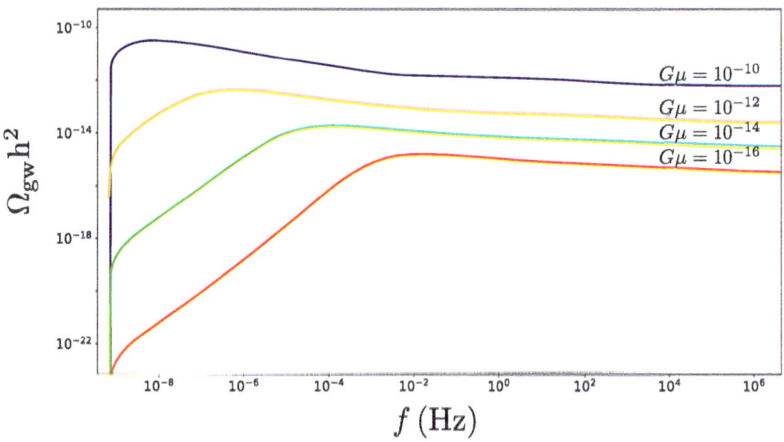

Fig. 3 Stochastic gravitational wave background generated by cosmic string networks for different values of string tension $G\mu$. Here, the emission of loops is assumed to be dominated by cusps and $\alpha = 10^{-8}$, $\Gamma = 50$ and $\mathcal{F} = 1$

strings with $G\mu > 1.5 \times 10^{-11}$ at a 95% confidence level [21]—more conservative, α-independent, constraints are much less stringent and set $G\mu < 1.3 \times 10^{-7}$ [55]. However, this dependence also opens the possibility of using a future detection of the stochastic gravitational wave background to probe the properties of cosmic strings and cosmic string loops. The upcoming Laser Interferometer Space Antenna (LISA) is particularly promising as to this regard. It shall be able to probe a wide range

Goto cosmic string network with $G\mu \sim 10^{-11} - 10^{-10}$ was proposed as a possible explanation [24, 40], but there is still no confirmation that this is an actual physical signal.

of scenarios—including more than 16 orders of magnitude in loop size—and to probe the tension of Nambu-Goto strings down to values of $G\mu \sim \mathcal{O}(10^{-17})$ [9], (an improvement of about 6 orders of magnitude over current probes). For scenarios with smaller loop sizes the improvement is also significant: LISA may potentially increase the constraints on these scenarios by at least four orders of magnitude.

1.4 Going Beyond the Standard

Cosmic strings may be formed in a variety of scenarios and several aspects affecting their stochastic gravitational wave background may depend on the underlying particle physics and/or cosmological model. There is, therefore, the prospect of probing cosmology and particle physics through this spectrum, but, in order to do so, one needs to be able to accurately characterize the impact that these modifications may have on the loop number density.

So far, numerical simulations have been of very limited use in this endeavour. As previously discussed, studying the production of loops was only possible for the simplest models: the Abelian-Higgs string or the Nambu-Goto string (that neglects the underlying particle physics model). Simulations have to be performed for a particular model and in a particular cosmological background, which makes the study of a large variety of non-standard scenarios time-intensive and computationally costly. The semi-analytical model in [74,75], however, allows us to depart easily and quickly from these standard scenarios and to study other relevant models proposed in the literature.

On one hand, the fact that the simplifying assumption of linear scaling was dropped enables the study of strings in non-standard cosmological scenarios—such as early matter domination, kination and models with additional massless degrees of freedom [35]—and the study of models in which strings do not undergo the standard cosmological evolution—for instance, strings created during an inflationary era [45]. In all of these cases it was possible to identify differences in the predicted spectra that may identify these non-standard scenarios in case of a detection (although the signatures of early matter domination and of strings created during inflation are very similar).

On the other hand, the VOS model—although derived from the Nambu-Goto action—is very versatile and can be extended, through the inclusion of additional physically- motivated phenomenological terms, to describe the dynamics of a variety of cosmic string models. A recent such extension was the inclusion of the effect of currents propagating along the strings [60,61], which resulted in a model to describe the large-scale dynamics of superconducting cosmic strings. Current-carrying strings are predicted in several particle physics scenarios (see e.g. [16,41,63,82])—with some even arguing that all strings should carry currents [39]—which motivates the interest in probing them through their gravitational wave background. Currents may have a significant impact on the evolution of a string network during the radiation era [61] and, as shown in [66], this translates into changes in the number of loops produced that may also be significant (particularly if currents are large). Besides this, currents also lead to a suppression of the gravitational wave emission efficiency and change the spectrum of emission of loops [66]. The combination of these effects results in a change in the amplitude of the background generated by the loops created in the radiation era, with potential changes to the shape of the peak of the spectra (in some limits) that may signal the existence of this type of string.

The computation of the stochastic gravitational wave background in these non-standard scenarios was enabled by the semi-analytical approach in [74,75] and herein lays its main advantage: this approach actually allows us to find signatures that may help us uncover the nature of cosmic strings.

2 Probing Cosmic Superstring with Gravitational Waves

To illustrate the potential of the stochastic gravitational wave background to help us discriminate between different string-forming scenarios, let us consider the case of cosmic superstrings.

Cosmic strings and String Theory's superstrings where initially thought to be unrelated [83]. First of all, macroscopic superstrings were expected to be unstable. But even if this problem was somehow overcome, superstrings are expected to form at energy scales close to the Planck scale and, thus, to have very high tensions, which would either mean that they would generate very strong observational signatures—that had not been detected—or that they would to have been diluted away during an inflationary era. In both cases, the possibility of superstrings playing a cosmological role similar to that of ordinary strings seemed to be ruled out.

This picture, however, was changed by the introduction of the brane-world realization of string theory. In this scenario, our universe is a 3-dimensional Dirichlet brane, or D3-brane, embedded in a $9 + 1$ dimensional universe. An inflationary stage would naturally arise in this paradigm if the universe had an additional pair of D-brane and anti-D-brane, since the accelerated expansion could be driven by their interactions as they move towards each other [2,26,50–52,57]. Inflation would then end when the pair of branes eventually collided and a part of their energy would be trapped into branes of lower dimensionality, especially D1 branes. In many such models, extra-dimensions are compactified and, if the collision of branes happens at highly curved regions known as throats, the daughter branes would appear to us 4-dimensional observers to have a much smaller tension (see e.g. [52]). Brane-inflation then provides a way to create strings after inflation with a tension that is compatible with our bounds. Moreover, in many cases, these were shown to be stable [50,72], which indicates that superstrings can indeed grow to macroscopic sizes and play the cosmological role of cosmic strings. These cosmic superstrings arose a lot of interest in the literature, since they provide us with a rare observational window into String Theory.

2.1 Intercommutation and Loop Production in Cosmic Superstring Networks

Although cosmic superstrings share some of the properties of ordinary strings—namely that they may also be regarded effectively as $1+1$-dimensional objects moving under the effect of their tension—there are fundamental differences that lead to them forming networks that are significantly more complex. These differences are mostly related to collisions and interactions between strings, which play a major role in loop formation, and may therefore have an impact on their gravitational wave signatures.

Brane-inflationary scenarios predict the production of two types of strings at the end of inflation: Fundamental strings, or *F-strings*, and 1-dimensional D-branes, usually dubbed *D-strings* [31]. These two types of strings have, in general, different tensions that are related by the dimensionless string coupling g_s:

$$\mu_F = g_s \mu_D\,, \tag{11}$$

where $g_s < 1$ and μ_F and μ_D are respectively the tensions of F- and D-strings.

A key difference between ordinary strings and superstrings arises from their quantum nature: their intercommutation probability P is—unlike that of ordinary strings—expected to be smaller than unity. This means that when a collision occurs, there is a non-vanishing probability that the strings will simply pass through each other without interaction. As a matter of fact, the probability of intercommutation in interactions between D-strings is expected to be such that $10^{-1} \lesssim P_{DD} \lesssim 1$, while $10^{-3} \lesssim P_{FF} \lesssim 1$ in F-string crossings [48]. The fact that cosmic superstrings are less likely to reconnect necessarily affects loop production and these are, in fact, less likely to be created when an intersection occurs. However, the effect that this has on the number density of loops produced is not what one would naively expect: if loop production is reduced, the network of strings loses energy less efficiently and becomes, as a result, denser. Since there are many more strings per unit volume when the intercommutation probability is reduced, we actually expect to end up having a higher loop number density too.

A decrease in the intercommutation probability can be introduced in the VOS model by reducing the energy-loss parameter $\tilde{c}$. For a weakly-interacting cosmic string network, with $\tilde{c} \ll 1$, one has, roughly, $L \propto \tilde{c}t$ and $\bar{v}^2 \sim 1/2$ [10]. As a result, the number of loops produced indeed increases and so does the amplitude of the stochastic gravitational wave background [76]:

$$\frac{dn_\ell}{dt} \propto \frac{1}{\tilde{c}^3} \quad \text{and} \quad \Omega_{\rm gw} \propto \frac{1}{\tilde{c}^2}\,. \tag{12}$$

A key question then is how $\tilde{c}$ is affected by the change in intercommutation probability. Although one may naively expect that the energy-loss parameter decreases proportionally with P [51], Nambu-Goto simulations of string networks with reduced intercommutation probability indicate that this decrease is, in fact, significantly slower [11][9]:

$$\tilde{c} \propto P^{1/3}. \tag{13}$$

This is explained by the fact that there is a build up of small-scale structure on the strings as P is reduced, which counteracts the impact that this reduction would have on the production of small loops. The amplitude of the stochastic gravitational wave background is thus expected to increase as

$$\Omega_{\rm gw} \propto P^{-2/3}, \tag{14}$$

which makes gravitational wave detectors particularly promising as a probe of cosmic superstrings.

Moreover, the fact that there are two distinct types of cosmic superstrings is also expected to have an impact on the gravitational wave spectrum. When two strings of different types collide, they cannot intercommute. Instead, they coalesce along their length and bind to create a new type of string. This process, illustrated in Fig. 4, may happen recursively, leading to the formation of (p, q)-strings—bound states of p F-strings and q D-strings—with a tension

$$\mu(p, q) = \mu_F \sqrt{p^2 + q^2/g_s^2}, \tag{15}$$

where p and q are coprime. Cosmic superstring networks then should be composed of several string species with a hierarchy of tensions that are connected by Y-type junctions, where three different strings meet. This means that, in general, one should expect cosmic string networks to have several loop populations with different tensions, all of which may contribute to the background of gravitational radiation.[10] One then needs to understand the cosmological evolution of each of the cosmic superstring species to compute the total loop number density.

In [12, 13], the VOS model was extended to allow for the description of multitension networks with junctions. This was done by considering several string species with different characteristic lengths L_i and RMS velocities $\bar{v}_i$, where the subscript i labels the ith type of string. The starting point of this model was the standard evolution equations of ordinary cosmic string networks—Eqs. (4) and (5)—as, after all, the main aspects determining their dynamics are expected to be similar: superstrings are also expected to be accelerated as a result of curvature and damped and stretched by expansion and loop chopping is also expected to play a key role. However the creation of bound states is also expected to have an impact on the dynamics and had to be included in these equations. When two different strings i and j interact, part of

---

[9] Note that the exponent 1/3 in (13) was observed both in radiation and matter era simulations. In Minkowski space, the simulations of [70] found that $\tilde{c} \propto P^{1/2}$.

[10] Note that loops with Y-junctions may also form, however these are expected to unzip and to separate into loops of different species [42].

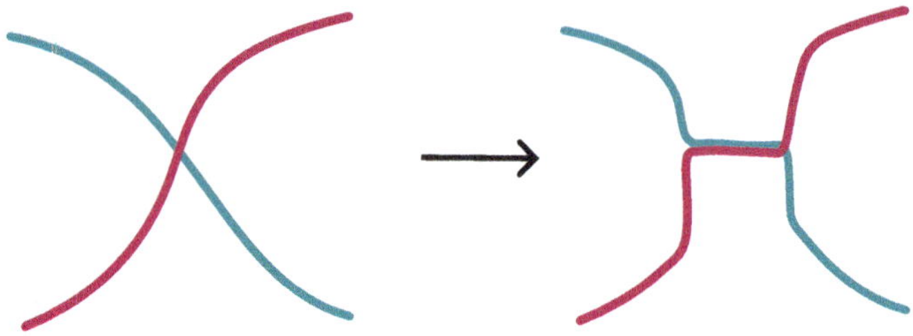

Fig. 4 Creation of bound states in collisions between two cosmic superstrings of different species. Here, different colours were used to represent different superstring types

their length is lost and used to produce a new segment k (which is a bound state of i and j). There is then a transfer of energy density from the i and j string networks into the network of k strings. This may be described by a term of the form [12, 13]

$$\mu_k \tilde{d}^k_{ij} \frac{\bar{v}_{ij}}{L_i^2} \frac{l_{ij}}{L_j} \equiv \mu_k D^k_{ij} \,, \tag{16}$$

where $\tilde{d}^k_{ij}$ is a phenomenological parameter describing the efficiency of the junction creation mechanism in collisions of i and j strings, $\bar{v}^2_{ij} = \bar{v}^2_i + \bar{v}^2_j$ and $l^{-1}_{ij} = L^{-1}_i + L^{-1}_j$. The self-interaction coefficients $\tilde{c}_i$ of each string species are, as we discussed, determined by their respective intercommutation probabilities P_{ii} and (13) can be used to determine their values. The cross-interaction coefficients $\tilde{d}^k_{ij}$ have been computed in [64]. These depend on the microphysical intercommutation probabilities P_{ij} describing interactions between i and j strings and they are also expected to scale as $\tilde{d}^k_{ij} \propto P^{1/3}_{ij}$. The formation of these bound states is also subject to kinematic constraints [13, 32, 33, 71] that further reduce the value of these coefficients, since for certain collision parameters their production is forbidden. This aspect was also taken into consideration in [64]. Moreover, there is an additional aspect that may affect the values of the coefficients $\tilde{c}_i$ and $\tilde{d}^k_{ij}$: since superstrings are embedded in a higher dimensional space, this would seem to make string collisions less likely. Cosmic superstrings, however, may not be free to move in these compact extra-dimensions and may be to some extent "pinned" to our D3-brane [48]. They may, however, due to their quantum nature fluctuate around it, thus exploring a fraction of the compact dimensions. In [64], they introduce a volume suppression parameter w, with $0 < w \leq 1$, to parameterize this effect (with $w = 1$ corresponding to the limit in which strings are stuck to our brane and $w < 1$ if strings can penetrate the compact dimensions). Since, in the latter case, strings are more likely to miss each other, both self- and cross-interactions parameters are expected to be smaller. Moreover, there is some excess energy left behind in the junction formation process because $\mu_k \neq \mu_i + \mu_j$. It is not yet clear whether this energy is absorbed by the new segment, as kinetic energy, or radiated away. The possibility that at least part of this energy is acquired by the segment may be taken into account by adding the following acceleration term to the evolution equation of $\bar{v}_k$:

$$\left.\frac{d\bar{v}_k}{dt}\right|_{\text{junction}} = \left(1 - \bar{v}_k^2\right) B D_{ij}^k \frac{\mu_i + \mu_j - \mu_k}{\mu_k} \frac{L_k^2}{v_k^2}, \tag{17}$$

where $0 \leq B \leq 1$ is a parameter that sets the fraction of the leftover energy that is absorbed by the segment.

With the addition of the terms describing the energy transfer between string species in (16) to the evolution equations for L_i and of the acceleration term in (17) to the $\bar{v}_i$ equations, and with the values of $\tilde{c}_i$ and $\tilde{d}_{ij}^k$ computed in [64], it is now possible to have a description of the cosmological evolution of cosmic superstring networks and infer from it the number density of loops for each species following the approach described in Sect. 1.2.

No simulations of cosmic superstring networks currently exist. Most numerical studies of this type of string either resort to Nambu-Goto strings with reduced intercommutation probability (as in [11,70]) or to field theory analogues (see e.g. [34,54,78]). All these seem to paint a similar picture to that of the VOS model for cosmic superstrings developed in [12,13]. There is clear evidence that the existence of junctions does not lead to a freezing of the network and that, deep in the radiation and matter eras, cosmic superstring networks also evolve towards linear scaling regimes. As discussed before, as a result of the reduced intercommutation probability, networks of F-strings are expected to be denser than networks of ordinary strings (or, in other words, to have smaller characteristic lengths), but the energy density is expected to decrease quickly as we consider string species with higher tensions. As a matter of fact, it was shown in [13,64] that it is generally enough to consider the three lightest type of strings—F- and D-strings and bound states of the two (hereafter referred to as *FD-strings*). It is then also enough to consider only three distinct loop populations when computing the stochastic gravitational wave background.

2.2 What to Expect for Cosmic Superstring Loops

Numerical and analytical studies of the evolution of cosmic superstring networks, unfortunately, tell us little about the sizes of loops that are created in these networks, which leaves us with a huge parameter space to explore. There are, however, several studies of cosmic superstrings that may help us shorten it a bit.

The first is related to the Nambu-Goto simulations of string networks with reduced intercommutation probability in [11]. Therein, as previously discussed, a build up of small-scale structure is observed as P is reduced, which mitigates the impact of this reduction on the production of small loops. The production of large loops, however, is indeed suppressed and, since for $P = 1$, only about 10% of the energy lost by the network is in the form of large loops, we should not expect to have a significant contribution from large loops for small P. It then seems reasonable to assume that all loops are created at the gravitational backreaction scale, with $\alpha = \Gamma G\mu$. Moreover, the formation of true cusps is also expected to be suppressed for cosmic superstrings [14,62]. Near-cusp events, with a subluminal velocity, may still occur but, in this case, the gravitational wave bursts are expected to be significantly weaker. Kinks, on the other hand, are not suppressed at all—since they are created by the intercommutation process—and they are, in fact, expected to proliferate due to reflection and transmission when, as they propagate along the string, they collide with the Y-junctions [17,18]. We may then expect the natural evolution of cosmic superstring networks to result in three populations of small and kinky loops, whose length is determined by the gravitational backreaction scale of each string species.

Figure 5 shows the stochastic gravitational wave background generated by cosmic superstring loops in this "theoretically-favoured" scenario. Therein, we may see that most of the spectrum looks very similar to that of ordinary cosmic strings, since it is dominated by the contribution of loops of F-string, the lightest (and densest) type of string. There are, however, distinct signatures of the loops of the heavier string species in the low-frequency range. Their contribution leads to significant changes to the shape of the peak of the spectrum that are prominent enough, in principle, to be detected directly. These signatures allow us to distinguish the spectrum of cosmic superstrings from that of ordinary cosmic strings and may then be considered tell-tale signs of their existence. Compellingly, the signature of D- and FD-strings is actually dependent on the relative tension of the different species (or, in other words, on the string coupling g_s), thus opening the possibility of probing the details of the underlying superstring scenario. For $g_s \sim 1$, F- and D-string have similar tensions and characteristic lengths and, as a result, the loops they produce have similar lengths. They then contribute to the gravitational wave background in the same frequency range and the effect of D-strings is, in some sense, to enhance the contribution of F-strings. The bound states, however, have a larger tension and characteristic length and, for this reason, contribute at smaller frequencies, leading to the small bump that may be seen in the spectrum in the bottom panel of Fig. 5. If $g_s \ll 1$, the tension of D-strings and FD-strings are both significantly larger than that of F-strings and so their loop populations contribute to the spectra at significantly lower frequencies, leading to a much more prominent signature that may be regarded as a secondary peak (upper panel of Fig. 5). For intermediate values of g_s, there may be situations in which each contribution is clearly distinguishable, causing a significant broadening of the peak (as seen in the middle panel of this figure). Figure 5 also shows that, if superstrings are allowed to explore the compact extra dimensions, there is an increase of the amplitude of the spectra. This increase is particularly relevant for F-string since heavier strings are in general expected to have smaller fluctuations [64]. Changing w then changes the relative contribution of the different loop populations, which may also allow us to probe w and to study observationally if cosmic superstrings can explore the compact dimensions or not. Finally, allowing the bound segments to acquire the energy left over in their creation does not seem to have a significant impact on the spectrum. This actually leads to an increase of the RMS velocity of the string network and to the consequent increase of its characteristic length, leading to a decrease in the number of loops produced. This decreases the amplitude of the signatures of heavier string types slightly, but not enough to erase them.

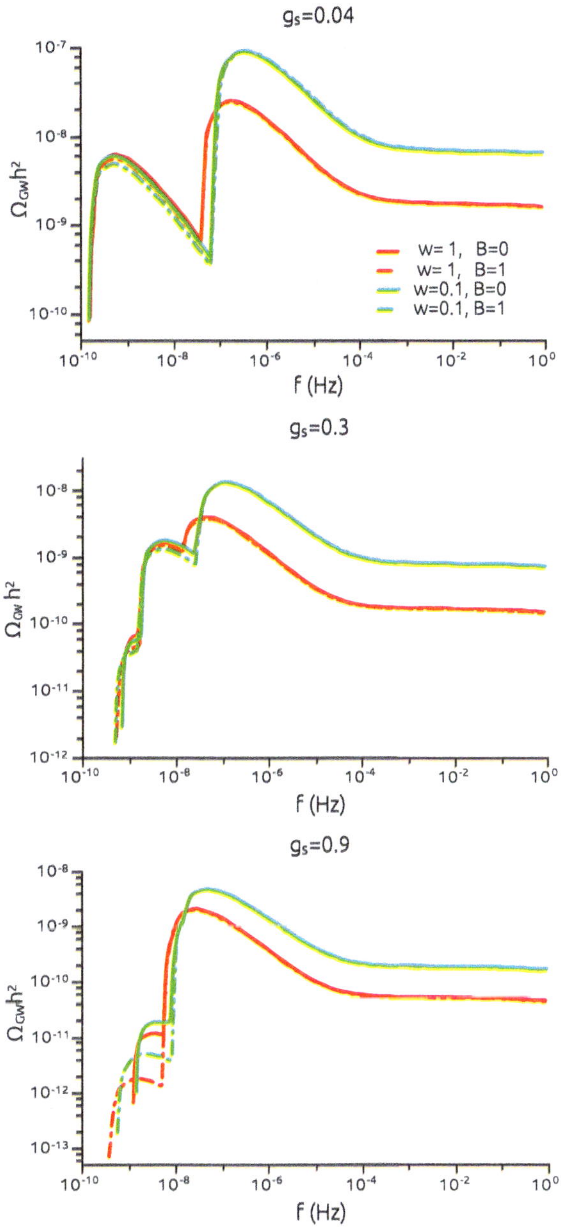

Fig. 5 Stochastic gravitational wave background generated by cosmic superstring networks with $\alpha_i = \Gamma G\mu_i$ and assuming that loops have kinks, for different values of the string coupling g_s. Each spectrum is plotted for $w = 1$ (red lines) and $w = 0.1$ (green lines) and for $B = 0$ (solid lines) and $B = 1$ (dashed lines) Here, $G\mu_F = 10^{-9}$, $\Gamma = 50$ and $\mathcal{F} = 1$. These plotted have been originally published in [76]

Overall, the main conclusion that can be drawn from these results is that the stochastic gravitational wave background generated by cosmic superstring networks may come in different shapes, depending on the underlying superstring model, and that this may allows us, in case of a detection, to extract its details and learn more about cosmic superstrings. Naturally, one may ask whether this is the case for any loop size, since we have assumed that the length is determined by the gravitational backreaction scale for every loop population, but in [76] it was shown that these signatures are present regardless of loop size (although they may be weaker if the loops are assumed to be significantly larger). Therein, however, it was assumed that the loop-size parameters are the same for all species, but relaxing this assumption and assuming significantly different α_i should, except for very fine-tuned choices, lead to visible signatures as well. In fact, this would mean that the length of loops can be significantly different between populations, thus contributing to different frequency ranges to the gravitational wave background.

2.3 Observational Constraints

Although the signatures of heavier string species are prominent and despite the fact that this is where the superstring physics is, most computations of the stochastic gravitational wave background generated by cosmic superstrings do not take this contribution into account. As a matter of fact, this spectrum is usually approximated using the spectrum of ordinary strings with reduced intercommutation probability, by simply using the relation in (13). This necessarily leads to a loss of information about the cosmic superstrings themselves as, at best, this is simply an approximation to describe the contribution of F-strings and, at most, this would allow us to reconstruct the intercommutation probability of this type of string. Note however that P_{FF} does not unequivocally translate into a value of the string coupling g_s, as it is also affected by how much of the compact dimensions the strings can explore. This is well illustrated by the spectra displayed in Fig. 6, where the predictions for two models with F-strings with the same intercommutation probability, but different g_s and w, are plotted alongside this simplified prediction obtained by rescaling the spectra of ordinary strings according to (13). Obviously, in both cases, the simplification fails to predict the spectrum accurately in the low-frequency range as it neglects the potential impact of D- and FD-strings. However, we may see therein that, although for $g_s = 0.3$ and $w = 1$, it does provide a good fit for most of the rest of the spectra, this is no longer the case for $g_s = 0.9$ and $w = 0.1$. In the latter case, this simplified prediction leads to an underestimation over the whole range of the spectrum because the tensions of F- and D-strings are similar and, thus, the corresponding loop populations contribute to the spectrum in a similar frequency range. Moreover, the energy lost by F- and D-strings (caused by the creation of FD-strings) has an impact on the energy density of these networks and on the number of loops produced, that is not taken into account when one uses this approximation. This means that inaccuracies are introduced by approximating only the contribution of the lightest type of string, even when probing the radiation era plateau, and these inaccuracies should necessarily translate into

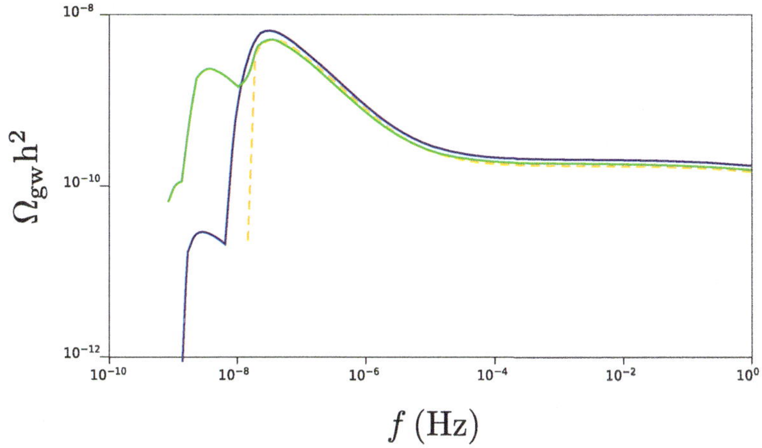

Fig. 6 Stochastic gravitational wave background generated by cosmic superstring networks—with $\alpha_i = \Gamma G\mu_i$ and assuming that loops have kinks—for different scenarios in which F-strings have the same intercommutation probability $P_{FF} = 0.03$. The green line corresponds to a model with $g_s = 0.3$ and $w = 1$, while the purple line represents the stochastic gravitational wave spectra for $g_s = 0.9$ and $w = 0.1$. The dashed yellow line corresponds to the spectra of ordinary strings with an intercommutation probability of $P = 0.3$. Here, $G\mu_F = 10^{-9}$, $\Gamma = 50$ and $\mathcal{F} = 1$

uncertainties in the reconstruction of the values of P_{FF} using observational data. Adding to this the degeneracy between g_s and w introduced by considering only, and approximately, the contribution of F-strings, there is significant loss of information about the underlying particle physics.

Pulsar timing arrays are currently the gravitational wave experiments that are most suitable to constrain the tension of cosmic strings and superstrings using their gravitational wave background, since they probe the low-frequency range of this spectrum, where the peak is expected to be located. They are also, and for the same reason, the best candidates to detect the signatures of heavier superstring species and to distinguish between ordinary strings and cosmic superstrings. In [76], NANOGrav 9-year data [6] was used to derive observational constraints on the tension of fundamental strings μ_F. This data sets an upper limit to the amplitude of the stochastic gravitational wave background at a reference frequency of $f_* = 1\text{yr}^{-1} \simeq 32n\text{Hz}$ of

$$\Omega_{\text{gw}}(f_*)h^2 < 4.15 \times 10^{-10+0.8s}, \tag{18}$$

which depends on the slope of the spectrum s at f_* (defined by assuming that locally $\Omega_{gw} \propto f^s$).[11] Assuming the theoretically-motivated model in which loops are kinky and created with a length similar to the gravitational backreaction scale, this bound on

[11] Usually, it is assumed that $s = -1/3$ for cosmic strings and superstrings motivated by the results of [36] for the gravitational have bursts emitted by cusps. However, as pointed out in [69], the slope of the spectra clearly cannot be assumed to be constant and, since $s < -1/3$ in large portions of the spectrum, this may lead to an overestimation of the constraints.

the background was translated into a conservative limit on the tension of fundamental strings of

$$G\mu_F < 3.2 \times 10^{-9}\,, \tag{19}$$

which is an improvement of about an order of magnitude over those resulting from primary cosmic microwave background anisotropies [29]. This constraint is conservative in the sense that it corresponds to the values of tension above which the stochastic gravitational wave background would have an amplitude above the NANOGrav bound at f_* for all values of g_s considered, assuming $w = 1$. It is also conservative in the sense that it is assumed that loops are created with a small length: for larger values of α, the amplitude of the gravitational wave background may, as we have seen, be significantly larger, which means that these would result in more stringent constraints.

As Fig. 7—where the exclusion regions of the $(g_s, G\mu_F)$-parameter space derived using NANOGrav 9-year data are plotted for $w = 1$ and $w = 0.1$—illustrates, these constraints go significantly beyond this conservative limit. Pulsar timing arrays may be used to detect not only the contribution of F-strings to the stochastic gravitational wave background but also the contribution of D-strings directly (sometimes in combination with that of FD-strings). There is, in fact, a secondary exclusion region for small enough g_S corresponding to the latter case. This shows that a detection of the signature of heavier string types with upcoming data is indeed conceivable in the future, as the sensitivity of the gravitational wave experiments increases. This figure also shows that the constraints derived using the simplified approximation, in which one only considers the effect of the intercommutation probability on F-strings, are safe in the sense that they always underestimate the signal—and thus do not lead to an exclusion of models that are allowed by the data—but lead to a significant loss of information about the underlying scenarios that are excluded. This highlights how important an accurate description of the underlying cosmic string model is for a precise prediction of their stochastic gravitational wave background and to use the current and upcoming data of gravitational wave detectors to its full potential to constrain and discriminate string-forming scenarios. Currently, this can only be achieved through a semi-analytical approach such as that presented here.

Here, it should be noted that there are several aspects of the dynamics and physical properties of cosmic superstrings that not entirely understood. A few of these where already mentioned and are taken into account in the VOS model for cosmic superstrings (for instance, it is not yet clear what happens to the excess energy left over in the formation of bound states or if strings can explore the compact dimensions). The abundance of FD-strings predicted in the VOS model [12, 13] is also higher than that observed in simulations [54], which hints at the possibility that they may decay as a result of the damping of velocities caused by expansion [15, 67]. This would result in a smaller contribution to the gravitational wave background from this type of string but, since this contribution is already subdominant when compared to that of F- and D-strings, this would not have a significant impact on the final shape of the spectrum. More importantly, however, superstrings could possibly be coupled to other fields, which would give then an additional decay channel besides gravitational waves.

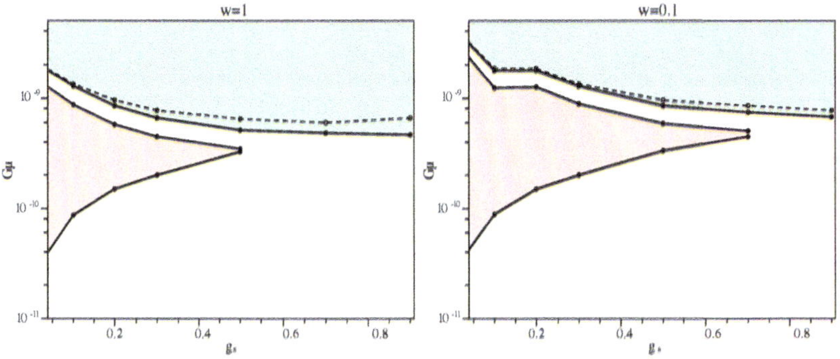

Fig. 7 Constraints on the tension of fundamental strings, $G\mu_F$, as a function of g_s, for cosmic superstrings with loops with kinks with $\alpha_i = \Gamma G\mu_i$, derived using NANOGrav 9-year data. The left panel corresponds to models with $w = 1$ and the right panel to $w = 0.1$. Solid lines represent the constraints derived taking into account the contribution of F-, D- and FD-strings, while dashed lines to those obtained using the simplified model. The blue shaded area represents the exclusion region that results from F-strings, while the pink shaded area represents that resulting from heavier string types. Here, $\Gamma = 50$ and $\mathcal{F} = 1$. This figure was originally published in [76]

3 Conclusions

A comparison of the stochastic gravitational wave background generated by ordinary cosmic strings and cosmic superstrings reveals that there could be tell-tale signs that may allows us to distinguish between these two scenarios. These signatures are, as a matter of fact, highly dependent on the superstring models considered and may then allow us to also discriminate between different superstring scenarios as well. This serves as a proof-of-concept of the possibility of using a detection of the gravitational wave background generated by cosmic strings to unveil the very early universe and probe the underlying particle physics. This work, however, shows that this will only be possible by making accurate predictions of the stochastic gravitational wave background for different string-forming scenarios, which in turn requires an understanding of the properties and phenomenology of the different types of strings that are formed. In the absence of results from numerical simulations, the semi-analytical approach proposed in [74, 75], due to its versatility, may prove to be a valuable tool in this endeavour as it enables the computation of the stochastic gravitational wave background in non-standard string-forming scenarios.

Acknowledgements L. S. is supported by FCT—Fundação para a Ciência e a Tecnologia through contract No. DL 57/2016/CP1364/CT0001. This work was also supported by FCT through the research grants UIDB/04434/2020 and UIDP/04434/2020 and through the R & D project 2022.03495. PTDC—*Uncovering the nature of cosmic strings*.

References

1. F.S. Accetta, L.M. Krauss, Nucl. Phys. B **319**, 747–764 (1989)
2. S.H.S. Alexander, Phys. Rev. D **65**, 023507 (2002)
3. B. Allen, E.P.S. Shellard, Phys. Rev. D **45**, 1898–1912 (1992)

4. B. Allen, P. Casper, A. Ottewill, Phys. Rev. D **50**, 3703–3712 (1994)
5. B. Allen, P. Casper, A. Ottewill, Phys. Rev. D **51**, 1546–1552 (1995)
6. Z. Arzoumanian et al. [NANOGrav], Astrophys. J. **821**(1), 13 (2016)
7. Z. Arzoumanian et al. [NANOGrav], Astrophys. J. Lett. **905**(2), L34 (2020)
8. P.G. Auclair, JCAP **11**, 050 (2020)
9. P. Auclair, J.J. Blanco-Pillado, D.G. Figueroa, A.C. Jenkins, M. Lewicki, M. Sakellariadou, S. Sanidas, L. Sousa, D.A. Steer, J.M. Wachter et al., JCAP **04**, 034 (2020)
10. P.P. Avelino, L. Sousa, Phys. Rev. D **85**, 083525 (2012)
11. A. Avgoustidis, E.P.S. Shellard, Phys. Rev. D **73**, 041301 (2006)
12. A. Avgoustidis, E.P.S. Shellard, Phys. Rev. D **78**, 103510 (2008)
13. A. Avgoustidis, E.J. Copeland, Phys. Rev. D **81**, 063517 (2010)
14. A. Avgoustidis, S. Chadburn, R. Gregory, Phys. Rev. D **86**, 063516 (2012)
15. A. Avgoustidis, A. Pourtsidou, M. Sakellariadou, Phys. Rev. D **91**(2), 025022 (2015)
16. A. Babul, T. Piran, D.N. Spergel, Phys. Lett. B **202**, 307–314 (1988)
17. P. Binetruy, A. Bohe, T. Hertog, D.A. Steer, Phys. Rev. D **82**, 083524 (2010)
18. P. Binetruy, A. Bohe, T. Hertog, D.A. Steer, Phys. Rev. D **82**, 126007 (2010)
19. J.J. Blanco-Pillado, K.D. Olum, B. Shlaer, Phys. Rev. D **89**(2), 023512 (2014)
20. J.J. Blanco-Pillado, K.D. Olum, B. Shlaer, Phys. Rev. D **92**(6), 063528 (2015)
21. J.J. Blanco-Pillado, K.D. Olum, X. Siemens, Phys. Lett. B **778**, 392–396 (2018)
22. J.J. Blanco-Pillado, K.D. Olum, J.M. Wachter, Phys. Rev. D **100**(12), 123526 (2019)
23. J.J. Blanco-Pillado, K.D. Olum, Phys. Rev. D **101**(10), 103018 (2020)
24. S. Blasi, V. Brdar, K. Schmitz, Phys. Rev. Lett. **126**(4), 041305 (2021)
25. C.J. Burden, Phys. Lett. B **164**, 277–281 (1985)
26. C.P. Burgess, M. Majumdar, D. Nolte, F. Quevedo, G. Rajesh, R.J. Zhang, JHEP **07**, 047 (2001)
27. R.R. Caldwell, B. Allen, Phys. Rev. D **45**, 3447–3468 (1992)
28. P. Casper, B. Allen, Phys. Rev. D **52**, 4337–4348 (1995)
29. T. Charnock, A. Avgoustidis, E.J. Copeland, A. Moss, Phys. Rev. D **93**(12), 123503 (2016)
30. S. Chen, R.N. Caballero, Y.J. Guo, A. Chalumeau, K. Liu, G. Shaifullah, K.J. Lee, S. Babak, G. Desvignes, A. Parthasarathy et al., Mon. Not. R. Astron. Soc. **508**(4), 4970–4993 (2021)
31. E.J. Copeland, R.C. Myers, J. Polchinski, JHEP **06**, 013 (2004)
32. E.J. Copeland, T.W.B. Kibble, D.A. Steer, Phys. Rev. Lett. **97**, 021602 (2006)
33. E.J. Copeland, H. Firouzjahi, T.W.B. Kibble, D.A. Steer, Phys. Rev. D **77**, 063521 (2008)
34. J.R.C.C.C. Correia, C.J.A.P. Martins, Phys. Rev. D **106**(4), 043521 (2022)
35. Y. Cui, M. Lewicki, D.E. Morrissey, J.D. Wells, Phys. Rev. D **97**(12), 123505 (2018)
36. T. Damour, A. Vilenkin, Phys. Rev. Lett. **85**, 3761–3764 (2000)
37. T. Damour, A. Vilenkin, Phys. Rev. D **64**, 064008 (2001)
38. R.L. Davis, Phys. Lett. B **180**, 225–230 (1986)
39. A.C. Davis, P. Peter, Phys. Lett. B **358**, 197–202 (1995)
40. J. Ellis, M. Lewicki, Phys. Rev. Lett. **126**(4), 041304 (2021)
41. A.E. Everett, Phys. Rev. Lett. **61**, 1807–1810 (1988)
42. H. Firouzjahi, J. Karouby, S. Khosravi, R. Brandenberger, Phys. Rev. D **80**, 083508 (2009)
43. D. Garfinkle, T. Vachaspati, Phys. Rev. D **36**, 2229 (1987)
44. B. Goncharov, R.M. Shannon, D.J. Reardon, G. Hobbs, A. Zic, M. Bailes, M. Curylo, S. Dai, M. Kerr, M.E. Lower et al., Astrophys. J. Lett. **917**(2), L19 (2021)
45. G.S.F. Guedes, P.P. Avelino, L. Sousa, Phys. Rev. D **98**(12), 123505 (2018)
46. M. Hindmarsh, J. Lizarraga, J. Urrestilla, D. Daverio, M. Kunz, Phys. Rev. D **96**(2), 023525 (2017)
47. M. Hindmarsh, J. Lizarraga, A. Urio, J. Urrestilla, Phys. Rev. D **104**(4), 043519 (2021)
48. M.G. Jackson, N.T. Jones, J. Polchinski, JHEP **10**, 013 (2005)
49. R. Jeannerot, J. Rocher, M. Sakellariadou, Phys. Rev. D **68**, 103514 (2003)
50. N.T. Jones, H. Stoica, S.H.H. Tye, JHEP **07**, 051 (2002)
51. N.T. Jones, H. Stoica, S.H.H. Tye, Phys. Lett. B **563**, 6–14 (2003)
52. S. Kachru, R. Kallosh, A.D. Linde, J.M. Maldacena, L.P. McAllister, S.P. Trivedi, JCAP **10**, 013 (2003)

53. T.W.B. Kibble, Nucl. Phys. B **252**, 227 (1985)
54. J. Lizarraga, J. Urrestilla, JCAP **04**, 053 (2016)
55. L. Lentati, S.R. Taylor, C.M.F. Mingarelli, A. Sesana, S.A. Sanidas, A. Vecchio, R.N. Caballero, K.J. Lee, R. van Haasteren, S. Babak et al., Mon. Not. R. Astron. Soc. **453**(3), 2576–2598 (2015)
56. L. Lorenz, C. Ringeval, M. Sakellariadou, JCAP **10**, 003 (2010)
57. M. Majumdar, A. Christine-Davis, JHEP **03**, 056 (2002)
58. C.J.A.P. Martins, E.P.S. Shellard, Phys. Rev. D **54**, 2535–2556 (1996)
59. C.J.A.P. Martins, E.P.S. Shellard, Phys. Rev. D **65**, 043514 (2002)
60. C.J.A.P. Martins, P. Peter, I.Y. Rybak, E.P.S. Shellard, Phys. Rev. D **103**(4), 043538 (2021)
61. C.J.A.P. Martins, P. Peter, I.Y. Rybak, E.P.S. Shellard, Phys. Rev. D **104**(10), 103506 (2021)
62. E. O'Callaghan, S. Chadburn, G. Geshnizjani, R. Gregory, I. Zavala, JCAP **09**, 013 (2010)
63. P. Peter, Phys. Rev. D **49**, 5052–5062 (1994)
64. A. Pourtsidou, A. Avgoustidis, E.J. Copeland, L. Pogosian, D.A. Steer, Phys. Rev. D **83**, 063525 (2011)
65. J.M. Quashnock, D.N. Spergel, Phys. Rev. D **42**, 2505–2520 (1990)
66. I.Y. Rybak, L. Sousa, JCAP **11**, 024 (2022)
67. I.Y. Rybak, A. Avgoustidis, C.J.A.P. Martins, Phys. Rev. D **99**(6), 063516 (2019)
68. I.Y. Rybak, L. Sousa, Phys. Rev. D **104**(2), 023507 (2021)
69. S.A. Sanidas, R.A. Battye, B.W. Stappers, Phys. Rev. D **85**, 122003 (2012)
70. M. Sakellariadou, JCAP **04**, 003 (2005)
71. P. Salmi, A. Achucarro, E.J. Copeland, T.W.B. Kibble, R. de Putter, D.A. Steer, Phys. Rev. D **77**, 041701 (2008)
72. S. Sarangi, S.H.H. Tye, Phys. Lett. B **536**, 185–192 (2002)
73. R.J. Scherrer, J.M. Quashnock, D.N. Spergel, W.H. Press, Phys. Rev. D **42**, 1908–1914 (1990)
74. L. Sousa, P.P. Avelino, Phys. Rev. D **88**(2), 023516 (2013)
75. L. Sousa, P.P. Avelino, Phys. Rev. D **89**(8), 083503 (2014)
76. L. Sousa, P.P. Avelino, Phys. Rev. D **94**(6), 063529 (2016)
77. L. Sousa, P.P. Avelino, G.S.F. Guedes, Phys. Rev. D **101**(10), 103508 (2020)
78. J. Urrestilla, A. Vilenkin, JHEP **02**, 037 (2008)
79. T. Vachaspati, A. Vilenkin, Phys. Rev. D **31**, 3052 (1985)
80. A. Vilenkin, Phys. Lett. B **107**, 47–50 (1981)
81. A. Vilenkin, E.P.S. Shellard, *Cosmic Strings and Other Topological Defects* (Cambridge University Press, 2000)
82. E. Witten, Nucl. Phys. B **249**, 557–592 (1985)
83. E. Witten, Phys. Lett. B **153**, 243–246 (1985)

Quantum Gravity and Mathematical Relativity

Entanglement in Conformal Field Theory and Holography

Nele Callebaut

Abstract

In these notes we give a pedagogical account of the replica trick derivation of CFT entanglement and its holographic counterpart, i.e. the Lewkowycz Maldacena derivation of the Ryu-Takayanagi formula. The application to an 'island set-up' for the calculation of black hole radiation entropy is briefly discussed. Further topics focused on are the relation to thermal entropy, thermofield double constructions and statements about the emergence of gravity from entanglement through reinterpretations of gravitational first laws.

1 CFT Entanglement

Entanglement entropy in a conformal field theory (CFT) can be defined as the von Neumann entropy of the reduced density matrix. We review in this chapter the replica trick derivation of CFT entanglement, following the work of Cardy and Calabrese [17,18,24] in Sect. 1.2 and of Holzhey, Larsen and Wilczek [46] in Sect. 1.3. This follows an introductory Sect. 1.1 covering the pictorial notation of wave-functionals and density matrices in the path integral formalism, which is also heavily used in Sect. 1.4 on the thermofield double construction.

1.1 Wave-Functionals and Density Matrices

We start with reviewing[1] some basic pictorial notation for wave-functionals and density matrices that will be used throughout.

[1]For more details, see Appendix A of [66] and section 4 of [43].

N. Callebaut (✉)
Institute for Theoretical Physics, University of Cologne, Zülpicherstraße 77, 50937 Köln, Germany
e-mail: nele.callebaut@thp.uni-koeln.de

B. Hartmann and J. Kunz (eds.), *Gravity, Cosmology, and Astrophysics*,
Lecture Notes in Physics 1022, https://doi.org/10.1007/978-3-031-42096-2_10

In a quantum mechanical system with Hamiltonian H and corresponding Euclidean action I, the transition amplitude from a position eigenstate at Euclidean time $t_E = -T$ to another position eigenstate at $t_E = 0$ can be written as a Euclidean path integral

$$\langle q_f | e^{-HT} | q_i \rangle = \int_{q_i,-T}^{q_f,0} \mathcal{D}q e^{-I}. \tag{1}$$

Inserting a complete set of energy eigenstates on the left hand side and taking the limit $T \to \infty$ picks out the vacuum state contribution $|\psi\rangle$, such that the wavefunction $\psi(q_f)$ is given by path integral evolution from past Euclidean infinity (at a fixed and unimportant initial position q_i)

$$\psi(q_f) = \int_{q_i,t_E=-\infty}^{q_f,t_E=0} \mathcal{D}q\, e^{-I}, \tag{2}$$

where q_f-independent factors have been dropped. The ket $|\psi\rangle$ is the function

$$|\psi\rangle = \int_{q_i,t_E=-\infty}^{\cdot,t_E=0} \mathcal{D}q\, e^{-I} \tag{3}$$

that takes a position q_f (at the dot $\cdot$) and gives back a complex number $\psi(q_f) = \langle q_f | \psi \rangle$. On slices of the path integral one recovers the Hilbert space of the theory.

Similarly, in a quantum field theory with field content $\phi(\mathbf{x}, t)$, the (unnormalized) vacuum state is prepared by Euclidean evolution

$$|\psi\rangle = \int_{\phi(t_E=-\infty)=\phi_i}^{\phi(t_E=0)=\cdot} \mathcal{D}\phi\, e^{-I} \quad = $$

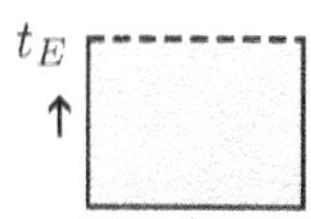

, (4)

where we have already for concreteness restricted the pictorial representation to field theories with one spatial direction x. The state can then further be evolved in Lorentzian time t, with the wave-functional

$$\psi(\phi_f) = \int_{\phi(t_i)=\phi_i}^{\phi(t_f)=\phi_f} \mathcal{D}\phi\, e^{i\mathcal{I}} = \int_{\phi(t_i)=\phi_i}^{\phi(t_f)=\phi_f} \mathcal{D}\phi\, e^{i\int_{t_i}^{t_f} L[\phi]} \tag{5}$$

shown in [55] to satisfy the Schrödinger evolution equation. Again, the ket for the vacuum state (4) is a functional with open boundary condition (at the dot $\cdot$) where it can take on a given field configuration ϕ_f and give back

$$\langle \phi_f | \psi \rangle = \int_{\phi(t_E=-\infty)=\phi_i}^{\phi(t_E=0)=\phi_f} \mathcal{D}\phi \, e^{-I} =$$ 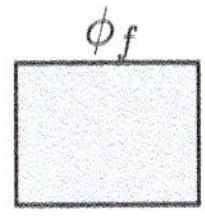

, (6)

obtained pictorially by gluing in the bra $\langle \phi_f |$.

The corresponding bra $\langle \psi |$ and (unnormalized) density matrix $\rho = |\psi\rangle\langle\psi|$ for the pure vacuum state can then pictorially be presented as

$$\langle \psi | =$$  , $$\rho =$$ 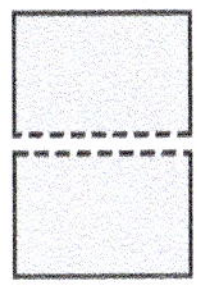 . (7)

An upper dashed line can take on a bra, a lower dashed line a ket, and the matrix ρ both, with matrix elements

$$\langle \phi | \rho | \phi' \rangle \equiv (\rho)_{\phi\phi'} =$$ 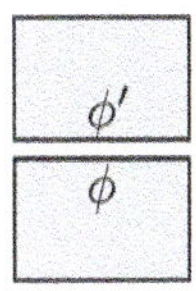

. (8)

Its trace $\operatorname{tr} \rho = \int \mathcal{D}\phi \langle \phi | \rho | \phi \rangle$, obtained by gluing along the previously dashed lines to identify and integrate out ϕ, gives the partition function Z,

$$Z = \operatorname{tr} \rho = \int \mathcal{D}\phi$$ 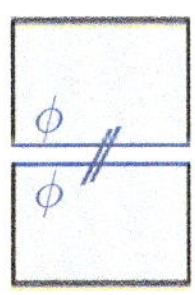

$=$ 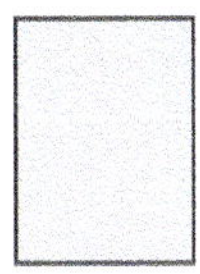 . (9)

In some figures of states (4) and density matrices (7) throughout in the text we will include field configurations in brackets to indicate how to read off the corresponding wave-functional (6) and density matrix elements (8).

1.2 CFT Entanglement from Replica Trick

For a given statistical ensemble described by ρ, where ρ is the probability distribution classically or density matrix quantum mechanically, the von Neumann entropy is defined in terms of the normalized density matrix $\hat{\rho} = \rho / \operatorname{tr} \rho = \rho / Z$ as

$$S = -\operatorname{tr}(\hat{\rho} \log \hat{\rho}). \tag{10}$$

It provides a fundamental, observer-dependent measure for the indeterminacy or lack of resolution of the system, e.g. $S = k_B \log \Omega(E)$ in the microcanonical ensemble, for an observer in a closed system, in which every microstate is equally probable $\hat{\rho} = 1/\Omega(E)$, or $S = (1 - \beta\partial_\beta) \log Z(\beta)$ for an open system observer in the canonical ensemble with Boltzmann probability distribution $\hat{\rho} = e^{-\beta H}/Z(\beta)$.

The von Neumann entropy can be applied in the context of a conformal field theory to define the concept of 'geometric entropy' or entanglement entropy.

The set-up we consider is a $(1 + 1)$-dimensional CFT with Euclidean path integral

$$Z = \int \mathcal{D}\phi \, e^{-I[\phi]}, \qquad I[\phi] = \int_{\mathbb{C}} dx dt_E \mathcal{L}[\phi(x, t_E)], \tag{11}$$

prepared in a pure state $|\psi\rangle$ as pictured in (4). The corresponding density matrix $\rho = |\psi\rangle\langle\psi|$ is pictured in (7). We consider a constant time slice and geometrically bipartition the system, assuming the Hilbert space can be factorized, into a spatial region A and its complement $\bar{A}$. An observer that only has access to region A will measure a different density matrix, called the reduced density matrix

$$\rho_A = \operatorname{tr}_{\bar{A}} \rho. \tag{12}$$

It is obtained from ρ by tracing out degrees of freedom in $\bar{A}$, and in general takes the form of a density matrix for a mixed state. The observer's lack of information about the full system can be quantified by the von Neumann entropy of the normalized reduced density matrix $\hat{\rho}_A = \rho_A / Z$,

$$S_A = -\operatorname{tr}(\hat{\rho}_A \log \hat{\rho}_A). \tag{13}$$

This is by definition the *geometric entropy* or *entanglement entropy* associated with region A. It is a measure for the amount of missing information from the point of view of the observer in A, vanishing in the limit that the observer has access to the full system since the von Neumann entropy for a pure density matrix is zero, and a measure for the amount of entanglement between degrees of freedom in A and degrees of freedom in $\bar{A}$, vanishing when the pure state of the CFT is separable, $|\psi\rangle = |\psi_A\rangle|\psi_{\bar{A}}\rangle$ and $\rho_A = |\psi_A\rangle\langle\psi_A|$, and there is thus no such entanglement.

From applying l'Hôpital's rule, the definition for the entanglement entropy S_A in (13) can be rewritten as

$$S_A = (1 - n\partial_n) \log Z(n)|_{n\to 1}, \tag{14}$$

$$Z(n) = \operatorname{tr}(\rho_A^n). \tag{15}$$

A positive integer n factors of ρ_A construct $Z(n)$, an n-fold replicated description of the system. Then n needs to be analytically continued to non-integer values of n to be able to take the derivative and limit in the definition (14) of S_A. This is called the replica method.

Now consider region A to be an interval $x = x_1..x_2$ at $t_E = 0$, with corresponding reduced density matrix ρ_A and its square ρ_A^2 given by

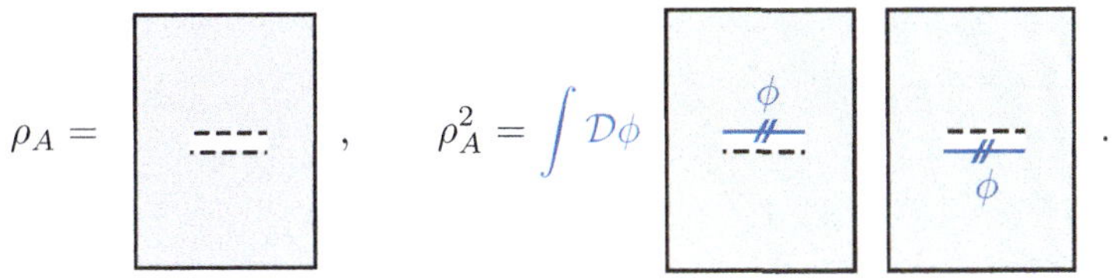

(16)

To calculate the corresponding entanglement S_A in (14), we need to construct $Z(n) = \mathrm{tr}\,\rho_A^n$ for integer $n > 1$. We can think of the gluing condition (referring to the identification and integrating out of the field configuration) in the matrix multiplication ρ_A^2 in two ways: (1) as continuing the *coordinates* (t_E, x) to live on a connected manifold consisting of two copies of the complex plane glued along the region A slit, or (2) as a condition on the *field content*, connecting the fields of two separate copies of the theory $\mathcal{L}(\phi_1)$ and $\mathcal{L}(\phi_2)$ along A. The first is a 'worldsheet' and the second a 'target space' perspective, with (t_E, x) running over $\mathcal{R}_{n,A}$ and $\mathbb{C}$ respectively. Constructing $\mathrm{tr}\,\rho_A^n$ in the first way gives rise to the replicated worldsheet $\mathcal{R}_{n,A}$, path integral integration over which gives $Z(n)$

$$\mathrm{tr}\,\rho_A^n = Z(n) = \int [\mathcal{D}\phi]_{\mathcal{R}_{n,A}}\, e^{-\int_{\mathcal{R}_{n,A}} dx dt_E \mathcal{L}[\phi(x,t_E)]}. \tag{17}$$

$\mathcal{R}_{n,A}$ is called the replica manifold and is pictured in the left figure of Fig. 1. In the $Z(1)$ manifold a rotation over 2π will bring you back to the same location, but in the $Z(n)$ manifold it takes a rotation of $2\pi n$ around the boundary points ∂A of region A to get back to the same location. That is, there are branch points and conical singularities at ∂A, with a conical excess of $2\pi n - 2\pi$.

In the second perspective we write (here $i = 1 \cdots n$)

$$\mathrm{tr}\,\rho_A^n = Z(n) = \int [\mathcal{D}\phi_i]_{\mathbb{C},\,bc}\, e^{-\int_{\mathbb{C}} dx dt_E \mathcal{L}^{(n)}[\phi_i(x,t_E)]} \tag{18}$$

with

$$\mathcal{L}^{(n)}[\phi_1, ..., \phi_n] = \mathcal{L}[\phi_1(x, t_E)] + \cdots + \mathcal{L}[\phi_n(x, t_E)] \tag{19}$$

$$bc = \begin{cases} \phi_1(t_E = 0^+, x \in A) = \phi_2(t_E = 0^-, x \in A) \\ \phi_2(t_E = 0^+, x \in A) = \phi_3(t_E = 0^-, x \in A) \\ \cdots \\ \phi_n(t_E = 0^+, x \in A) = \phi_1(t_E = 0^-, x \in A) \end{cases}. \tag{20}$$

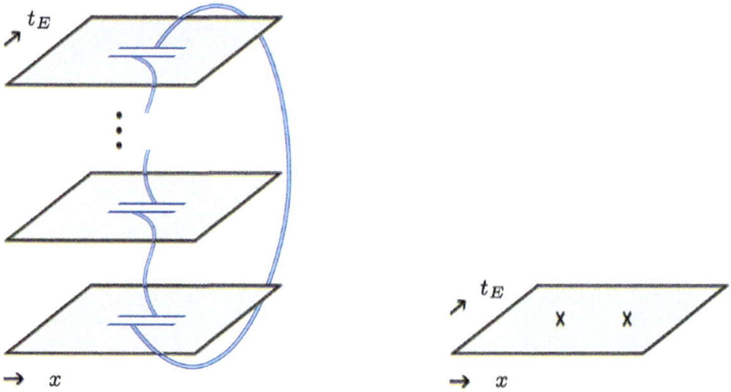

Fig. 1 $Z(n) = \text{tr}\,\rho_A^n$ from the WS perspective (17) (left) and the TS perspective (18) (right). It is the path integral of the theory $\mathcal{L}[\phi]$ on the $\mathbb{Z}_n$ symmetric replica manifold $\mathcal{R}_{n,A}$, or equivalently the path integral of the theory $\mathcal{L}^{(n)}[\phi_i]$ over the orbifold manifold $\mathbb{C} \equiv \mathcal{R}_{n,A}/\mathbb{Z}_n$ in the presence of twist fields. On the right, $Z(n) = \langle T_n(x_1)\tilde{T}_n(x_2)\rangle$

The boundary conditions bc express a global symmetry of the theory $\mathcal{L}^{(n)}[\phi_1, ..., \phi_n]$ under exchange of the fields $\phi_i \to \phi_{i+1}$, the $\mathbb{Z}_n$ permutation symmetry. The conditions can be implicitly implemented by placing twist fields at ∂A. These have the property that when circling a twist field T_n resp. anti twist field $\tilde{T}_n$, a field $\phi_{i \text{ mod } n}$ is transformed into $\phi_{i+1 \text{ mod } n}$, resp. to $\phi_{i-1 \text{ mod } n}$. Then,

$$\begin{aligned} Z(n) &= \int [\mathcal{D}\phi_i]_{\mathbb{C}}\; T_n(x_1)\, \tilde{T}_n(x_2)\, e^{-\int_{\mathbb{C}} dx dt_E \mathcal{L}^{(n)}[\phi_i(x,t_E)]} \\ &= \langle T_n(x_1)\, \tilde{T}_n(x_2)\rangle_{\mathcal{L}^{(n)},\mathbb{C}} \end{aligned} \tag{21}$$

where we used a condensed notation for the locations of the twist fields $T_n(x = x_1, t_E = 0)$ and $\tilde{T}_n(x = x_2, t_E = 0)$, and in the second line we rewrite the $Z(n)$ partition function as a 2-point function of twist fields. This interpretation of $Z(n)$ is pictured on the right of Fig. 1.

1.2.1 Replica Manifold

We will proceed first with calculating $Z(n)$ in the replicated worldsheet point of view of Fig. 1a, following [24], and comment on the twist field correlator perspective later.

For a theory with stress tensor defined via $\delta I = \frac{1}{4\pi}\int T_{\mu\nu}\delta g^{\mu\nu}\sqrt{g}\, d^d x$, the partition function satisfies $\delta \log Z = -\frac{1}{4\pi}\int \langle T_{\mu\nu}\rangle \delta g^{\mu\nu}\sqrt{g}\, d^d x$. This allows us to write down the behavior of $Z(n)$ under a coordinate transformation $x^\mu \to x'^\mu = x^\mu + \delta x^\mu$ (which induces a metric transformation $\partial_\mu \delta x_\nu + \partial_\nu \delta x_\mu$)

$$\delta \log Z(n) = -\frac{1}{2\pi}\int \langle T^\mu{}_\nu\rangle \frac{\partial \delta x^\nu}{\partial x^\mu} d^2 x. \tag{22}$$

This is for the theory $\mathcal{L}[\phi(x, t_E)]$ on the replica manifold $\mathcal{R}_{n,A}$ spanned by $x^\mu = (x, t_E)$, which is everywhere flat except at the branch points $\partial A = (x_1, x_2)$.

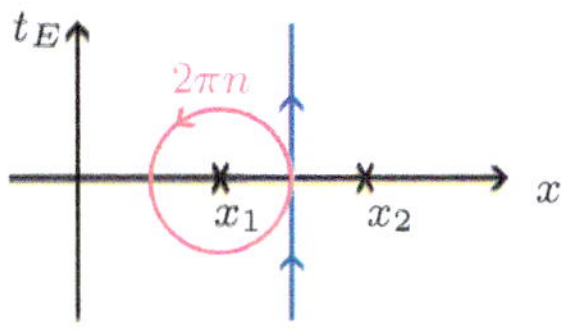

Fig. 2 Contour for calculating $\delta \log Z(n)$ in (24)

We can now consider a coordinate transformation that induces a change δL of the length $L = x_2 - x_1$ of the interval A, and then integrate $\delta \log Z(n)$ to find the dependence of $Z(n)$ on x_1 and x_2. We can choose for example the non-conformal transformation $x' = x + \theta(x - x_0)\delta L$, where $x_1 < x_0 < x_2$. This indeed has the effect $L' = x_2' - x_1' = L + \delta L$. Then

$$\delta \log Z(n) = -\frac{\delta L}{2\pi} \int \langle T_{xx} \rangle \delta(x - x_0) dx dt_E = -\frac{\delta L}{2\pi} \int \langle T_{xx}(x_0, t_E) \rangle dt_E \tag{23}$$

$$= -\delta L \left(\oint_{x_1} \frac{d\zeta}{2\pi i} \langle T_{\zeta\zeta}(\zeta) \rangle - \oint_{x_1} \frac{d\bar{\zeta}}{2\pi i} \langle T_{\bar{\zeta}\bar{\zeta}}(\bar{\zeta}) \rangle \right) \tag{24}$$

In the second line we moved to complex coordinates $\zeta = x + it_E$, deforming the contour along the full infinite range of t_E at $x = x_0$ to ζ encircling x_1 (x_2 is an equally good choice), see Fig. 2. Now we still need to know $\langle T_{\zeta\zeta} \rangle$. On a regular complex plane with coordinate z, the stress tensor expectation value is zero because of rotational and translation invariance. The replica manifold can be conformally transformed into the complex plane, i.e. uniformized, by the conformal transformation

$$z = \left(\frac{\zeta - x_1}{x_2 - \zeta} \right)^{1/n} \tag{25}$$

which consists of first mapping the branch points to 0 and infinity, and then going to the $\mathbb{Z}_n$ orbifold, pictured as 'pizza slice' in Fig. 3. Under this conformal transformation, the stress tensor of the CFT transforms anomalously, $T_{\zeta\zeta}(\zeta) = (\partial z / \partial \zeta)^2 \, T_{zz}(z) + \frac{c}{12}\{z, \zeta\}$, with c the central charge of the CFT. It follows that $\langle T_{\zeta\zeta} \rangle$ is determined by the Schwarzian derivative of $z(\zeta)$ to be

$$\langle T_{\zeta\zeta}(\zeta) \rangle = \frac{c}{24} \left(1 - \frac{1}{n^2} \right) \frac{(x_2 - x_1)^2}{(\zeta - x_1)^2 (\zeta - x_2)^2}. \tag{26}$$

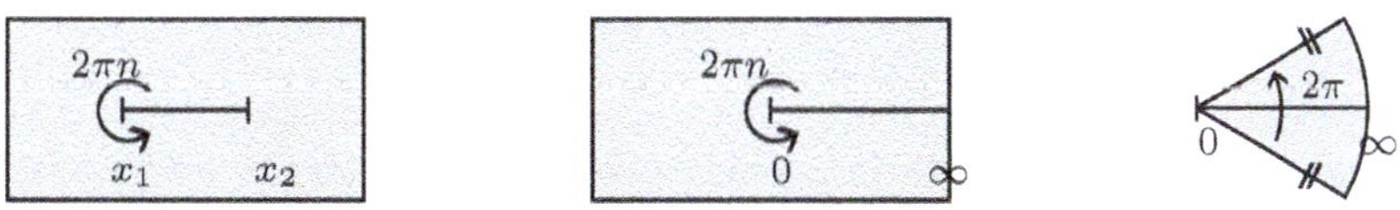

Fig. 3 Coordinate transformation (25) from ζ on the replica manifold (left) to $\frac{\zeta - x_1}{x_2 - \zeta}$ (middle) to z on the orbifold, i.e. the regular complex plane (right). The 'pizza slice' representing the orbifold is drawn here for $n = 6$

In evaluating the complex integrals in (24), the residue theorem picks up an extra n. This is because ζ is a complex coordinate with an argument of range $2\pi n$ around x_1 or x_2. It follows that

$$\frac{\delta \log Z(n)}{\delta L} = -\frac{c}{6}\frac{n - 1/n}{L} \tag{27}$$

or $\log Z(n) = -\frac{c}{6}(n - 1/n)\log L + ...$, where the dots are L-independent terms. We have obtained an expression for the L-dependence of $Z(n) = \mathrm{tr}\, \rho_A^\alpha|_{\alpha=n}$ for integer values of $\alpha = n \geq 1$. The full partition function for complex α is then of the form $Z(\alpha) + \sin(\pi\alpha)g(\alpha)$ in the Re $\alpha > 1$ region with g an analytic function. It can be shown based on the fact that $\hat{\rho}_A$ has eigenvalues $\lambda \in [0, 1]$ and using Carlson's theorem [69] that $g(\alpha) \equiv 0$ and therefore the obtained $Z(n)$ is valid beyond integer n. After substitution in (14), this leads to the famous formula for the vacuum entanglement of an interval A in a 2-dimensional CFT

$$S_A = \frac{c}{3}\log\frac{L}{\epsilon}. \tag{28}$$

Its scaling with c shows the Weyl anomaly of the CFT crucially enters the derivation—in (26) in this case, as the Schwarzian derivative transformation rule is implied by the Weyl anomaly. The UV cutoff ϵ has to be introduced for dimensional reasons, and regulates the arbitrarily large contributions to the entropy from UV degrees of freedom arbitrarily close to ∂A.

For the choice of δL-inducing coordinate transformation we could have also followed [46] and chosen a scale transformation.

1.2.2 Replicated Target Space

Let us also comment on the twist field correlator derivation of $\log Z(n)$, i.e. the perspective of Fig. 1b. Details can be found in [17, 18].

The stress tensor expectation value $\langle T_{\zeta\zeta}(\zeta)\rangle$ in (26) corresponds to the insertion of one stress tensor operator on the replica manifold in Fig. 1a, let's say on the i-th sheet. From the point of view of Fig. 1b, it is the insertion of the stress tensor of the i-th copy of the theory in the presence of twist field operators and thus (normalizing left and right with the insertion of the unit operator)

$$\left\langle T_{\zeta\zeta}(\zeta)\right\rangle_{\mathcal{L},\mathcal{R}_{n,A}} = \frac{\left\langle T^{(i)}_{\zeta\zeta}(\zeta)\, \mathcal{T}_n(x_1)\, \tilde{\mathcal{T}}_n(x_2)\right\rangle_{\mathcal{L}^{(n)},\mathbb{C}}}{\left\langle \mathcal{T}_n(x_1)\, \tilde{\mathcal{T}}_n(x_2)\right\rangle_{\mathcal{L}^{(n)},\mathbb{C}}}, \tag{29}$$

where x_1, x_2 are short for the twist field locations $(\zeta_1, \bar{\zeta}_1)$, $(\zeta_2, \bar{\zeta}_2)$ more generally, and where we have included for clarity subscripts specifying the theory in which the expectation values are taken. The denominator of the right hand side is the $Z(n)$ we are to determine. For primary twist fields [23] it takes the form $1/|x_1 - x_2|^{2d_n}$, with

d_n the unknown scaling dimension of T_n and $\tilde{T}_n$. The numerator of the right hand side can then be rewritten in terms of the twist field two-point function,

$$\left\langle T^{(i)}_{\zeta\zeta}(\zeta)\, T_n(x_1)\, \tilde{T}_n(x_2)\right\rangle = \frac{1}{n}\frac{d_n}{2}\frac{(x_1-x_2)^2}{(\zeta-x_1)^2(x_2-\zeta)^2}\left\langle T_n(x_1)\, \tilde{T}_n(x_2)\right\rangle, \tag{30}$$

by applying the Ward identity for the $(\mathcal{L}^{(n)}, \mathbb{C})$ theory with stress tensor $T^{(n)}_{\zeta\zeta} = nT^{(i)}_{\zeta\zeta}$ (by extensivity of the Lagrangian and thus Hamiltonian density). Then (26), (29) and (30) impose $d_n = \frac{c}{12}(n - 1/n)$, leading to the same $\log Z(n)$ as in the previous section, and the same result for S_A in (28).

One main advantage of this perspective is that the known transformation behavior of the conformal two-point function (21) under conformal transformations immediately gives us the formula for $\log Z(n)$ and thus the entanglement S_A in a conformally related geometry, as we will now briefly discuss.

In the notation $f = x + it_E$, $\bar{f} = x - it_E$ for our current set-up, with metric $ds^2 = df d\bar{f} = dt_E^2 + dx^2$ and state the vacuum state $|0\rangle_f$ as measured in f-coordinates, we have

$$S_A = \frac{c}{3}\log\left|\frac{f_1 - f_2}{\epsilon_f}\right| \tag{31}$$

for

$$\langle T_n(f_1, \bar{f}_1)\tilde{T}_n(f_2, \bar{f}_2)\rangle = |f_1 - f_2|^{-\frac{c}{6}(n-1/n)}. \tag{32}$$

Under a conformal transformation $f = f(z)$, $\bar{f} = \bar{f}(\bar{z})$, the metric transforms $ds^2 = |\partial f/\partial z|^2 dz d\bar{z} \equiv \Omega(z, \bar{z}) dz d\bar{z}$ with a Weyl factor Ω, and the entanglement in z-coordinates becomes

$$S_A = \frac{c}{6}\log\frac{(f(z_1) - f(z_2))(\bar{f}(\bar{z}_1) - \bar{f}(\bar{z}_2))}{\sqrt{f'(z_1)f'(z_2)\bar{f}'(\bar{z}_1)\bar{f}'(\bar{z}_2)\epsilon_z\epsilon_{\bar{z}}}}. \tag{33}$$

This follows from

$$\langle T_n(z_1, \bar{z}_1)\tilde{T}_n(z_2, \bar{z}_2)\rangle = \left|\frac{f(z_1) - f(z_2)}{\sqrt{f'(z_1)f'(z_2)}}\right|^{-\frac{c}{6}(n-1/n)} \tag{34}$$

$$= |f'(z_1)f'(z_2)|^{d_n}\langle T_n(f_1, \bar{f}_1)\tilde{T}_n(f_2, \bar{f}_2)\rangle \tag{35}$$

where we used the transformation behavior of a primary field $\mathcal{O}(z, \bar{z}) = \left(\partial z'/\partial z\right)^h \left(\partial \bar{z}'/\partial \bar{z}\right)^{\bar{h}} \mathcal{O}'(z', \bar{z}')$ for the twist fields of dimension $h = \bar{h} = d_n/2$ under an $f(z)$ transformation.

As can be seen from comparison of (31) and (33), the UV cutoffs as measured in f or z coordinates are related by [46]

$$\epsilon_f = f'(z)\epsilon_z \tag{36}$$

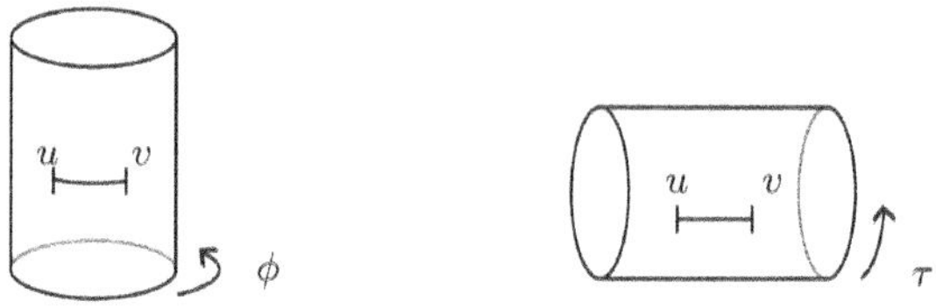

Fig. 4 Left: Finite size formula $S_A = \frac{c}{3}\log\left(\frac{\Sigma}{\pi\epsilon}\sin\frac{\pi(v-u)}{\Sigma}\right)$ for an interval $z_1 = \bar{z}_1 = u$, $z_2 = \bar{z}_2 = v$ on the cylinder of size $\Delta\phi = \Sigma$ follows from (33) with $f(z) = \exp\left(\frac{2\pi}{\Sigma}iz\right)$ relating the f-coordinate of the plane to the z-coordinate of the cylinder with compact $\phi = \text{Re}\, z$. Right: Finite temperature formula $S_A = \frac{c}{3}\log\left(\frac{\beta}{\pi\epsilon}\sinh\frac{\pi(v-u)}{\beta}\right)$ for an interval $z_1 = \bar{z}_1 = u$, $z_2 = \bar{z}_2 = v$ on the cylinder of size $\Delta\tau = \beta$ follows from (33) with $f(z) = \exp\left(\frac{2\pi}{\beta}z\right)$ relating the f-coordinate of the plane to the z-coordinate of the cylinder with compact $\tau = \text{Im}\, z$

(and writing ϵ as $\sqrt{\epsilon_1\epsilon_2}$). Indeed, from $f(z+\epsilon_z) \approx f(z) + \epsilon_z f'(z)$ with $\epsilon_z \equiv \delta z$, it follows that $\epsilon_f \equiv \delta f$ is given by the above.

The formula (33) can be applied immediately to the cases pictured in Fig. 4 of intervals in the z-cylinder.

1.3 Relation to Thermal Entropy

There is a 'short-cut' for deriving the interval entanglement S_A in (28) which will also be important for the holographic interpretation. It consists of mapping the set-up of Sect. 1.2 to a thermal system. This section follows [46].

A conformal change of coordinates induces a change of basis among the operators of the theory and affects the density matrix through a unitary transformation, to which the trace in (13) is insensitive. Moreover, the anomalous ($\sim c$) contribution to the transformed Hamiltonian only affects the unnormalized density matrix (and thus e.g. $\log(Z)$), but not the normalized density matrix that appears in the von Neumann entropy (13). This is to say that the entanglement entropy is conformal invariant, and hence its calculation can be simplified by considering well-chosen conformal mappings.

We consider the same theory $Z \equiv Z(1)$ of Sect. 1.2, in complex coordinates $\zeta = x + it_E$, $\bar{\zeta} = x - it_E$. First we translate the interval A from $x = [x_1, x_2]$ to $x = [0, L]$. The vacuum state of the system is pictured in Fig. 5a. Then we consider the conformal transformation to $w = (\zeta - x_1)/(x_2 - \zeta)$ or

$$w = \frac{\zeta}{L - \zeta}, \tag{37}$$

which maps the interval to the positive half-line. Keeping track of the UV cutoff, the more precise statement is that the $x = [\epsilon, L - \epsilon]$ interval is mapped to $w = \text{Re}\, w = [\frac{\epsilon}{L}, \frac{L}{\epsilon}]$. Then, we further transform to

$$z = \frac{1}{\kappa}\log w, \tag{38}$$

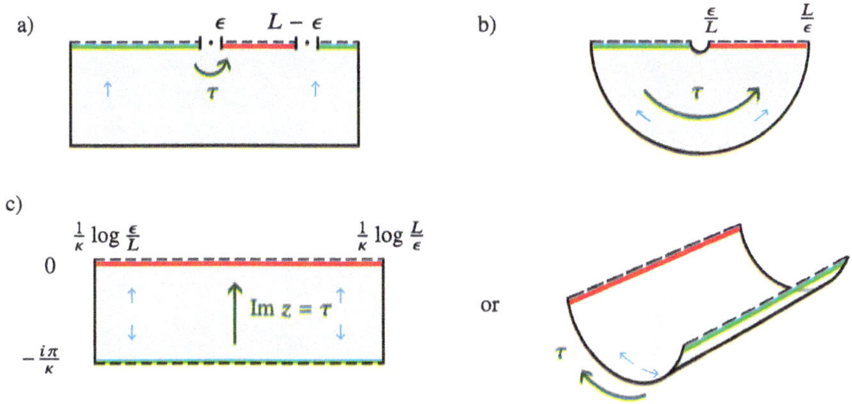

Fig. 5 Prepared state, with region A in red and complementary region $\bar{A}$ in green, under conformal mappings between the ζ, w- and z-geometry, in respectively **a**, **b** and **c**. In terms of metrics (for $\kappa = 1$ choice), $d\tau^2 + \frac{dR^2}{R^2} = dzd\bar{z}$ (cylinder) $= d(\log w)d(\log \bar{w}) = \frac{1}{w\bar{w}}dwd\bar{w} \to dwd\bar{w}$ (plane) $= R^2 d\tau^2 + dR^2$ with $w = R\exp(i\tau)$ and $z = x_z + i\tau = (\log R) + i\tau$. Under these mappings, a periodic coordinate τ around ∂A becomes the angle on the annulus or the Euclidean time coordinate of the cylinder. The small blue arrows point in the direction of the Euclidean evolution t_E

with κ an arbitrary real number, that should therefore not affect any physics. It maps the positive half-line to $z = \text{Re}\, z = [\frac{1}{\kappa}\log\frac{\epsilon}{L}, \frac{1}{\kappa}\log\frac{L}{\epsilon}]$ and the negative half-line to $z = [\frac{1}{\kappa}\log\frac{\epsilon}{L} - \frac{i\pi}{\kappa}, \frac{1}{\kappa}\log\frac{L}{\epsilon} - \frac{i\pi}{\kappa}]$, so that A is now one side of a strip of width π/κ. The partition function $Z(1) = \text{tr}\,\rho$ is mapped to the w-*annulus* and the z-*cylinder* respectively. The imaginary part of z

$$\text{Im}\, z \equiv \tau \tag{39}$$

brings forth a periodic coordinate; as the angle of the w-annulus with periodicity $\Delta\tau = 2\pi$, and as the compact direction of the z cylinder with periodicity $\Delta\tau = 2\pi/\kappa$. A direct consequence of this is that the reduced density matrix ρ_A in both pictures takes the form of, not just any mixed density matrix, but a *thermal* density matrix

$$\rho_A = e^{-(\Delta\tau)H_\tau} = \rho_{thermal} \tag{40}$$

as demonstrated in Fig. 6. On the annulus, H_τ is the generator of rotation in the Euclidean plane, and thus the boost generator in Lorentzian signature. On the cylinder, H_τ is the actual Hamiltonian or generator of cylinder (Euclidean) time translation. The corresponding inverse temperature is denoted β by choosing $\kappa = 2\pi/\beta$. It is a fictitious temperature measured by an observer that only has access to half of the w-plane or only one side of the z-strip. Figure 6b gives us the path integral derivation of the Bisognano Wichmann theorem (40) [13] (e.g. [45,73]).

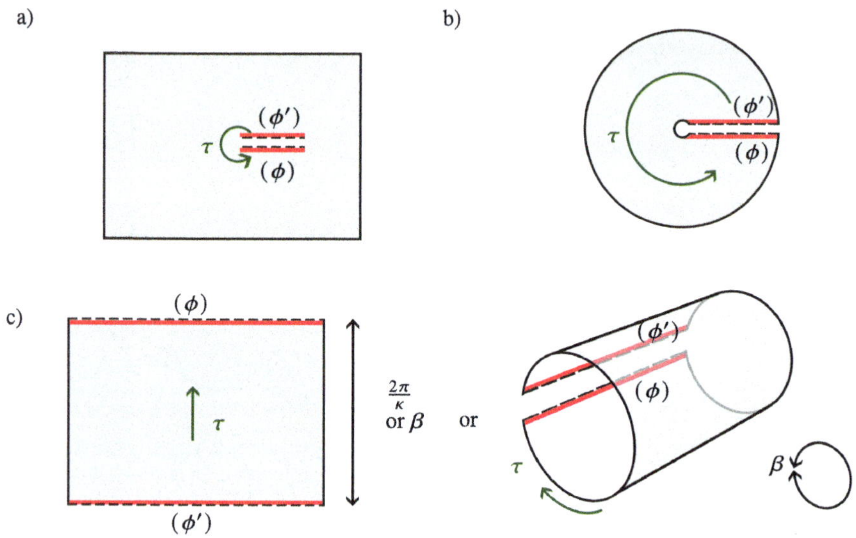

Fig. 6 Reduced density matrix ρ_A for the state in Fig. 5, under conformal mappings between the ζ, w- and z-geometry, in respectively **a**, **b** and **c**. In brackets are the field configurations to read off the matrix elements $(\rho_A)_{\phi\phi'} = \langle\phi|\rho_A|\phi'\rangle$

In the cylinder picture, S_A is now equal to the thermal entropy $S_{thermal}$ of the strip of width β, which is simply given by the thermodynamic formula $S_{thermal} = \beta E + \log Z(\beta)$ in terms of the energy and free energy, or

$$S_{thermal} = (1 - \beta\partial_\beta) \log Z(\beta). \tag{41}$$

The partition function of the CFT on the strip of width β and length $L_z \gg \beta$ was determined in [1, 14], again by making use of the Weyl anomaly, to be (to leading order in $1/\beta$)

$$\log Z(\beta) = \text{const}\, \beta\, L_z + \frac{\pi c}{6\beta} L_z\,. \tag{42}$$

The constant in the first term is non-universal, i.e. CFT-dependent, and drops out of the formula for $S_{thermal}$, which gives

$$S_{thermal} = \frac{c\pi}{3\beta} L_z = \frac{c}{3} \log \frac{L}{\epsilon}\,. \tag{43}$$

L_z is the full (divergent) length A of the cylinder in Fig. 5c, which is given by $\frac{\beta}{\pi} \log \frac{L}{\epsilon} \equiv \frac{\beta}{\pi} \log \frac{L_\zeta}{\epsilon_\zeta}$, so that in the second equality we find, as we should, $S_A = \frac{c}{3} \log \frac{L}{\epsilon}$ in (28).

As pointed out in [25], there is an IR/UV quality to the relation $L_z \sim \log L/\epsilon$ between the cylinder regulator (large L_z) and the plane regulator (small ϵ), which

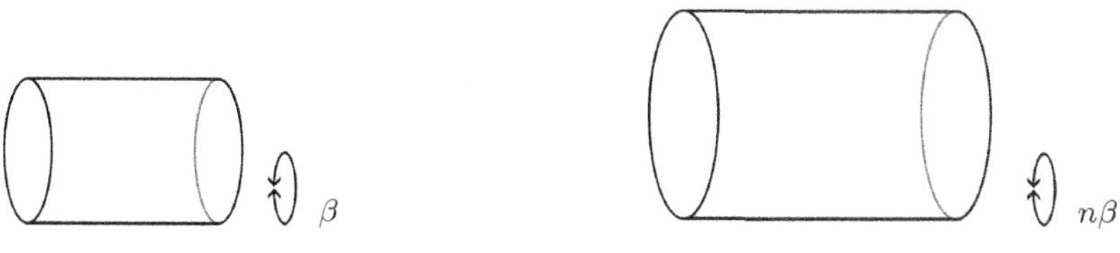

Fig. 7 $Z(1)$ and $Z(n)$

you typically find in a holographic setting.[2] A last comment about the derivation is that the formula for the entanglement of a finite interval on the thermal cylinder in Fig. 4b should become the thermal entropy for the state in Fig. 5c in the limit of $L_z \gg \beta$. This can be easily checked: $\lim_{L_z\to\infty} \log(\beta/(\pi\epsilon_z)\sinh(\pi L_z/\beta)) = \pi L_z/\beta + \log(\beta/(2\pi\epsilon_z))$, with the second term vanishing for $\epsilon_z = \beta/(2\pi) \ll L_z$ following from (36) with $z = \beta/(2\pi)\log(\zeta/(L-\zeta))$.

In the above paragraph, $Z(\beta)$ is the $Z(1)$ on the thermal cylinder (where β could also be taken to be 2π by choosing $\kappa = 1$). The replicated manifold $\mathrm{tr}\,\rho_A^n$, obtained by gluing n copies of the $Z(1)$ strip along A, is again a cylinder, $Z(n\beta) \equiv Z(n)$, only this time with periodicity $n\beta$ (Fig. 7). By the conformal mapping to the cylinder, the conical singularities at ∂A have been removed from the replica manifold, and the analytic continuation to non-integer n is immediate, since the periodicity of the cylinder can be varied continuously. The replica trick (14) indeed gives $S_A = S_{thermal}$

$$S_{thermal} = (1 - n\partial_n)\log Z(n)|_{n\to 1} = \frac{c\pi}{3\beta}L_z \tag{44}$$

from the trivial n-dependence in $Z(n) \equiv Z(n\beta) = \frac{\pi c}{6n\beta}L_z + \text{const}\; n\beta L_z$. The triviality of the thermal replica trick on the cylinder, combined with the argument that S_A is conformal invariant, signals that the replica derivation of the interval entanglement S_A and in particular the question of analytic continuation of n should be well-defined (Sect. 1.2), as should its holographic interpretation (Sect. 2.2).

The replica manifold in the annulus picture does have a conical singularity at the origin $\epsilon \to 0$ with conical excess $2\pi(n-1)$. The length of the region A can be varied by a scale transformation $x^\mu \to x'^\mu = (1 - 2\frac{\delta\epsilon}{\epsilon})x^\mu$, such that $Z(n)$ in (22) has to satisfy

$$\delta \log Z(n) = \frac{\delta \log \epsilon}{\pi} \int \langle T^\mu{}_\mu \rangle d^2x. \tag{45}$$

Here, the trace of the stress tensor can be obtained from applying the argument leading to (26), or by directly making use of the Weyl anomaly $\langle T^\mu{}_\mu\rangle = \frac{c}{12}R$ which relates it to the curvature $R \sim (1-n)(\delta^{(2)}(x_1) + \delta^{(2)}(x_2))$ of the manifold and its boundaries (see also [68] e.g.). This procedure for deriving (28), detailed in [46],

[2] The difference is in the contractability of the τ-circle (non-contractible on the cylinder, contractible on the plane), which is also reflected in the holographic description (non-contractible on the cylinder, contractible in the bulk description, see Fig. 13).

gives an alternative to the replica trick derivation of Sect. 1.2.1 that makes the coarse graining physics behind the geometric entropy S_A apparent.

1.4 Thermofield Double

In the previous section we encountered the possibility of an observer restricted to a part of spacetime A measuring a state ρ_A that is thermal even though the full system is in the vacuum state. This fits in the general concept of the thermofield double [60,70], which will be discussed in this section.

A thermofield double (TFD) is a particular vacuum state that is constructed to reproduce thermal physics of a given QFT. We will focus on a conformal QFT in particular, with a Hamiltonian H_R and Hilbert space $\mathcal{H}_R$, and on $1+1$ dimensions. The Euclidean path integral for compactified Euclidean time with period $\Delta\tau = \beta$ gives the thermal partition function $Z(\beta)$ of the theory. In path integral visualization, $Z(\beta) = \text{tr}\,\rho_{thermal}$ is a cylinder $\mathcal{C}(\beta)$ when the spacelike direction of the theory is non-compact, and thus $\rho_{thermal}$ is the cylinder before tracing, i.e. with open cut

$\rho_{thermal} =$ 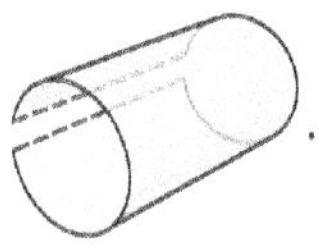 . (46)

The statement is that the same physics can be described by constructing a state $|\psi\rangle$, called the TFD state, as the state

$|\psi\rangle =$ 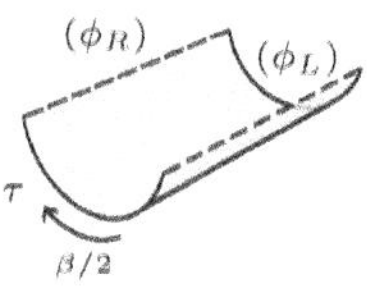

(47)

living in the doubled Hilbert space $\mathcal{H}_L \times \mathcal{H}_R$, with total Hamiltonian $H = H_R - H_L$ (as time τ runs downwards in the left copy and upwards in the right copy). We refer to the left copy as the one where τ starts and the right copy where τ arrives. This state is constructed such that it satisfies the property that its density matrix

$\rho = |\psi\rangle\langle\psi| =$ 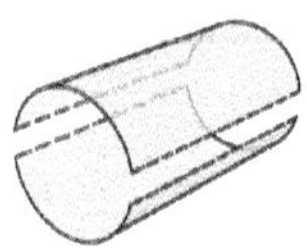 (48)

reduces to the thermal density matrix of the original ‘right’ system when the left copy is traced out,[3]

$$\rho_R \equiv \mathrm{tr}_L\, \rho =$$

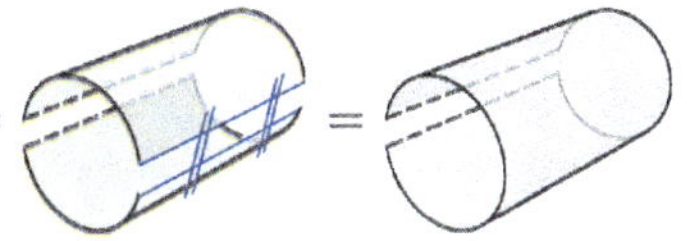

$$= \rho_{thermal}. \tag{49}$$

Said otherwise, the TFD state is the answer to the question [54] “is there a state $|\psi\rangle \equiv |0(\beta)\rangle$ for which the QFT vacuum expectation value $\langle\psi|\mathcal{O}|\psi\rangle = \mathrm{tr}(\mathcal{O}\rho)$ reproduces the statistical average $\mathrm{tr}(\mathcal{O}\rho_{thermal})$ of an operator $\mathcal{O}$?",

$$\langle\psi|\mathcal{O}|\psi\rangle = \mathrm{tr}(\mathcal{O}\rho_{thermal})$$

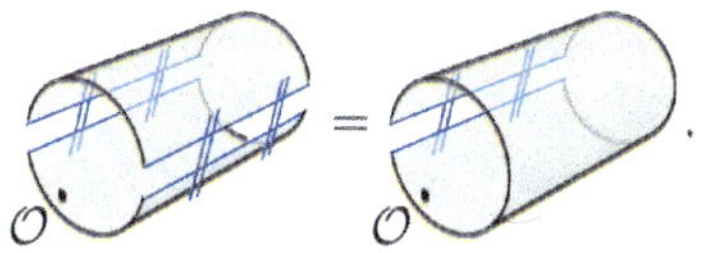

(50)

Of course, for $\mathcal{O}$ the identity operator, this is just the statement that

$$\langle\psi|\psi\rangle = \mathrm{tr}(\rho_{thermal})$$

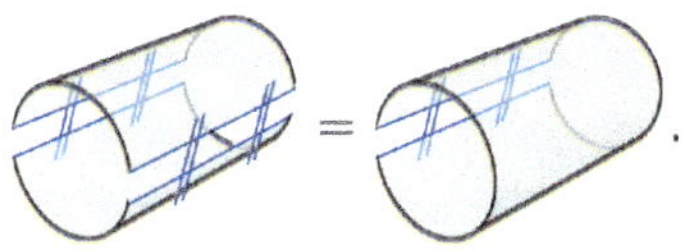

(51)

For a compact space dimension,

$$|\psi\rangle =$$

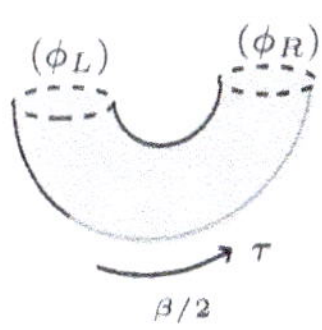

(52)

[3] In Eq. (49) and in the rest of the section, the pictorial representation of tracing out fields is short for the notation introduced in (9), leaving out the explicit integration over the fields for conciseness.

and

$$\rho_L \equiv \mathrm{tr}_R\, \rho =$$ 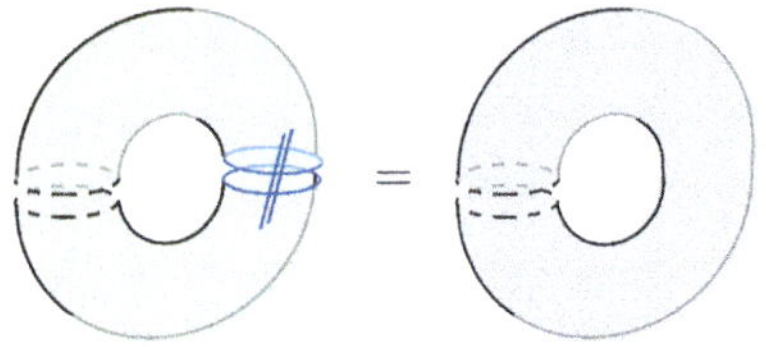 .

(53)

The above statements are given in path integral language $|\psi\rangle = \int^{\cdot} \mathcal{D}\phi\, e^{-I[\phi]}$ or

$$\psi(\phi_L, \phi_R) = \int^{\substack{\phi(\tau_0) = \phi_L \\ \phi(\tau_0 + \beta/2) = \phi_R}} \mathcal{D}\phi\, e^{-I[\phi]}, \tag{54}$$

as this gives an immediate visual derivation of (49)–(50). It is a simple exercise to derive (49)–(50) from the explicit formula for the TFD state (47) given by

$$|\psi\rangle = \sum_n \sqrt{p_n}\, |E_n\rangle_L |E_n\rangle_R, \qquad p_n = \frac{e^{-\beta E_n}}{Z(\beta)} \tag{55}$$

with $|E_n\rangle$ the energy eigenstates of H_L and H_R. It is a pure state per construction (using (48)), and the construction is therefore referred to as purification. That is, $\mathrm{tr}(\rho^2) = \mathrm{tr}\, \rho$ and hence its von Neumann entropy is zero, while the von Neumann entropy of one copy $S_A = -\,\mathrm{tr}(\hat{\rho}_R \log \hat{\rho}_R)$ is the thermal entropy

$$S_A = S_{thermal}, \tag{56}$$

where we used the notation A for the region the observer of the 'right' theory has access to. The TFD construction thus provides a set-up where the reduced density matrix is thermal (49), and the entanglement entropy S_A is exactly given by a thermal entropy. As discussed in Sect. 1.3, this can be applied to the calculation of CFT entanglement of an interval A by conformally mapping the interval to one copy of a TFD in Fig. 5.

As is apparent from the path integral pictures above, a thermofield double description arises for Euclidean manifolds that can be divided into two disconnected parts by the specification of *two* values of a periodic coordinate τ (the values τ_0 and $\tau_0 + \beta/2$ in (54)), such as the cylinder or torus.

Let us discuss two more examples. (For more details, see e.g. [40].) The first is the Euclidean disk with metric $ds^2 = dR^2 + R^2 d\tau^2$. In polar coordinates with polar angle τ, one can consider the TFD state

$$|\psi\rangle =$$ 

(57)

with

$\mathrm{tr}_L\, \rho =$ 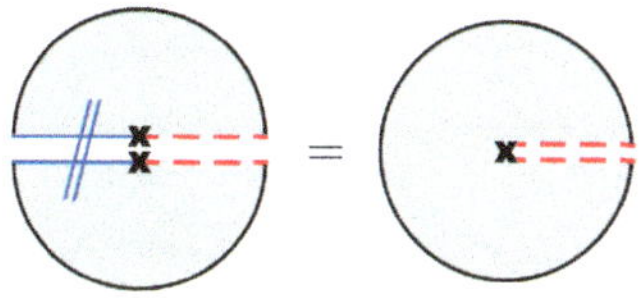 $= \rho_{thermal}$ (58)

the thermal density matrix encountered in Fig. 6b, for $\beta = 2\pi$ the inverse Unruh temperature [72] detected by a Rindler observer in Lorentzian signature (Fig. 8a). Euclidean evolution prepares the vacuum state $|\psi\rangle$ on the Minkowski plane $ds^2 = -dt^2 + dx^2$, at the spacelike slice located at $t = 0 = t_E$, i.e. the intersection of zero Lorentzian and Euclidean time $t_E = it$. A Rindler observer is a boosted observer (with acceleration set to one below) who only has access to the positive half-line $x > 0$ at $t = 0$, with domain of dependence the Rindler wedge

$$ds^2_{Rindler} = -dt^2 + dx^2, \qquad (x \geq 0,\ |t| < x) \tag{59}$$

$$= dR^2 - R^2 dt_S^2, \qquad (R \geq 0,\ \text{all } t_S) \tag{60}$$

which is covered by Rindler coordinates (R, t_S) or (x_z, t_S) with $x_z = \log R$. These are related to Minkowski coordinates by $x = R\cosh t_S$, $t = R\sinh t_S$. The half-line A $(x > 0)$ is effectively separated from the half-line $\bar{A}$ $(x < 0)$ by the Rindler horizon $R = 0$.

A last example is the cigar manifold, which interpolates between the Euclidean disk geometry (at the tip of the cigar) and the Euclidean cylinder. It appears in the Wick-rotated metric of black hole backgrounds, e.g. the Schwarzschild black hole in Fig. 8b. The associated TFD state $|\psi\rangle$ provides the Hartle-Hawking state [41] of quantum fields on the black hole background, defined by doing the path integral over half the Euclidean geometry, and $\mathrm{tr}_L\, |\psi\rangle\langle\psi| = \rho_{thermal}$ describes a thermal state at the Hawking temperature of the black hole. The prepared state at the intersection of zero Euclidean and Lorentzian time can then be further evolved in Lorentzian time t.

The holographic dual of the TFD

The extended $(2+1)$-dimensional AdS-Schwarzschild black hole or BTZ solution [9] is shown in Fig. 8c, with a Hartle-Hawking state prepared by evolution over the Poincaré disk $ds^2 = 4dwd\bar{w}/(1 - w\bar{w})^2$. This $(2+1)$-dimensional geometry has two asymptotic boundaries where two dual CFT copies live. It is the holographic dual of the TFD state of $(1+1)$-dimensional CFT [59]. Indeed, when the suppressed spacelike coordinate x_z of the CFT is added back to the Euclidean half of the Penrose diagram in Fig. 8c, the AdS TFD state at the conformal boundary becomes the CFT TFD state of Fig. (47) or (52), depending on whether x_z has an infinite or compact range.

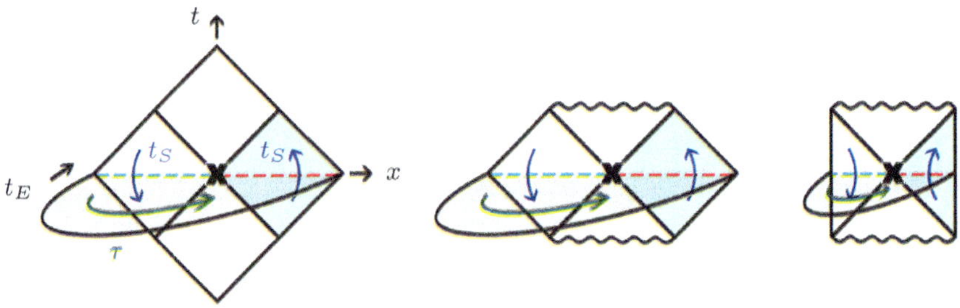

Fig. 8 Penrose diagrams of **a** Minkowski spacetime, **b** the extended Schwarzschild black hole, and **c** the extended AdS-Schwarzschild or BTZ black hole, covered by the full range of (t, x) coordinates. In the case of the black holes, (t, x) are the Kruskal coordinates. Superimposed is the Euclidean preparation of the TFD state, which provides the Hartle-Hawking state. The state can then be further evolved in Lorentzian time. To an observer in resp. the Rindler wedge, the (1-sided) Schwarzschild geometry and the (1-sided) BTZ geometry (all marked in light blue), the state appears thermal. The (Lorentzian) time coordinate t_S that covers these regions is Wick rotated, $t_S = i\tau$, to the periodic coordinate τ of the thermofield double construction

Fig. 9 Region A in red, with entanglement S_A given in (62)

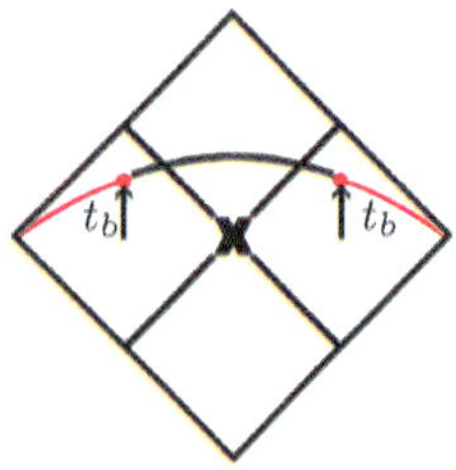

An application of Eq. (33)

To end this section, we focus on the Minkowski geometry of Fig. 8a and determine the Lorentzian time t_b-dependence of the entanglement of the region A pictured in Fig. 9, following Hartman and Maldacena [42]. The region consists of the half-line $x_z > 0$ in each Rindler wedge, in cylinder coordinates $dx_z^2 - dt_S^2$.

At time $t_b = 0$, the Cauchy slice on which A is defined is simply at $t = 0$. Then, point P_1 in the left Rindler wedge has Rindler time coordinate $\tau = 0$ and $x_z = 0$, while point P_2 in the right Rindler wedge has coordinates $\tau = \beta/2$ and $x_z = 0$ or $(z_2, \bar{z}_2) = (i\beta/2, -i\beta/2)$. Next, the Cauchy slice can be pushed upwards, with time coordinate t_b equal to t_S in the right Rindler wedge and equal to $-t_S$ in the left one. By analytic continuation, the locations of the interval endpoints ∂A are

$$(z_1, \bar{z}_1) = (-t_b, t_b), \qquad (z_2, \bar{z}_2) = (t_b + \frac{i\beta}{2}, -t_b - \frac{i\beta}{2}). \tag{61}$$

Applying Eq. (33) with these values, and with $f(z) = \exp(2\pi z/\beta)$ the planar coordinate, one obtains

$$S_A = \frac{c}{3} \log\left(\frac{\beta}{\pi\epsilon_z} \cosh(\frac{2\pi}{\beta} t_b)\right) \tag{62}$$

for the time-dependence. At large times, the behavior is linear $S_A \sim t_b$.

This is indeed $S_A = c/3 \log(\Delta x/\epsilon_x)$ with the relation $x = e^{2\pi x_z/\beta} \cosh(2\pi t_S/\beta)$ between planar and Rindler coordinates evaluated at $x_z = 0$ at the interval endpoints, and $\epsilon_x = \frac{2\pi}{\beta}\epsilon_z$ by Eq. (36) with (61). The holographically dual calculation of S_A is also performed in [42].

2 Holographic Entanglement

Holography, in a general sense, posits there is a physically dual description of CFT in terms of a higher-dimensional theory of gravity. The CFT sets asymptotic boundary conditions for the gravitational theory, and the duality can be most succinctly stated as an equality of bulk (gravity) and boundary (CFT) partition functions

$$Z_{grav} = Z_{CFT}. \tag{63}$$

A subscript is included to refer to the partition function of the previous chapter Z as the *CFT* theory. No further knowledge of holography or AdS/CFT [56] will be assumed.

In this chapter we will discuss the holographic interpretation of the $\log L$ formula for the CFT entanglement (28). It famously has a geometric interpretation as a minimal area, according to the 'Ryu-Takayanagi' (RT) formula [67,68]. The proof of the RT formula by Lewkowycz and Maldacena [57], involving a bulk replica trick, is summarized in the first section. It is henceforth referred to as the LM derivation. In recent years, extensions of their proof have been used to calculate the entropy of Hawking radiation with the so-called 'island rule' and address the black hole information paradox. We comment on this briefly in Sect. 2.1.1. Then returning to holography, we discuss in Sect. 2.2 how the RT formula was initially proven for a special, U(1)-symmetric case by first mapping the CFT entanglement to a thermal entropy, for which a standard holographic dual interpretation is known. This discussion constitutes the holographic dual of the discussion on CFT entanglement as thermal entropy in Sect. 1.3. Finally, in Sect. 2.3, the intuition gained from the U(1) case proof of the RT prescription is used to discuss how gravitational first laws have interpretations in terms of CFT entanglement, leading to statements on the emergence of gravity from entanglement. Section 2.3.2 comments briefly on similar statements about emergent gravity from entanglement in non-holographic set-ups.

2.1 Lewkowycz Maldacena Derivation of Ryu-Takayanagi Formula

Here the set-up is a gravitational theory (e.g. [38])

$$Z_{grav} = \int \mathcal{D}g\mathcal{D}\phi e^{-I[g,\phi]}, \tag{64}$$

$$I = -\frac{1}{16\pi G}\int d^{d+1}x\sqrt{g}(R-2\Lambda) + I_{GHY} + I_m[g,\phi]. \tag{65}$$

The dynamics of the metric field g with curvature R is described by the Euclidean action I of Einstein gravity, with negative cosmological constant Λ, Gibbons-Hawking-York boundary term I_{GHY}, and the action I_m for the matter fields ϕ. We will focus, to begin with, on the $d = 2$ case, interpreting the gravitational theory as the 3-dimensional, asymptotically AdS dual of the 2-dimensional CFT of chapter 1. By the AdS/CFT duality (63), the CFT entanglement $S_A = (1 - n\partial_n)\log Z_{CFT}(n)|_{n\to 1}$ discussed in the previous chapter has a dual interpretation as gravitational entropy

$$S_A = (1 - n\partial_n)\log Z_{grav}(n)|_{n\to 1}. \tag{66}$$

It is this object we are interested in calculating in this section, by employing a bulk version of the replica trick.

The full, formal quantum gravity path integral Z_{grav} has a semi-classical approximation $Z_{grav} \approx \exp(-I[g_*, \phi_*])$ in terms of the on-shell gravitational action, evaluated on a classical solution (g_*, ϕ_*). In this saddle point approximation, the gravitational entropy becomes calculable as

$$S_A = -(1 - n\partial_n)I[g_*, \phi_*](n)|_{n\to 1}. \tag{67}$$

The solution has to satisfy boundary conditions set by the CFT replica manifold partition function $Z_{CFT}(n)$, with g_* extending asymptotically to the manifold over which $Z_{CFT}(n)$ path integrates. From here on we will consider pure gravity in the bulk for simplicity, and at the end of the section mention results on extensions, such as the inclusion of matter fields ϕ, higher derivative gravity, etc.

Several incarnations of $Z_{CFT}(n)$ (with integer $n > 1$) were discussed in the previous chapter: as the path integral of a replicated theory on the non-replicated orbifold manifold $\mathbb{C}$ (Fig. 1b), as the path integral of the non-replicated theory on the replicated manifold $\mathcal{R}_{n,A}$ (Fig. 1a) or as the path integral of the non-replicated theory on the replicated cylinder (Fig. 7b). The last one provides the right starting point for constructing the semi-classical $Z_{grav}(n)$, because it is a smooth replica manifold, with a smooth corresponding bulk solution. A bulk geometry that extends the conical singularity of $\mathcal{R}_{n,A}$ into the bulk is not an acceptable solution of the sourceless Einstein equations that allows for a saddle point evaluation [44].[4] Boundary condi-

[4] An interpretation of the bulk configurations considered in [35] is discussed in [30].

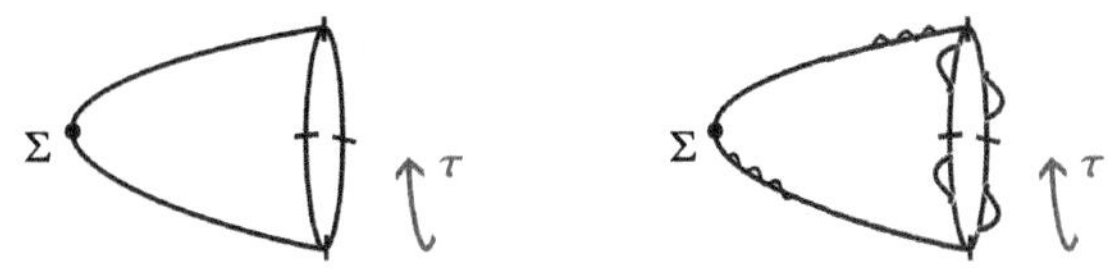

Fig. 10 **a** Bulk replica manifold $\mathcal{M}_n$ (here pictured for $n = 4$) that on the conformal boundary $\partial\mathcal{M}_n$ reduces to the cylinder replica manifold of Fig. 7b (with the spacelike direction of the CFT suppressed), for the density matrix $\rho_A = \exp(-\beta H_\tau)$ of (40). The uncontractible circle in the boundary becomes contractible in the bulk at Σ. **b** Bulk replica manifold $\mathcal{M}_n$ for more general, non-$U(1)$ symmetric case $\rho_A = \mathcal{P}\exp\int_\beta^0 d\tau\mathcal{H}(\tau)$

tions are imposed at the conformal boundary of the asymptotically AdS theory, thus allowing to use the replicated cylinder, which is conformally related to $\mathcal{R}_{n,A}$.

The replica cylinder of Fig. 7b extends into the bulk in a (by ansatz) $\mathbb{Z}_n$ symmetric solution $\mathcal{M}_n$ that smoothly ends in a $\mathbb{Z}_n$ fixed point, as illustrated in Fig. 10a. With the boundary replica manifold describing a thermal state, $\operatorname{tr}\rho_A^n = \operatorname{tr}\exp(-n\beta H_\tau)$, $\mathcal{M}_n$ is a black hole solution and the bulk replica method will reduce to the Gibbons-Hawking path integral derivation of its Bekenstein-Hawking entropy (see Sect. 2.2), providing the dual of the CFT entanglement S_A for τ an angular coordinate going around the boundary ∂A of the interval A, as in Fig. 6. The relevant conformal mapping employed there

$$ds^2 = R^2 d\tau^2 + dR^2 \rightarrow d\tau^2 + \frac{dR^2}{R^2} \tag{68}$$

maps the τ-direction on the plane to the uncontractible circle S^1 of a hyperbolic cylinder $S^1(\beta) \times \mathbb{H}_1$ (easily extended to $S^1(\beta) \times \mathbb{H}_{d-1}$ in the $d > 2$ case, with corresponding replica cylinder $S^1(n\beta) \times \mathbb{H}_{d-1}$). The hyperbolic cylinder has a U(1) isometry generated by the Killing vector ∂_τ, reflecting the conservation of the Hamiltonian H_τ. However, the present discussion crucially is more generally valid: one can take as boundary conditions a CFT state

$$\rho_A = \mathcal{P}e^{-\int_0^\beta d\tau\mathcal{H}(\tau)} \tag{69}$$

with Euclidean Hamiltonian density $\mathcal{H}(\tau)$ that is not necessarily U(1) invariant, but instead in general depends on τ. The more general set-up is the following. Consider a spatial region A in a d-dimensional boundary CFT at time equal to zero. Going to Euclidean signature, one can *locally* at the boundary ∂A of A choose an angular coordinate τ that has the property that it encircles ∂A. For this coordinate, locally the mapping to a hyperbolic cylinder will be valid, but globally τ-dependence can appear in the conformally scaled boundary metric, i.e. U(1) invariance is lost. Still, the τ-dependence has to be consistent with the 2π-periodicity of the circle that τ parametrizes in the original geometry. That is, the (conformal) boundary metric can depend on τ through powers of $\exp(\pm i\tau)$. We follow [57] in setting $\beta = 2\pi$.

The more general set-up is still characterized by a non-contractible τ-circle $S^1(2\pi)$ in the boundary metric. The corresponding replica manifold is obtained by gluing together n circles to $S^1(2\pi n)$, and is thus $\mathbb{Z}_n$ symmetric by construction, as a consequence of the periodicity in τ. The bulk replica manifold $\mathcal{M}_n$ is then constructed as the (smooth) solution of the sourceless Einstein's equations that extends the replica manifold into the bulk in a $\mathbb{Z}_n$ symmetric way. Its general form, with $\mathbb{Z}_n$ fixed point Σ where the boundary τ-circle $S^1(2\pi n)$ becomes contractible, is illustrated in Fig. 10b. To be more precise, Σ is a set of points of co-dimension 2 in the $(d+1)$-dimensional bulk.

Now we are ready to evaluate the entropy

$$S_A = -(1 - n\partial_n) I[\mathcal{M}_n]|_{n\to 1}. \tag{70}$$

This can be written as $S_A = n\partial_n \left(I[\mathcal{M}_n] - nI[\mathcal{M}_1]\right)|_{n\to 1}$ in terms of two smooth manifolds, $\mathcal{M}_n$ and $\mathcal{M}_1$, with different boundaries $\partial\mathcal{M}_n \neq \partial\mathcal{M}_1$. In order to compare them, they should be brought in a form where each has the same boundary. To achieve this, periodicity in τ (or said otherwise, $\mathbb{Z}_n$ symmetry) can be used to rewrite $I[\mathcal{M}_n]$ in the first term[5]

$$I[\mathcal{M}_n] = \int_0^{2\pi n} \mathcal{L}[g(n)]d\tau \tag{71}$$

as an integer n times an auxiliary action $I[\mathcal{O}_n]_{\text{smooth}}$ defined as

$$I[\mathcal{O}_n]_{\text{smooth}} \equiv \int_0^{2\pi} \mathcal{L}[g(n)]d\tau. \tag{72}$$

Here, $\mathcal{L}$ is the Lagrangian density after integrating over all but the τ coordinate, and $\mathcal{O}_n$ the orbifold manifold $\mathcal{O}_n \equiv \mathcal{M}_n/\mathbb{Z}_n$ obtained from the replica manifold by quotienting the replica symmetry, with boundary $\partial\mathcal{O}_n = \partial\mathcal{M}_1$. The relation

$$I[\mathcal{M}_n] = n\, I[\mathcal{O}_n]_{\text{smooth}} \tag{73}$$

is well-defined also for non-integer n, and it will be used below to obtain S_A in the $n \to 1$ limit, or $\epsilon \to 0$ for $n = 1 + \epsilon$.

Let us discuss the orbifold for different values of n. For $n = 1$, it is equal to the original dual of the CFT, $\mathcal{O}_1 = \mathcal{M}_1$, which we know is a solution of the $I[g]$ equations of motion. As it extremizes the action, this implies

$$I[\mathcal{O}_{1+\epsilon}] = I[\mathcal{O}_1]. \tag{74}$$

[5] Alternatively, one can proceed by rewriting $nI[\mathcal{M}_1]$ in the second term, into an action evaluated on a manifold with boundary $\partial\mathcal{M}_n$ and conical excess $2\pi(n-1)$ [22,57]. After adding and subtracting the action evaluated on the manifold with the same boundary but a regulated, rounded off conical singularity, and using the fact that $\mathcal{M}_n$ is a solution of the equations of motion, the remaining contribution to S_A comes from the regulated conical tip and is proportional to its area [34,36].

For $n \neq 1$, the orbifold has a conical singularity at the location of the $\mathbb{Z}_n$ fixed point in $\mathcal{M}_n$, with a deficit angle of $2\pi(n-1)/n$ or opening angle $2\pi/n$. The corresponding curvature singularity, which is proportional to the deficit angle, contributes a term $I_{\text{sing}} \sim \int_{\text{sing}} \sqrt{g} R$ to the gravitational action evaluated on $\mathcal{O}_n$,

$$I[\mathcal{O}_n] = I[\mathcal{O}_n]_{\text{smooth}} + I_{\text{sing}}. \tag{75}$$

The first term contains the regular contribution $I[\mathcal{O}_n]_{\text{smooth}} \sim \int_{\text{cone}\setminus\text{sing}} \sqrt{g} R$. It is this regulated contribution that appears in (73) as a rewriting of the *smooth* left hand side.

Because of the conical singularity, $\mathcal{O}_n$ is not a solution of the $I[g]$ equations of motion, but of Einstein's equations in the presence of a source localized at the singularity. The required source is a co-dimension 2 brane (in the $(d+1)$-dimensional bulk) of tension $T = (n-1)/(4nG)$ and action $I_{\text{brane}} = T \int d^{d-1}y\sqrt{h}$, for y and h the induced coordinates and metric on the brane [27]. This means that on-shell,

$$I[\mathcal{O}_n] + I_{\text{brane}} = I[\mathcal{O}_n]_{\text{smooth}} \tag{76}$$

or said otherwise, that $I_{\text{sing}} = -I_{\text{brane}}$, which can be checked explicitly. In evaluating $I[\mathcal{O}_n]_{\text{smooth}}$ one can thus think of it either as integrating the gravitational Lagrangian over the regulated cone with no contribution of the singularity (75), or alternatively as *adding* the brane (76): $\int_{\text{cone}\setminus\text{sing}} \mathcal{L} d\tau dr d\mathbf{x} = \int_{\text{cone}} \mathcal{L} d\tau dr d\mathbf{x} + \int_{\Sigma} \mathcal{L}_{brane} d\mathbf{x}$ (with here $\mathcal{L}$ the gravitational Lagrangian density).

Next we can write $I[\mathcal{M}_{1+\epsilon}]$ in terms of $I[\mathcal{O}_n]_{\text{smooth}}$ by (73), and then use (76) and finally (74) to obtain

$$I[\mathcal{M}_{1+\epsilon}] = (1+\epsilon) I[\mathcal{M}_1] + \epsilon \frac{\mathcal{A}[\Sigma_{min}]}{4G}. \tag{77}$$

The last term is I_{brane} in the limit of a tensionless brane $\epsilon \to 0$, which does not back react and settles in the location Σ that minimizes its Nambu-Goto 'worldvolume' or area $\mathcal{A}[\Sigma_{min}]$. The entropy S_A, with the limit reexpressed for $n = 1+\epsilon$,

$$S_A = \left. \frac{I[\mathcal{M}_{1+\epsilon}] - I[\mathcal{M}_1]}{\epsilon} - I[\mathcal{M}_1] \right|_{\epsilon \to 0} \tag{78}$$

finally becomes

$$S_A = \frac{\mathcal{A}[\Sigma_{min}]}{4G}. \tag{79}$$

This is the Ryu-Takayanagi (RT) formula for the holographic entanglement of a spacelike region A. It also includes a homology condition, which has been discussed in the context of the LM derivation in [39].

For general time-dependent states, the area should be extremized according to the HRT formula [47]. A derivation of this covariant prescription was given in [28].

Fig. 11 Left: The set-up for the calculation of radiation entropy, in Lorentzian signature. It consists of an AdS_2 black hole (blue) with a bath region that collects black hole radiation attached (green). For visualization one can think of the bath regions as extending backwards. Right: Radiation region R with entropy given by the formula (81), which contains contributions from a possible island region I

Fig. 12 Left: The state is the combined TFD state of the black hole and the bath region, consisting of the Poincaré disk evolution of Fig. 8c in blue and $x_z > 0$ cylinder evolution of (47) in green. Right: Orbifold with twist fields (crosses) at boundary ∂R of radiation region R (at $t_b = 0$) and twist fields as well as branes (blue dots) at boundary ∂I of possible island region I

The generalization to higher derivative gravity involving Wald entropy [48,75] was derived in [29].

The inclusion of matter fields—in a semi-classical treatment of the bulk theory—lead to a first-order correction $\mathcal{O}(G^0)$ to the extremal area of bulk matter entropy S_m [32,57]. It is consistent at that order with the later proposed Engelhardt-Wall formula [31], at arbitrary orders in the bulk Planck constant,

$$S_A = S_{gen}[\Sigma_{QES}] \tag{80}$$

where the 'quantum extremal surface' Σ_{QES} is the minimal co-dimension 2 surface that extremizes the bulk generalized entropy $S_{gen} \equiv \mathcal{A}/4G + S_m(EW)$. Here, EW is short for entanglement wedge, defined as the bulk region confined by A and its RT surface Σ, and $S_m(EW)$ is the von Neumann entropy of bulk matter fields in EW. The QES prescription was derived in [29] and in [5] by inserting twist fields at Σ for the replicated matter sector in the orbifold $\mathcal{O}_n$, see also [64].

2.1.1 Islands

The gravitational entropy formula (80) has also been applied beyond the context of holography in set-ups that exhibit (a version of) a black hole information paradox to calculate the entropy of Hawking radiation. Without going into detail, we will sketch the idea.

The set-up considered in [5] is given by a 2-dimensional AdS-black hole solution of Jackiw-Teitelboim gravity [4,49,71] connected to a non-gravitational bath region that collects black hole radiation. There is a matter CFT of large central charge $c \gg 1$ that extends over both the gravitational and non-gravitational regions. The bath consists of an $x_z > 0$ Minkowski geometry attached to the right asymptotic boundary of AdS and another copy to the left (Fig. 11a). One can prepare the state at time zero by Euclidean evolution over half the Euclidean AdS_2-black hole solution and half the $x_z > 0$ cylinder, resulting in the combined TFD state of the black hole and the bath region (Fig. 12a). The set-up is entirely 2-dimensional and non-holographic, but does have a holographic interpretation [2,26] where the AdS_2 region forms an end of the world brane of a 3-dimensional AdS bulk.

Next, one can consider a 'radiation' region R extending from spacelike infinity to a value of x_z greater than zero in each copy of the bath. The entanglement S_m of CFT fields in R can be calculated by the replica trick of Sect. 1.2. Step one is to construct the replica manifold consisting of n copies of the system with cuts along R that are glued cyclically, as in Fig. 1a. Upon analytic continuation, the dependence of S_m on time t_b going forward in each copy of the theory can be obtained, as in the derivation of Eq. (62). The result is a continuously rising $S_m(t_b) = \frac{c}{3}\log(2\cosh t_b)$, the Hawking entropy for black hole radiation collected in R. The absence of a Page curve for this entropy constitutes a black hole information paradox [3].

The gravitational region in the replica manifold is left to dynamically determine its geometry, given the boundary conditions fixed at its edge. The replica manifold giving rise to the Hawking entropy only has connections between the n sheets along R, and disconnected gravitational regions. Now the insight of [5] was that connections between the n sheets in the gravitational region also can arise, when a gravitational solution is taken into account that represents a (replica symmetric) wormhole between the replicas.[6] The connection between the sheets within the gravitational region is along a region I called the island. In the orbifold $\mathcal{O}_n = \mathcal{M}_n/\mathbb{Z}_n$ of the replica manifold $\mathcal{M}_n$, pictured in Fig. 12b, the endpoints ∂R of the radiation region are smooth and carry twist fields (as in Fig. 1b), while the endpoints ∂I of the island region have conical deficits (as in (75)–(76)) and carry both twist fields and branes. An extension of the Lewkowycz Maldacena argument in Sect. 2.1 leads to the QES formula (80) for the entropy S_R as the minimal generalized entropy [5]

$$S_{gen} = \frac{\Phi(\partial I)}{4G} + S_m(R \cup I), \tag{81}$$

with Φ the Jackiw-Teitelboim dilaton and ∂I the location extremizing S_{gen}. The replica manifold with the island produces a radiation entropy expression S_R that becomes minimal for large t_b, giving rise to the sought-after Page curve.

6 For the role of wormholes in replica trick computations, see also [65], reviewed in [61].

2.2 Holographic Entanglement and Thermal Entropy

In this section, we focus on the U(1) symmetric case of the holographic entanglement of Sect. 2.1, or the Casini Huerta Myers derivation [25] of the Ryu-Takayanagi formula for the CFT entanglement (28) of an interval. This will provide useful intuition for the discussion of the entanglement first law and its gravitational dual in the next section.

The simplification in the U(1) case is that the entanglement maps to a thermal entropy, and one can construct the direct holographic dual interpretation of the CFT discussion of Sect. 1.3. This will lead to Fig. 15 presenting the bulk equivalent of the mapping pictured in Fig. 5, and to the RT formula coinciding with the black hole entropy formula.

2.2.1 Holographic Thermal Entropy

Let us first consider the $(2+1)$-dimensional BTZ black string solution

$$ds^2 = -(r^2 - r_+^2)dt_S^2 + \frac{dr^2}{r^2 - r_+^2} + r^2 dx_z^2 \tag{82}$$

with horizon at $r = r_+$, AdS boundary at $r \to \infty$ and AdS radius $l = 1$. The ranges of the coordinates are $t_S = -\infty..\infty$, $r \geq r_+$, and $x_z = -\infty..\infty$. The conformal boundary in Euclidean signature ($t_S = i\tau$) is the Euclidean cylinder $ds_{CFT}^2 = d\tau^2 + dx_z^2$, with Euclidean time periodicially identified $\tau \sim \tau + \beta$. The uncontractible τ-circle in the dual CFT becomes contractible in the bulk at $r = r_+$ (where $(r^2 - r_+^2)d\tau^2 \to 0$). This typical feature of a solution with a horizon is represented in Fig. 13b by a cigar geometry in the (r, τ) directions. Smoothness of the geometry near the tip of the cigar requires the periodicity of τ to be fixed in terms of r_+ to (e.g. [16])

$$\beta = \frac{2\pi}{r_+}. \tag{83}$$

This is the inverse Hawking temperature of the black string, and the Euclidean gravitational path integral $Z_{grav}(\beta)$, shown in Fig. 13 as the solid cylinder, is interpreted as a thermal partition function tr ρ_{grav}. The corresponding entropy

$$S_{thermal} = (1 - \beta\partial_\beta) \log Z_{grav}(\beta) \tag{84}$$

is the holographic dual of the thermal entropy (41) of the CFT. It can be calculated in different ways, as is nicely summarized in e.g. [8]. The original Gibbons-Hawking method [37] took an 'on-shell approach': as β is varied, so is the mass (and thus the horizon radius r_+) of the black hole, in such a way that the on-shell relation (83) is maintained. Variation of the on-shell action ($-\log Z_{grav}(\beta)$) [7,43] then gives rise to the famous Bekenstein-Hawking entropy of the black hole

$$S_{thermal} = \frac{\mathcal{A}_\langle}{4G}, \tag{85}$$

where $\mathcal{A}_h$ is the horizon area. Another option is the 'off-shell approach': when only β is varied, the tip of the cigar turns into a conical singularity. As discussed at length in Sect. 2.1, such a singular geometry does not solve the (sourceless) gravitational equations of motion. Equation (84) can then be written in terms of the conical excess ϵ, with $\delta\beta/\beta = \delta\epsilon$ [8,62], as

$$S_{thermal} = (1-\partial_\epsilon)\log \operatorname{tr} \rho_{grav}^{1+\epsilon}|_{\epsilon\to 0} = (1-n\partial_n)\log \operatorname{tr} \rho_{grav}^{n}|_{n\to 1}, \tag{86}$$

reproducing for $n = 1+\epsilon$ the replica trick formula (66). Indeed the Gibbons-Hawking entropy derivation in this approach reduces to a special case of the gravitational replica trick[7] of Sect. 2.1: the result (79) for the entropy reduces to (85). It is a special case because the gravitational solution (82) has a U(1) Killing vector ∂_{t_S}, describing a system in thermal equilibrium.

2.2.2 Ryu-Takayanagi from Thermal Entropy Argument

The BTZ string spacetime (82) is just a different (namely hyperbolic) slicing of 3-dimensional AdS. The explicit coordinate transformation to AdS in Poincaré coordinates

$$ds^2 = \frac{1}{Z^2}\left(dZ^2 - dt^2 + dx^2\right) \tag{87}$$

is given e.g. in the appendix of [6] (via embedding coordinates). It reduces asymptotically to the transformation from Rindler to Minkowski coordinates, $x = e^{x_z}\cosh t_S$ and $t = e^{x_z}\sinh t_S$. In BTZ slicing, the constant time slice of AdS_3 is presented in Fig. 14, with the BTZ string metric (82) covering one half. Asymptotically, this means the half-space of the plane, which we can call region A as in Fig. 5b, is described by the full range $x_z = -\infty..\infty$ of the BTZ string coordinate. This is pictured in Fig. 15b and c. Indeed, each of the conformal transformations depicted in Fig. 5 that are needed to map an interval A in a 2-dimensional CFT to the full system size at an

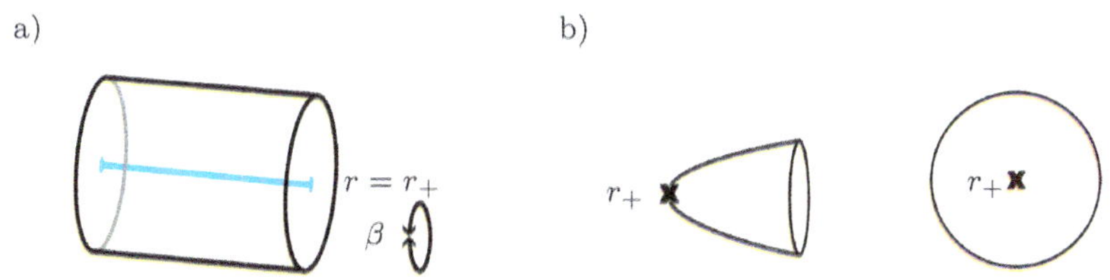

Fig. 13 **a** The thermal CFT partition function of Fig. 7a is equated, by AdS/CFT, to the thermal partition function $Z_{grav}(\beta)$ of a black string solution of AdS gravity, represented by the corresponding solid cylinder. **b** An (r,τ) slice of the solid cylinder pictured as a cigar, with U(1) fixed point at $r = r_+$, or a Poincaré disk $ds^2 = 4dw d\bar{w}/(1-w\bar{w})^2$ with $w = \sqrt{(r-r_+)/(r+r_+)}\exp(ir_+\tau)$

[7] It is more directly related to the Lewkowycz Maldacena derivation following the option in Footnote 5.

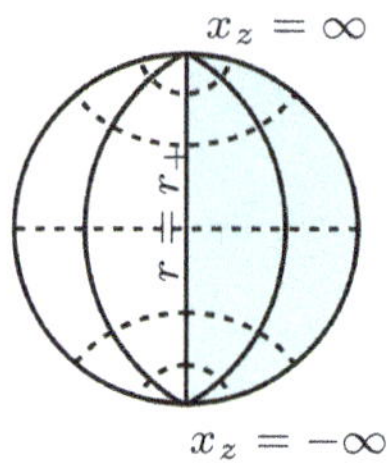

Fig. 14 A Poincaré disk constant time slice of AdS_3 in hyperbolic coordinates, with lines of constant r (solid) and constant x_z (dashed). The BTZ string metric (82) covers one half of the disk

auxiliary temperature T is dual to an AdS bulk coordinate transformation. This is illustrated in Fig. 15, which provides the holographic dual interpretation of Fig. 5, in Euclidean signature. The explicit bulk coordinate transformation from Fig. 15a to 15b, dual to conformally mapping the interval to the half-line, can also be found in [25].[8]

Having arrived at Fig. 15c, the entanglement S_A for an interval A is equal to the thermal entropy $S_{thermal}$ given by the Bekenstein-Hawking formula (85) for the BTZ string geometry with Hawking temperature T. That is, S_A is calculated holographically by the area of the BTZ horizon. Under the mappings from Fig. 15c back to the coordinates of Fig. 15a, the horizon is mapped to the RT minimal surface, and the RT prescription is proven by having found the bulk transformations dual to the conformal transformations of Holzhey et al. [46] combined with the holographic prescription (85) for calculating a thermal entropy. The length of the RT geodesic is divergent because it extends to the boundary, reflecting as it should the UV divergence of S_A. This is consistent with the area of the *planar* BTZ *string* geometry being divergent, as opposed to the finite area of the BTZ black hole (obtained from a quotient of AdS).

From (86), it is clear that the above sketched Casini Huerta Myers proof of the RT conjecture forms a special, U(1) symmetric case of the bulk replica proof of Sect. 2.1, with the RT surface provided by the U(1) fixed point or more generally the $\mathbb{Z}_n$ fixed point. By continuity, the RT surface ending on the $\mathbb{Z}_n$ fixed points ∂A in the boundary consists of the set of $\mathbb{Z}_n$ fixed points in the bulk.

For each of the Euclidean bulk pictures in Fig. 15, Fig. 16 shows the corresponding Lorentzian one. In Lorentzian signature, the Poincaré coordinates (Z, t, x) cover the region of AdS within the AdS-Poincaré horizon. This horizon intersects the AdS boundary at the null boundaries of the Minkowski spacetime Penrose diagram of Fig. 6a, with BTZ string coordinates describing the bulk dual of each of the Rindler wedges.

[8] While we have restricted to the AdS_3/CFT_2 case for concreteness, the discussion in [25] holds for any dimension.

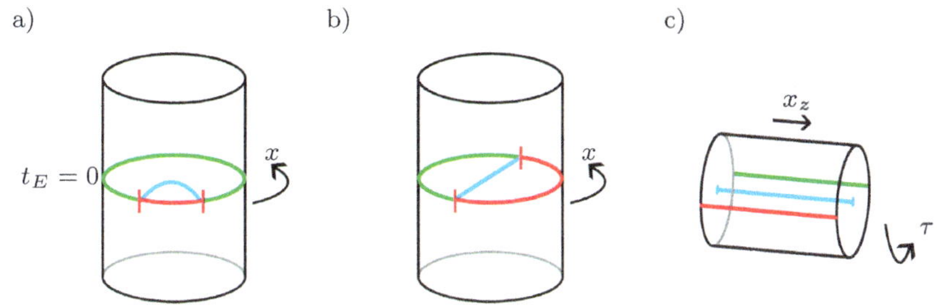

Fig. 15 The AdS (in **a** and **b**) and BTZ string dual (in **c**) of the 2-dimensional CFT set-ups in Fig. 5, in Euclidean signature. They are related by conformal mappings in the boundary or coordinate transformations in the bulk. The CFT region A is in red, the complementary region $\bar{A}$ in green, and the RT surface $r = r_+$ in blue

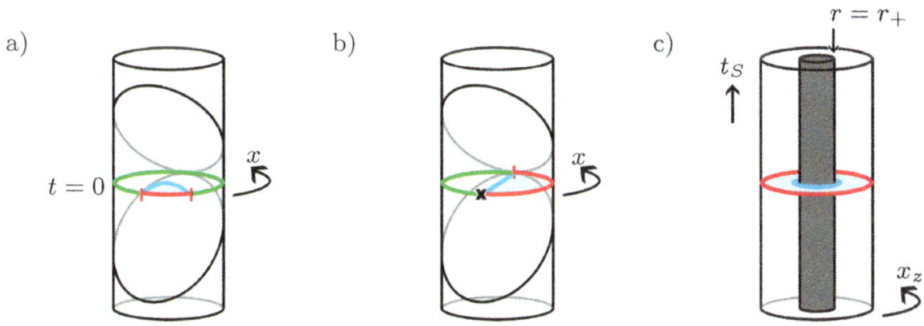

Fig. 16 The AdS (in **a** and **b**) and BTZ string dual (in **c**) of the 2-dimensional CFT set-ups in Fig. 5, in Lorentzian signature. The AdS-Poincaré coordinates (t, x) cover the region of the cylinder inside the Poincaré horizon. The entanglement wedge (light blue shaded) of A (red) maps to the outside-horizon region of the BTZ string geometry. In the right-most figure, the spatial coordinate has been compactified for reasons of presentation, because in this way it is most clear visually that the Bekenstein-Hawking entropy inspiration behind the RT prescription becomes explicit in the U(1) case. However, in this picture, the coordinate x_z should still be considered to run from $-\infty$ to ∞, covering the cylinder infinitely many times, and thus the horizon area in Fig. **c** is still divergent

2.3 Gravitational EOM from CFT Entanglement

The RT formula gives an intriguing geometric interpretation to CFT entanglement. Now the bulk geometry has to satisfy Einstein's equations and we present in this section the basic ideas behind entanglement interpretations of the gravitational dynamics. At first we focus on holography, in the last subsection we comment on similar ideas in non-holographic contexts.

2.3.1 Holographic Emergent Gravity

Bulk Metric Equation of Motion

We can consider the first law of black hole thermodynamics [11] for the BTZ string black hole of Fig. 16c,

$$\delta M = T\delta S_{thermal}, \tag{88}$$

with $T = 1/(2\pi)$ (corresponding to the choice of $\kappa = 1$ in (40)). Equation (88) expresses how a small change in mass of the black hole is related to its change in entropy. As it relates two solutions of the Einstein equations (one with mass M and one with mass $M + \delta M$), the first law is an on-shell expression that is equivalent to the linearized Einstein equations. This is made precise in the Iyer-Wald formalism [48, 75]. In essence, that formalism writes the black hole first law (88) as a Stokes equation $\delta E_\infty - \delta E_h = 0$, with E_∞ and E_h the energy at infinity and the horizon. However, it is valid far more broadly than the case above, allowing to write a gravitational first law expressing energy conservation in a chosen region of a solution of the equations of motion, for any gravitational theory with a diffeomorphism invariant Lagrangian. For example, the Iyer-Wald formalism can be applied directly to the entanglement wedge region in Fig. 16a. In fact, we will shortly interpret the black hole first law (88) as the gravitational first law for the entanglement wedge of A. We are focusing here for concreteness on the most intuitive U(1) symmetric case, where the entanglement wedge maps to a black hole outside-horizon region.

One can now ask what the dual CFT interpretation of the gravitational law (88) is. By the Bekenstein-Hawking formula (85) and the mappings in Fig. 16, i.e. by the RT formula of AdS/CFT, the right hand side of (88) is equal to $S_A/(2\pi)$, the vacuum entanglement entropy of region A. On the left hand side, the mass M is obtained from integrating the Brown-York stress tensor, which is identified in AdS/CFT with the expectation value of the CFT stress tensor [7], so that $\delta M = \delta\langle H_{t_S}\rangle$ in terms of the boundary Rindler Hamiltonian. Because of the Bisognano Wichmann result (40) (with τ the Euclidean Rindler time $\tau = -it_S$), the Rindler Hamiltonian is equal (up to a normalization constant) to $1/(2\pi)$ times the 'modular Hamiltonian' $H_{mod,A}$, which is an operator defined by writing the positive semi-definite reduced density matrix ρ_A of the vacuum CFT state in the form

$$\rho_A = \frac{e^{-H_{mod,A}}}{\operatorname{tr} e^{-H_{mod,A}}}. \tag{89}$$

Having used the AdS/CFT dictionary on each side of the black hole first law (88), it now reads, in terms of CFT entanglement concepts:

$$\delta\langle H_{mod,A}\rangle = \delta S_A. \tag{90}$$

This is a relation known as the 'first law of entanglement' [12,56,76], for its close resemblance, and in this case connection, to the first law of thermodynamics. The conclusion is that a gravitational first law in the bulk, namely the one associated with the entanglement wedge region of A, corresponds to an entanglement first law in the CFT. It follows that from the CFT perspective, the entanglement first law (90) imposes linearized Einstein equations in the dual bulk theory. In this sense gravity can be said to emerge from entanglement in AdS/CFT [33,56,74].

Bulk matter can be included in a semi-classical treatment of the bulk theory, leading to an additional matter energy contribution to the first law (88), $\delta M = T\delta S + \delta E_m$. The gravitational first law is then still consistent with the first law of entanglement *if* the RT prescription is corrected to include a bulk matter entropy contribution, which is indeed the proposal of the FLM [32] and QES [31] prescriptions.

Bulk Matter Equation of Motion

We discussed how equations of motion for the geometry in a gravitational theory can be thought of as having an entropic origin. One can also ask about the equations of motion for bulk *matter* degrees of freedom, and to what extent these can be described using statements about boundary entanglement. Considering for concreteness a free bulk scalar field ϕ, we can consider how it behaves under the action of the total CFT modular Hamiltonian, defined for a given CFT region A as the difference $H^{tot}_{mod,A} \equiv H_{mod,A} - H_{mod,\bar{A}}$. It is an operator that can be written as a linear combination of the CFT conformal generators L_0, L_1 and L_{-1} [77]. One obtains what is known as the JLMS result [52,53]

$$[H^{tot}_{mod,A}, \phi] = \mathcal{L}_\xi \phi \tag{91}$$

with $\mathcal{L}_\xi$ the Lie derivative in the direction of ξ, the AdS Killing vector acting in the entanglement wedge of A and vanishing on the RT surface. One can then argue [20] that the bulk equation of motion for ϕ, written in the form $EOM = 0$, should also satisfy

$$[H^{tot}_{mod,A}, EOM] = \mathcal{L}_\xi(EOM). \tag{92}$$

This gives a consistency condition between a modular Hamiltonian condition in the CFT and the bulk matter equation of motion, strongly constraining the latter. For electromagnetically or gravitationally interacting bulk scalar fields, the right hand side of the JLMS condition (91) gets corrected, reflecting the non-locality of such bulk objects. In particular, the bulk equation of motion for a gravitationally interacting scalar field in three dimensions satisfies (92) with ξ equal to ξ_{Ban}, the asymptotic Killing vector of the Bañados geometry [10].

2.3.2 Non-holographic Emergent Gravity

The idea of a connection between Einstein's equations and entropy laws goes back to the seminal work of Jacobson [50]. A modern entanglement entropy version of it is discussed in [51] and describes how a given CFT (in dimensions $d > 2$) can be coupled to a theory of gravity in the same number of dimensions by imposing the 'maximal vacuum entanglement condition'. This statement, similarly to the statements on holographic emergent gravity from entanglement in Sect. 2.3.1, can be traced back to a reinterpretation of a gravitational first law in terms of entanglement [15,51]. A 2-dimensional version of Jacobson's emergent gravity from maximal vacuum entanglement has been discussed in [19], see also [63,64], and a closely related account on entropic emergence of Jackiw-Teitelboim gravity for a CFT with boundary was presented in [21].

Acknowledgements I would like to thank my collaborators, and more broadly my friends and colleagues, in the fascinating world of entanglement and holography. Special thanks to Vincent Callebaut for the figures. Funded, in part, by the Deutsche Forschungsgemeinschaft (DFG, German Research Foundation) – Projektnummer 277101999 – TRR 183 (project A03 and B01).

References

1. I. Affleck, Phys. Rev. Lett. **56**, 746–748 (1986)
2. A. Almheiri, R. Mahajan, J. Maldacena, Y. Zhao, JHEP **3**, 149 (2020)
3. A. Almheiri, R. Mahajan, J. Maldacena, Islands outside the horizon (2019)
4. A. Almheiri, J. Polchinski, JHEP **11**, 014 (2015)
5. A. Almheiri, T. Hartman, J. Maldacena, E. Shaghoulian, A. Tajdini, JHEP **5**, 013 (2020)
6. C. Asplund, N. Callebaut, C. Zukowski, JHEP **9**, 154 (2016)
7. V. Balasubramanian, P. Kraus, Commun. Math. Phys. **208**, 413–428 (1999)
8. V. Balasubramanian, B. Czech, B. Chowdhury, J. Boer, JHEP **10**, 220 (2013)
9. M. Banados, C. Teitelboim, J. Zanelli, Phys. Rev. Lett. **69**, 1849–1851 (1992)
10. M. Banados, AIP Conf. Proc. **484**, 147–169 (1999)
11. J. Bardeen, B. Carter, S. Hawking, Commun. Math. Phys. **31**, 161–170 (1973)
12. E. Bianchi, Horizon entanglement entropy and universality of the graviton coupling (2012)
13. J. Bisognano, E. Wichmann, J. Math. Phys. **17**, 303–321 (1976)
14. H. Bloete, J. Cardy, M. Nightingale, Phys. Rev. Lett. **56**, 742–745 (1986)
15. P. Bueno, V. Min, A. Speranza, M. Visser, Phys. Rev. D **95**, 046003 (2017)
16. M. Cadoni, M. Melis, Found. Phys. **40**, 638–657 (2010)
17. P. Calabrese, J. Cardy, J. Stat. Mech. **406**, P06002 (2004)
18. P. Calabrese, J. Cardy, J. Phys. A **42**, 504005 (2009)
19. N. Callebaut, JHEP **2**, 153 (2019)
20. N. Callebaut, G. Lifschytz, JHEP **8**, 289 (2022)
21. N. Callebaut, H. Verlinde, JHEP **5**, 045 (2019)
22. J. Camps, JHEP **3**, 070 (2014)
23. J. Cardy, O. Castro-Alvaredo, B. Doyon, J. Statist. Phys. **130**, 129–168 (2008)
24. J. Cardy, *Conformal Field Theory and Statistical Mechanics, Les Houches Summer School: Session 89: Exacts Methods In Low-Dimensional Statistical Physics And Quantum Computing* (2008)
25. H. Casini, M. Huerta, R. Myers, JHEP **5**, 036 (2011)
26. H. Chen, R. Myers, D. Neuenfeld, I. Reyes, J. Sandor, JHEP **10**, 166 (2020)
27. X. Dong, Nature Commun. **7**, 12472 (2016)
28. X. Dong, A. Lewkowycz, M. Rangamani, JHEP **11**, 028 (2016)
29. X. Dong, A. Lewkowycz, JHEP **1**, 081 (2018)
30. X. Dong, D. Harlow, D. Marolf, JHEP **10**, 240 (2019)
31. N. Engelhardt, A. Wall, JHEP **1**, 073 (2015)
32. T. Faulkner, A. Lewkowycz, J. Maldacena, JHEP **11**, 074 (2013)
33. T. Faulkner, M. Guica, T. Hartman, R. Myers, M. Van Raamsdonk, JHEP **3**, 051 (2014)
34. D. Fursaev, S. Solodukhin, Phys. Rev. D **52**, 2133–2143 (1995)
35. D. Fursaev, JHEP **9**, 018 (2006)
36. D. Fursaev, A. Patrushev, S. Solodukhin, Phys. Rev. D **88**, 044054 (2013)
37. G. Gibbons, S. Hawking, Phys. Rev. D **15**, 2752–2756 (1977)
38. G. Gibbons, S. Hawking, M. Perry, Nucl. Phys. B **138**, 141–150 (1978)
39. F. Haehl, T. Hartman, D. Marolf, H. Maxfield, M. Rangamani, JHEP **5**, 023 (2015)
40. D. Harlow, Rev. Mod. Phys. **88**, 015002 (2016)
41. J. Hartle, S. Hawking, Phys. Rev. D **13**, 2188–2203 (1976)
42. T. Hartman, J. Maldacena, JHEP **5**, 014 (2013)
43. T. Hartman, *Lectures on Quantum Gravity and Black Holes*

44. M. Headrick, Phys. Rev. D **82**, 126010 (2010)
45. M. Headrick, *Lectures on Entanglement Entropy in Field Theory and Holography* (2019)
46. C. Holzhey, F. Larsen, F. Wilczek, Nucl. Phys. B **424**, 443–467 (1994)
47. V. Hubeny, M. Rangamani, T. Takayanagi, JHEP **7**, 062 (2007)
48. V. Iyer, R. Wald, Phys. Rev. D **50**, 846–864 (1994)
49. R. Jackiw, Nucl. Phys. B **252**, 343–356 (1985)
50. T. Jacobson, Phys. Rev. Lett. **75**, 1260–1263 (1995)
51. T. Jacobson, Phys. Rev. Lett. **116**, 201101 (2016)
52. D. Jafferis, A. Lewkowycz, J. Maldacena, S. Suh, JHEP **6**, 004 (2016)
53. D. Kabat, G. Lifschytz, JHEP **5**, 017 (2019)
54. R. Laflamme, Nucl. Phys. B **324**, 233–252 (1989)
55. F. Larsen, F. Wilczek, Ann. Phys. **243**, 280–298 (1995)
56. N. Lashkari, M. McDermott, M. Van Raamsdonk, JHEP **4**, 195 (2014)
57. A. Lewkowycz, J. Maldacena, JHEP **8**, 090 (2013)
58. J. Maldacena, Adv. Theor. Math. Phys. **2**, 231–252 (1998)
59. J. Maldacena, JHEP **4**, 021 (2003)
60. P. Martin, J. Schwinger, Phys. Rev. **115**, 1342–1373 (1959)
61. T. Mertens, G. Turiaci, *Solvable Models of Quantum Black Holes: A Review on Jackiw-Teitelboim Gravity* (2022)
62. W. Nelson, Phys. Rev. D **50**, 7400–7402 (1994)
63. J. Pedraza, A. Svesko, W. Sybesma, M. Visser, JHEP **12**, 134 (2021)
64. J. Pedraza, A. Svesko, W. Sybesma, M. Visser, Phys. Rev. D **105**, 126010 (2022)
65. G. Penington, S. Shenker, D. Stanford, Z. Yang, JHEP **3**, 205 (2022)
66. J. Polchinski, *String Theory. Vol. 1: An Introduction to the Bosonic String* (Cambridge University Press, 2007)
67. S. Ryu, T. Takayanagi, Phys. Rev. Lett. **96**, 181602 (2006)
68. S. Ryu, T. Takayanagi, JHEP **8**, 045 (2006)
69. S. Solodukhin, Living Rev. Rel. **14**, 8 (2011)
70. Y. Takahasi, H. Umezawa, Collect. Phenom. **2**, 55–80 (1975)
71. C. Teitelboim, Phys. Lett. B **126**, 41–45 (1983)
72. W. Unruh, Phys. Rev. D **14**, 870 (1976)
73. W. Unruh, N. Weiss, Phys. Rev. D **29**, 1656 (1984)
74. M. Van Raamsdonk, Gen. Rel. Grav. **42**, 2323–2329 (2010)
75. R. Wald, Phys. Rev. D **48**, R3427–R3431 (1993)
76. G. Wong, I. Klich, L. Pando Zayas, D. Vaman, JHEP **12**, 020 (2013)
77. J. Boer, F. Haehl, M. Heller, R. Myers, JHEP **8**, 162 (2016)

Coordinates are Messy—Not Only in General Relativity

Carla Cederbaum and Melanie Graf

Abstract

The coordinate freedom of General Relativity makes it challenging to find mathematically rigorous and physically sound definitions for physical quantities such as the center of mass of an isolated gravitating system. We will argue that a similar phenomenon occurs in Newtonian Gravity once one ahistorically drops the restriction that one should only work in Cartesian coordinates when studying Newtonian Gravity. This will also shed light on the nature of the challenge of defining the center of mass in General Relativity. Relatedly, we will give explicit examples of asymptotically Euclidean relativistic initial data sets which do not satisfy the Regge–Teitelboim parity conditions often used to achieve a satisfactory definition of center of mass. These originate in our joint work [4] with Jan Metzger. This will require appealing to Bartnik's asymptotic harmonic coordinates.

1 Preferred Systems of Coordinates (or not)

As we all know, Euclidean space—the stage of Newtonian Gravity—knows preferred systems of coordinates, called Cartesian coordinates. In such coordinates, the Euclidean metric δ takes its canonical form. Similarly, the Minkowski spacetime—the setting of Special Relativity—carries preferred systems of coordinates in which

C. Cederbaum (✉) · M. Graf
Universität Tübingen, Fachbereich Mathematik, Auf der Morgenstelle 10, 72076 Tübingen, Germany
e-mail: cederbaum@math.uni-tuebingen.de

M. Graf
e-mail: graf@math.uni-tuebingen.de; melanie.graf@uni-hamburg.de

M. Graf
Present Address:
Fachbereich Mathematik, Universität Hamburg, Bundesstraße 55, 20146 Hamburg, Germany

B. Hartmann and J. Kunz (eds.), *Gravity, Cosmology, and Astrophysics*,
Lecture Notes in Physics 1022, https://doi.org/10.1007/978-3-031-42096-2_11

the Minkowski metric η takes its canonical form. In contrast, curved spacetimes—the mathematical framework of General Relativity—and initial data sets therein are well-known not to admit any 'canonical' or 'preferred' coordinates in general. This freedom in the choice of coordinates makes it challenging to find mathematically rigorous and physically sound definitions for physical quantities such as the center of mass of an isolated gravitating system in General Relativity as is well-known and will be discussed in this article. We will argue that a similar phenomenon occurs in Newtonian Gravity once one ahistorically and somewhat unnecessarily drops the restriction that one should only work in Cartesian coordinates when studying Newtonian Gravity. This will also shed light on the nature of the challenge of defining the center of mass in General Relativity. Relatedly, we will give explicit examples of asymptotically Euclidean relativistic initial data sets which do not satisfy the "Regge–Teitelboim (parity) conditions" often used to achieve a satisfactory definition of center of mass. These originate in our joint work [4] with Jan Metzger.

2 Isolated Systems at a Given Instant of Time

Let's begin by recalling the standard definitions of an "isolated system at a given instant of time" in both Newtonian Gravity and General Relativity. We will also recall the standard definitions of (total) mass of such systems along the way and discuss the convergence of the involved integrals.

2.1 Isolated Systems at a Given Instant of Time in Newtonian Gravity

In Newtonian Gravity, we can think of an "isolated system at a given instant of time" as given by a matter density function $\rho\colon \mathbb{R}^3 \to [0,\infty)$ which has compact support or at least decays suitably fast towards infinity. For example, one could ask that $\rho = O(r^{-3-\varepsilon})$ as $r \to \infty$ for some (small) $\varepsilon > 0$, that is, ρ decays to zero at least as fast as $r^{-3+\varepsilon}$, where r denotes the radial coordinate on $\mathbb{R}^3$. Alternatively but not equivalently, one could ask that $\rho \in L^1(\mathbb{R}^3)$. Both assumptions are independent of the chosen Cartesian coordinates because any two systems of Cartesian coordinates on $\mathbb{R}^3$ differ only by a rigid motion. Either of these decay assumptions is sufficient for the total mass

$$m = \iiint_{\mathbb{R}^3} \rho(\mathbf{x})\, d\mathbf{x} \tag{1}$$

to be well-defined and finite. Anticipating the discussion below, let us point out that the O-assumption suggests computing the integral in (1) as an improper Riemann integral in polar coordinates, while the L^1-assumption suggests treating it as a Lebesgue integral. Of course, the resulting mass m will be the same whatever notion of integral one refers to, as long as it converges. However, taking the former viewpoint, we can take advantage of cancellations in the spherical integrals. Also note that the decay assumptions are of course *not* independent of arbitrary coordinate changes.

2.2 Isolated Systems at a Given Instant of Time in General Relativity

In General Relativity, an "isolated system at a given instant of time" is modelled as an asymptotically Euclidean (or asymptotically flat) relativistic initial data set (or time-slice): As usual, an initial data set (M, g, K) consists of a 3-dimensional Riemannian manifold (M, g) carrying a symmetric $(0, 2)$-tensor field K playing the role of second fundamental form (or extrinsic curvature) of the initial data set in the spacetime modelling the system.

In addition, a relativistic initial data set carries an energy density $\mu\colon M \to \mathbb{R}$ and a momentum density one-form J related to g and K via the well-known Einstein constraint equations

$$\mathrm{R} - |K|^2 + (\mathrm{tr}\, K)^2 = 2\mu \tag{2}$$
$$\mathrm{div}(K - (\mathrm{tr}\, K)g) = -J. \tag{3}$$

and derived from the energy-momentum tensor T of the spacetime. Here, R denotes the scalar curvature of (M, g), and $|\cdot|$, tr, and div denote the tensor norm, trace, and divergence with respect to g, respectively. We will adopt the following standard definition, see also Fig. 1.

Definition 1 (*Asymptotically Euclidean Relativistic Initial Data Set*) A relativistic initial data set (M, g, K) with energy density μ and momentum density J is called *asymptotically Euclidean* if there exists a compact set $C \subset M$, a radius $R > 0$, and an *asymptotic coordinate chart* $\mathbf{x}\colon M \setminus C \to \mathbb{R}^3 \setminus B_R(0)$ such that

$$g_{ij} - \delta_{ij} = O_2(r^{-\frac{1}{2}-\varepsilon}) \tag{4}$$
$$K_{ij} = O_1(r^{-\frac{3}{2}-\varepsilon}) \tag{5}$$
$$\mu, J_i = O_0(r^{-3-\varepsilon}) \tag{6}$$

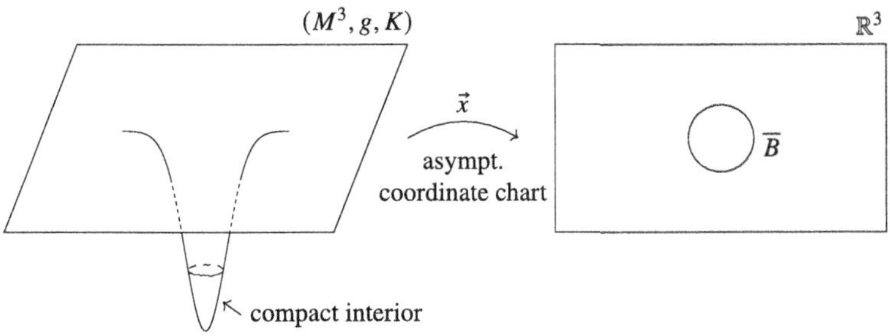

Fig. 1 An asymptotically Euclidean relativistic initial data set (M, g, K) and the image of its asymptotic end $M \setminus C$ in $\mathbb{R}^3$ under the asymptotic coordinate chart $\mathbf{x}$

as $r = |\mathbf{x}| \to \infty$ for some decay parameter $\varepsilon > 0$, where g_{ij}, K_{ij}, and J_i denote the components of g, K, and J in the coordinates $\mathbf{x}$, respectively. Alternatively but not equivalently, one can replace Assumption (6) by asking that μ, $J_i \in L^1(\mathbb{R}^3 \setminus B_R(0))$.

Here, the index $k \in \mathbb{N}_0 \cup \{\infty\}$ in $O_k(r^\alpha)$ for some $\alpha < 0$ is a shorthand for asking that derivatives of order up to k decay 'accordingly' as $r \to \infty$, that is, first derivatives decay to zero at least as fast as $r^{\alpha-1}$, second derivatives decay to zero at least as fast as $r^{\alpha-2}$, etc. In what follows, we will slightly abuse notation and extend this to the 'decay rate' $\alpha = 0$, so that $f = O_k(r^0)$ will mean that the function f stays bounded as $r \to \infty$, while its order l derivatives decay to zero at least as fast as r^{-l} as $r \to \infty$ whenever $l \leq k$.

The (total) mass m_{ADM} of an asymptotically Euclidean relativistic initial data set (M, g, K) was defined by Arnowitt, Deser, and Misner in [1] via the (total) energy E_{ADM} and (total) linear momentum $\mathbf{P}_{\mathrm{ADM}}$ and has become the standard definition, satisfying many desirable properties such as for example positivity ([13, 14]):

$$E_{\mathrm{ADM}} := \frac{1}{16\pi} \lim_{R\to\infty} \iint_{S_R(0)} \sum_{i,j=1}^{3} (\partial_i g_{ij} - \partial_j g_{ii}) \frac{x^j}{R} dA_\delta, \tag{7}$$

$$P^j_{\mathrm{ADM}} := \frac{1}{8\pi} \lim_{R\to\infty} \iint_{S_R(0)} \sum_{i=1}^{3} \left((\mathrm{tr}\, K) g_{ij} - K_{ij}\right) \frac{x^i}{R} dA_\delta, \tag{8}$$

$$m_{\mathrm{ADM}} := \sqrt{E^2_{\mathrm{ADM}} - |\mathbf{P}_{\mathrm{ADM}}|^2}, \tag{9}$$

where dA_δ denotes the Euclidean area measure on $S_R(0)$. In [2], Bartnik uses harmonic asymptotically Euclidean coordinate charts to prove that E_{ADM} is well-defined and independent of the choice of asymptotically Euclidean coordinate charts, see Sect. 3. Note that Bartnik uses the weaker L^1-decay condition on μ. From this, (8), (5) and (6), it follows directly that m_{ADM} is also well-defined and independent of the choice of asymptotically Euclidean coordinate chart via the Divergence Theorem. Again, this argument only uses the weaker L^1-condition on J.

You may wonder why one asks for such general decay rates in (4)–(6), and not just for, say,

$$g_{ij} - \delta_{ij} = O_2(r^{-1}) \tag{10}$$

etc.. This has two reasons: First, the theory of asymptotically Euclidean relativistic initial data sets does not become more complicated if one does so. Second, it actually becomes richer, i.e., allows for more examples, see [4] and the references cited therein.

Next, let us discuss how a single relativistic initial data set can carry different asymptotic coordinate charts and discuss the relationship between different such charts.

3 Comparing Different Asymptotic Coordinate Systems

Clearly, if a relativistic initial data set is asymptotically Euclidean for some asymptotic coordinate chart $\mathbf{x}$, it will also be asymptotically Euclidean for any asymptotic coordinate chart $\mathbf{y}$ arising from $\mathbf{x}$ by a rigid motion,[1] for the same decay parameter ε. Moreover, the class of possible transformations between two asymptotic coordinate charts $\mathbf{x}$ and $\mathbf{y}$ with respect to which a given relativistic initial data set (M, g, K) is asymptotically Euclidean is much richer than just rigid motions, even when fixing the decay parameter ε. Here is an example: Let (M, g, K) be a relativistic initial data set with energy and momentum densities μ and J which is asymptotically Euclidean with respect to some asymptotic coordinate chart $\mathbf{x}$ and for some decay parameter $\varepsilon > 0$. Then (M, g, K) will also be asymptotically Euclidean with decay parameter ε with respect to the asymptotic coordinate chart

$$\mathbf{y} := \mathbf{x} + \sin(\ln r)\,\mathbf{a} \tag{11}$$

for some non-vanishing $\mathbf{a} \in \mathbb{R}^3$, with $r = |\mathbf{x}|$ as before. This can be seen by a straightforward computation. In particular, note that (11) is very similar to a mere translation, differing only in the bounded factor $\sin(\ln r) = O_\infty(r^0)$ as $r \to \infty$.

On the other hand, it crucially depends on the choice of asymptotic coordinate chart whether a given relativistic initial data set "is" asymptotically Euclidean: For example, any relativistic initial data set which is asymptotically Euclidean with respect to an asymptotic coordinate chart $\mathbf{x}$ will not be asymptotically Euclidean with respect to the chart $\mathbf{y} := 2\mathbf{x}$ as one easily computes. We will hence refer to an asymptotic coordinate chart $\mathbf{x}$ as an *asymptotically Euclidean coordinate chart* for a given relativistic initial data set (M, g, K) (with energy and momentum densities μ and J) if (M, g, K) is asymptotically Euclidean with respect to $\mathbf{x}$.

Summarizing, the class of asymptotically Euclidean coordinate charts for a given relativistic initial data set is much richer than the class of Cartesian coordinate systems on Euclidean space. This applies in particular to the Euclidean relativistic initial data set $(M = \mathbb{R}^3, g = \delta, K = 0)$ sitting inside the Minkowski spacetime. Here, one sees that the Cartesian coordinate systems are asymptotically Euclidean coordinate charts, but by far not the only asymptotically Euclidean coordinate charts.

3.1 Divergence of Mass

At this point, it is instructive to recall that the decay condition (4) cannot be relaxed as was shown by a counter-example by Denissov and Solovyev [8]. Inspired by their example, let us consider the Euclidean relativistic initial data set $(M = \mathbb{R}^3, g = \delta, K = 0)$ in the coordinates

[1] The careful reader may note that one may need to change the compact set C and the radius R from 1 for the coordinate chart $\mathbf{y}$; we will ignore such subtleties in this article for the sake of readability.

$$\mathbf{y} := \left(1 + \frac{a}{\sqrt{r}}\right) \mathbf{x} \tag{12}$$

for some non-zero $a \in \mathbb{R}$ which leads to (4) with $\varepsilon = 0$ and the unphysical result $m_{\text{ADM}} = \frac{a^2}{8}$. One can argue similarly for (5) as we will discuss elsewhere; alternatively, one can compute in a lengthy but straightforward way that the decay conditions (4) and (5) transform equivariantly under coordinate boosts in the ambient spacetime. From this and the example by Denissov and Solovyev, one can conclude that (5) is necessary for physicality of the definition of m_{ADM}.

In summary, (convergence and coordinate independence of) mass is very well understood in both Newtonian Gravity and General Relativity and depends crucially on the decay of the matter variables, as well as, in General Relativity, on the asymptotics of the relativistic initial data set itself.

One of the main tools introduced by Bartnik for the study of mass and energy are the "harmonic asymptotically Euclidean coordinates" we will now explain.

4 A Canonical Choice: Harmonic Coordinates

Cartesian coordinates are not only canonical for the Euclidean metric, they are also *harmonic*, that is, they satisfy the system of partial differential equations

$$\triangle_\delta \mathbf{x} = 0, \tag{13}$$

a shorthand for the system of equations

$$\triangle_\delta x^i = 0 \text{ for } i = 1, 2, 3, \tag{14}$$

where $\triangle_\delta$ denotes the Euclidean Laplacian.

Exploiting this insight, Bartnik showed in [2] that asymptotically Euclidean relativistic initial data sets (M, g, K) always possess *harmonic* asymptotically Euclidean coordinate charts, that is, asymptotically Euclidean coordinate charts satisfying the geometric system of partial differential equations

$$\triangle \mathbf{x} = 0, \tag{15}$$

where $\triangle$ denotes the Laplacian with respect to g. Here, "geometric" means that the partial differential equations themselves do not depend on a choice of (local or asymptotic) coordinate chart.

Furthermore, Bartnik showed [2, Theorem 3.1] that two such harmonic asymptotically Euclidean coordinate charts $\mathbf{x}, \mathbf{y}$ are related by a rigid motion up to suitably lower order terms,

$$\mathbf{y} = Q\mathbf{x} + \mathbf{a} + O_0(r^{\frac{1}{2}-\varepsilon}), \tag{16}$$

for a special orthogonal matrix $Q \in \mathbb{R}^{3\times 3}$ and a vector $\mathbf{a} \in \mathbb{R}^3$. In particular, there are *more* harmonic asymptotically Euclidean coordinate charts on Euclidean space than just the Cartesian coordinate systems: for example, the coordinates $\mathbf{y} := \mathbf{x} + \frac{\mathbf{b}}{r}$ for some non-trivial vector $\mathbf{b} \in \mathbb{R}^3$ are also harmonic.

5 On the Center of Mass of Isolated Systems at a Given Instant of Time

Let us now move on to the definition of (total) center of mass, where the situation is somewhat drastically different than for energy, linear momentum, and mass. Again, we will first take a look at the (total) center of mass of an isolated system at a given instant of time in Newtonian Gravity.

5.1 On the Center of Mass in Newtonian Gravity

Provided $m \neq 0$, the center of mass in Newtonian Gravity is naturally defined as the averaged weighted integral of the position vector $\mathbf{x}$,

$$\mathbf{C} := \frac{1}{m} \iiint_{\mathbb{R}^3} \rho(\mathbf{x})\, \mathbf{x}\, d\mathbf{x}. \tag{17}$$

Looking at (17) as a Lebesgue integral, it suggests itself that one should ask that $\rho r \in L^1(\mathbb{R}^3)$. Instead, for this to be well-defined and finite as an improper Riemann integral,

$$\mathbf{C} = \lim_{R\to\infty} \int_0^R \iint_{S_r(0)} \rho(\mathbf{x})\, \mathbf{x}\, dA_\delta\, dr, \tag{18}$$

it suggests itself to assume $\rho = O_0(r^{-4-\varepsilon})$ for some $\varepsilon > 0$, in analogy with the choice of decay rate used for ensuring that the mass is well-defined; this of course settles the convergence issue for Newtonian gravitating systems.

However, let us—ahistorically—take a different approach in analogy with the standard approach taken to resolve the corresponding issue in General Relativity. To this end, let us instead make a further parity-based decay assumption, namely

$$\rho^{\text{odd}} = O_0(r^{-4-\varepsilon}), \tag{19}$$

where

$$\rho^{\text{odd}}(\mathbf{x}) := \frac{1}{2}(\rho(\mathbf{x}) - \rho(-\mathbf{x})) \tag{20}$$

is the odd part of ρ. This approach relies on the insight that the contribution to (18) of the even part

$$\rho^{\text{even}} := \rho - \rho^{\text{odd}} \tag{21}$$

vanishes by parity on each sphere $S_r(0)$. In view of the analogous approach taken in General Relativity, let us take the time to consider the *parity condition*

$$\rho^{\mathrm{odd}} = O_0(r^{-4-\varepsilon}) \tag{22}$$

and its properties in more detail. Importantly, we would like to bring to the reader's attention that the parity condition is *not* independent of the choice of Cartesian coordinate systems because the reflection $\mathbf{x} \mapsto -\mathbf{x}$ involved in the definition of ρ^{odd} does not interact well with translations. However, the desirable invariance under choice of Cartesian coordinate systems can be restored if one assumes that in some, and hence all, Cartesian coordinate systems, one has $\rho = O_1(r^{-3-\varepsilon})$, by appealing to the Mean Value Theorem.

5.1.1 Transformation Behavior of the Center of Mass in Newtonian Gravity

Of course, when well-defined by asking that $\rho \in L^1(\mathbb{R}^3)$ or $\rho = O_0(r^{-4-\varepsilon})$ for some $\varepsilon > 0$, the center of mass $\mathbf{C}$ transforms as expected under changes of Cartesian coordinate systems, which can suggestively be written as

$$\mathbf{C}_{\mathbf{y}} = Q\mathbf{C}_{\mathbf{x}} + \mathbf{a}. \tag{23}$$

But what happens if one—ahistorically—allows asymptotically Euclidean coordinate charts on the Euclidean stage of Newtonian Gravity? It will be instructive to study this in an explicit example similar to (11), i.e., $\mathbf{y} := \mathbf{x} + \sin(\ln r)\,\mathbf{a}$, but modified to obtain a global coordinate chart $\mathbf{z}$ on $\mathbb{R}^3$,

$$\mathbf{z} := \mathbf{x} + \sigma(r)\sin(\ln r)\,\mathbf{a}, \tag{24}$$

where $\sigma\colon [0,\infty) \to \mathbb{R}$ is a strictly increasing cut-off function satisfying $\sigma(r) = 0$ for $r < 2$ and $\sigma(r) = 1$ for $r > 3$. Computing the center of mass $\mathbf{C}_{\mathbf{z}}$ according to (17) with respect to the asymptotically Euclidean coordinate chart $\mathbf{z}$, for a point particle matter density $\rho(\mathbf{x}) = m\delta(\mathbf{x})$, one finds $\mathbf{C}_{\mathbf{x}} = \mathbf{0}$ but $\mathbf{C}_{\mathbf{z}}$ diverges like $\sin(\ln s)\mathbf{a}$ for $s = |\mathbf{z}| \to \infty$.

This can be made mathematically more precise by using a surface integral approach via the Divergence Theorem and the Poisson equation for the Newtonian potential as elaborated by Cederbaum and Nerz [5].

Briefly put, once one ahistorically allows more general asymptotically Euclidean coordinate charts in Newtonian Gravity, the center of mass is *not* generically a well-defined quantity even if $\rho = O_0(r^{-4-\varepsilon})$ as $r \to \infty$. From the perspective of Newtonian Gravity arising as the Newtonian limit of General Relativity for slow speeds and small masses, it thus becomes reasonable to expect a similar phenomenon to occur in General Relativity. We will now turn to this.

5.2 On the Center of Mass in General Relativity

In General Relativity, a (total) notion of center of mass $\mathbf{C}_{\text{BÓRT}}$ of an isolated system at a given instant of time was put forward by Beig and Ó Murchadha in [3], based on previous work by Regge and Teitelboim [12] and similar in spirit and derivation to the ADM-quantities. For an asymptotically Euclidean relativistic initial data set (M, g, K) with $E_{\text{ADM}} \neq 0$, its components are *formally* defined by

$$C^{\ell}_{\text{BOM}} := \frac{1}{16\pi E_{ADM}} \lim_{R\to\infty} \Big\{ \iint_{S_R(0)} x^{\ell} \sum_{i,j=1}^{3} (\partial_i g_{ij} - \partial_j g_{ii}) \tfrac{x^j}{R} \, dA_\delta - \iint_{S_R(0)} \sum_{i=1}^{3} \big(g_{i\ell} \tfrac{x^i}{R} - g_{ii} \tfrac{x^{\ell}}{R} \big) dA_\delta \Big\} \tag{25}$$

with respect to the given asymptotically Euclidean coordinate chart $\mathbf{x}$. Just as in the Newtonian case, this is a formal definition in the sense that it need not and does not always converge.

One instance where it diverges[2] is the canonical Schwarzschild relativistic initial data set $(\mathbb{R}^3 \setminus \overline{B_{2m}(0)}, \frac{1}{1-\frac{2m}{r}} dr^2 + r^2 d\Omega^2, K = 0)$ of mass $m \neq 0$, when considered with respect to the asymptotic coordinate chart $\mathbf{y}$ arising from the Cartesian coordinates $\mathbf{x}$ computed from the spherical polar Schwarzschild coordinates via (11). As in the Newtonian Gravity case discussed above, one finds via a lengthy computation that $(\mathbf{C}_{\text{BÓRT}})_{\mathbf{y}}$ diverges like $\sin(\ln s)\mathbf{a}$ for $s = |\mathbf{y}| \to \infty$, while of course $(\mathbf{C}_{\text{BÓRT}})_{\mathbf{x}}=0$ and $E_{\text{ADM}} = m$. We would like to draw the reader's attention to the fact that this initial data set has $\mu = 0$, $J = 0$, so the divergence problem clearly does not arise from poor decay of the matter.

A first idea one might have to remedy the divergence problem of $\mathbf{C}_{\text{BÓRT}}$ could be to assume the stronger decay condition $g_{ij} - \delta_{ij} = O_1(r^{-2-\varepsilon})$, thereby enforcing convergence in a way similar to remedying the convergence issue of E_{ADM} discovered by Denissov and Solovyev, see Sect. 3.1. However, this implies $E_{\text{ADM}} = 0$ which is undesirable when interested in the center of mass.

Instead, one usually resorts to parity assumptions. Before we do so in Sect. 5.2.2, let us briefly take a look at the transformation behavior of the center of mass under changes of asymptotic coordinates.

5.2.1 Transformation Behavior of the Center of Mass in General Relativity

As in the Newtonian case discussed in Sect. 5.1.1, when $\mathbf{C}_{\text{BÓRT}}$ is well-defined (see below), the center of mass $\mathbf{C}_{\text{BÓRT}}$ transforms as expected under "asymptotic Euclidean motions", i.e., under changes of asymptotic coordinate systems that can be written as

$$\mathbf{y} = Q\mathbf{x} + \mathbf{a} \tag{26}$$

with Q and $\mathbf{a}$ as before. That is to say that (23) holds also in the relativistic case.

[2] For further examples of divergence of the center of mass, see [5] and the references cited therein.

5.2.2 Introducing the Regge–Teitelboim Parity Conditions

As hinted to in the Newtonian Gravity discussion above, the standard way out of the divergence dilemma is to assume parity conditions as suggested by Regge and Teitelboim in [12].

Definition 2 (*Regge–Teitelboim Conditions*) An initial data set (M, g, K) with an asymptotically Euclidean coordinate chart $\mathbf{x}$ is said to satisfy the *weak (strong) Regge–Teitelboim conditions* if there exists $\varepsilon > 0$ such that, for $\eta = \frac{1}{2}$ $(\eta = 1)$ and

$$g_{ij}^{\text{odd}} = O_2(r^{-\frac{1}{2}-\eta-\varepsilon}) \tag{27}$$

$$K_{ij}^{\text{even}} = O_1(r^{-\frac{3}{2}-\eta-\varepsilon}) \tag{28}$$

$$\mu^{\text{odd}}, J_i^{\text{odd}} = O_0(r^{-3-\eta-\varepsilon}) \tag{29}$$

as $r = |\mathbf{x}| \to \infty$, where the even and odd parts are taken with respect to $\mathbf{x}$.

It was shown by Beig and Ó Murchadha in [3] that the strong Regge–Teitelboim conditions indeed suffice to ensure convergence of $\mathbf{C}_{\text{BÓRT}}$. Consistently, the above Schwarzschild example does not satisfy any Regge–Teitelboim conditions in the asymptotically Euclidean coordinate chart $\mathbf{y}$ introduced in (11), as can be seen by a tedious computation for which we refer the interested reader to [4].

It is well-known (see [5] and the references cited therein) that the weak Regge–Teitelboim conditions do not suffice to ensure convergence of $\mathbf{C}_{\text{BÓRT}}$; yet, as we will see at the end of this article, they are very relevant for analyzing $\mathbf{C}_{\text{BÓRT}}$.

Moreover, as in the Newtonian Gravity case, neither the strong nor the weak Regge–Teitelboim conditions are invariant under changes between different asymptotically Euclidean coordinate charts because of the same conflict between reflections and translations. But they suffer from even more fundamental issues.

6 (In-)Existence of Coordinate Systems Satisfying the Regge–Teitelboim Conditions

We have just seen that the class of coordinate systems satisfying the Regge–Teitelboim conditions is not closed under translations. But, more fundamentally, do all asymptotically Euclidean relativistic initial data sets even possess *any* asymptotically Euclidean coordinate charts in which the (strong) Regge–Teitelboim conditions hold? As we have investigated with Jan Metzger in [4], this turns out not to be the case; indeed, we will soon give explicit counter-examples.

In order to prove *in*existence of such asymptotically Euclidean coordinate charts on a given relativistic initial data set, we utilize Bartnik's harmonic asymptotically Euclidean coordinate charts, see Sect. 4, and methods from [2, 10] as well as a bootstrapping argument to show the following result. We refer the interested reader to our joint work with Jan Metzger [4] for more details and the proofs of the following theorems.

Theorem 1 *Let (M, g, K) be an asymptotically Euclidean relativistic initial data set and assume it satisfies the weak (strong) Regge–Teitelboim conditions with respect to an asymptotically Euclidean coordinate chart* $\mathbf{x}$. *Then there exists a smooth harmonic asymptotically Euclidean coordinate chart* $\mathbf{y}$ *such that* $\mathbf{x}-\mathbf{y}=O_3(|\mathbf{x}|^{\frac{1}{2}-\overline{\varepsilon}})$ *as* $|\mathbf{x}|\to\infty$ *and*

$$\overline{g}_{ij}^{\text{odd}} = O_1(|\mathbf{y}|^{-\frac{1}{2}-\eta-\overline{\varepsilon}}) \tag{30}$$
$$\overline{K}_{ij}^{\text{even}} = O_0(|\mathbf{y}|^{-\frac{3}{2}-\eta-\overline{\varepsilon}}) \tag{31}$$

as $|\mathbf{y}|\to\infty$ *for some* $\overline{\varepsilon}>0$, *where* $\eta=\frac{1}{2}$ *(respectively* $\eta=1$*), and where the components* $\overline{g}_{ij}$ *and* $\overline{K}_{ij}$ *of* g *and* K *as well as their odd and even parts are computed with respect to* $\mathbf{y}$.

In other words, the Regge–Teitelboim conditions are inherited by harmonic asymptotically Euclidean coordinate charts up to a potential loss of derivatives. As a corollary of this analysis, the reduced derivative weak (respectively strong) Regge–Teitelboim conditions (30), (31) are satisfied for one set of harmonic asymptotically Euclidean coordinate charts if and only if they are satisfied for all such charts.

We also get the following "converses", which readily follow from a more careful analysis of decay rates.

Theorem 2 *Let (M, g, K) be an asymptotically Euclidean relativistic initial data set with respect to an asymptotically Euclidean coordinate chart* $\mathbf{x}$, *but assume that*

$$K_{ij}^{\text{even}} \neq O_0(|\mathbf{x}|^{-2-\varepsilon}) \tag{32}$$

as $|\mathbf{x}|\to\infty$ *for some decay parameter* $\varepsilon>0$.

Then the harmonic coordinate chart constructed in 1 is asymptotically Euclidean but cannot satisfy the weak Regge–Teitelboim conditions. More precisely, we get

$$\overline{K}_{ij}^{\text{even}} \neq O_0(|\mathbf{y}|^{-2-\varepsilon}) \tag{33}$$

as $|\mathbf{y}|\to\infty$.

If, in addition, g satisfies additional decay assumptions such as for example

$$g_{ij}^{\text{odd}} = O_2(|\mathbf{x}|^{-3/2-\varepsilon}), \tag{34}$$

while

$$K_{ij}^{\text{even}} \neq O_0(|\mathbf{x}|^{-\frac{5}{2}-\varepsilon}) \tag{35}$$

as $|\mathbf{x}|\to\infty$ *then the harmonic coordinate chart constructed in 1 is asymptotically Euclidean but cannot satisfy the strong Regge–Teitelboim conditions. More precisely, we get*

$$\overline{K}_{ij}^{\text{even}} \neq O_0(|\mathbf{y}|^{-\frac{5}{2}-\varepsilon}) \tag{36}$$

as $|\mathbf{y}|\to\infty$.

In a nutshell, we have seen that ruling out the existence of asymptotically Euclidean coordinate charts in which a given relativistic initial data set satisfies the strong or the weak Regge–Teitelboim conditions can be reduced to asking (almost) the same question only about harmonic asymptotically Euclidean coordinate charts. This allows us to give a number of explicit examples of relativistic initial data sets not allowing for any asymptotically Euclidean coordinate charts satisfying the strong (respectively weak) Regge–Teitelboim conditions.

6.1 Graphical Counter-Examples to Existence of Regge–Teitelboim Coordinates

All examples discussed in this section originate from our joint work with Jan Metzger [4]. Following ideas by Cederbaum and Nerz [5], we focus on relativistic initial data sets in the Schwarzschild spacetime of mass $m \in \mathbb{R}$ in the Cartesian coordinates $\mathbf{x}$ associated with the Schwarzschild coordinates. These can and will be described as graphs over the canonical relativistic initial data set $\{t = 0\}$ of suitable graph functions $T \colon \mathbb{R}^3 \setminus C \to \mathbb{R}$ for a suitable compact set $C \subset \mathbb{R}^3$. Writing the Schwarzschild spacetime as

$$N(r) = \sqrt{1 - \frac{2m}{r}}, \tag{37}$$

$$h = \frac{1}{N^2} dr^2 + r^2 d\Omega^2, \tag{38}$$

on $\mathbb{R} \times (r_m, \infty) \times \mathbb{S}^2$ with $d\Omega^2$ denoting the canonical metric on the sphere $\mathbb{S}^2$, one finds

$$g^T_{ij} = h_{ij} - N^2 \partial_i T \partial_j T \tag{39}$$

$$K^T_{ij} = \frac{\partial_i T \partial_j N + \partial_j T \partial_i N + N \mathrm{Hess}_h(T)_{ij} - N^2 \partial_i T \partial_j T \, dN(\mathrm{grad}_h(T))}{\sqrt{1 - N^2 |dT|^2_h}} \tag{40}$$

on the graph $M_T = \{t = T(\mathbf{x}) : \mathbf{x} \in \mathbb{R}^3 \setminus C\}$, see Fig. 2.

Choosing

$$T_1(\mathbf{x}) = \sin(\ln r) + \frac{\mathbf{u} \cdot \mathbf{x}}{r} \tag{41}$$

as in [5] for non-trivial $\mathbf{u} \in \mathbb{R}^3$, one obtains a relativistic initial data set $(M_{T_1}, g^{T_1}, K^{T_1})$ which satisfies neither the weak nor the strong Regge–Teitelboim conditions with respect to $\mathbf{x}$; in fact, $\mathbf{C}_{\text{BÓRT}}$ diverges like $\sin(\ln r)\mathbf{u}$ in this example, see [4,5]. It is worth noting that the metric g^{T_1} in fact *does* satisfy the weak (but not the strong) Regge–Teitelboim conditions with respect to $\mathbf{x}$ (see [5]); they fail to hold only for K^{T_1}.

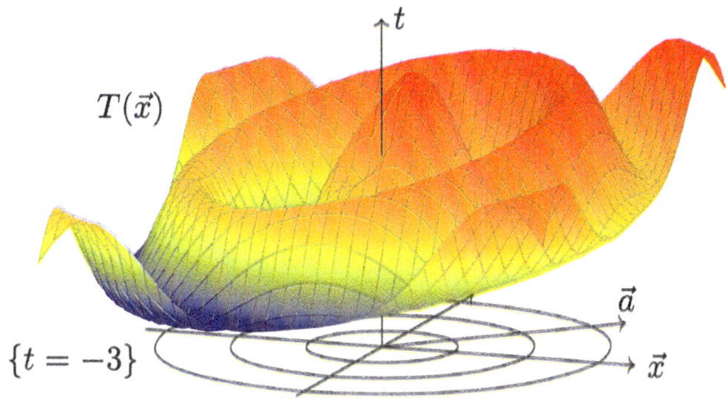

Fig. 2 Graphical example in Schwarzschild spacetime for $\{t = T_1\}$, logarithmic plot

Suitably exploiting Theorems 1 and 2 and the decay of T_1, N, and h, one can assert that $(M_{T_1}, g^{T_1}, K^{T_1})$ does not satisfy the weak nor the strong Regge–Teitelboim conditions in *any* asymptotically Euclidean coordinate chart.

Similarly, choosing

$$T_2(\mathbf{x}) = \frac{\sin(\ln r)}{r^\beta} \tag{42}$$

for $0 < \beta < \frac{1}{2}$, one finds that $(M_{T_2}, g^{T_2}, K^{T_2})$ *does* satisfy the weak Regge–Teitelboim conditions (for $\varepsilon < \frac{1}{2}$) but does not possess *any* asymptotically Euclidean coordinate chart in which the strong Regge–Teitelboim conditions hold. Again, the problematic (non-)decay occurs in K^{T_2}.

6.2 Why the Weak Regge–Teitelboim Conditions are Relevant for the Center of Mass

Finally, we still owe the reader a justification of why the weak Regge–Teitelboim conditions are relevant for the study of the center of mass $\mathbf{C}_{\text{BÓRT}}$: Indeed, Huisken and Yau in [9] developed an alternative definition of center of mass, called $\mathbf{C}_{\text{CMC}}$, via asymptotic Constant Mean Curvature (CMC) foliations. In a series of works culminating in a paper by Nerz [11], it was shown that, for asymptotically Euclidean relativistic initial data sets satisfying the weak Regge–Teitelboim conditions, one has

$$\mathbf{C}_{\text{BÓRT}} = \mathbf{C}_{\text{CMC}} \tag{43}$$

in the sense that either both centers diverge or both converge to the same limit.

Roughly, Huisken and Yau in [9] and Nerz in [11] prove existence and uniqueness of a foliation (that is, a smoothly parametrized partition into smooth 2-spheres parametrized by $\sigma \in (0, \sigma_0)$) of the asymptotic end of an asymptotically Euclidean relativistic initial data set, such that the leaves have constant mean (i.e., average extrinsic) curvature $H(\Sigma_\sigma) = \sigma$. The leaves (i.e., the 2-spheres) Σ_σ of this foliation are indicated as colored curves in Fig. 3. Pushing forward the leaves via the

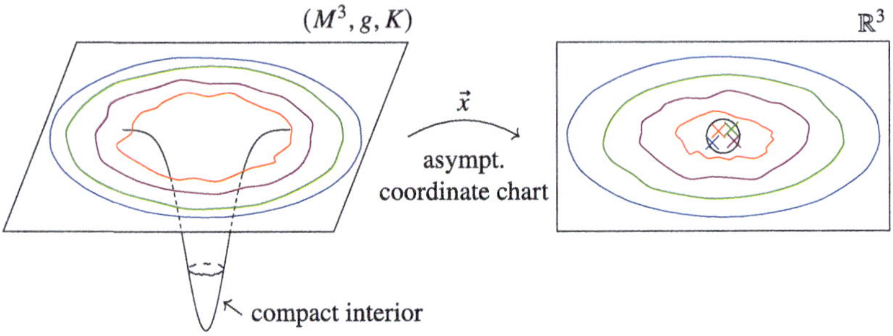

Fig. 3 The leaves of a foliation near infinity and their images in $\mathbb{R}^3$ under the asymptotic coordinates $\mathbf{x}$. The crossed positions indicate the Euclidean coordinate centers $\mathbf{c}(\Sigma_\sigma)$ of the surfaces Σ_σ of the same color (representing the same parameter σ)

asymptotic coordinates $\mathbf{x}$, $\mathbf{x}(\Sigma_\sigma)$, gives rise to a foliation of a neighborhood of infinity in $\mathbb{R}^3$ and one computes the average position of a point on $\mathbf{x}(\Sigma_\sigma)$ in $\mathbb{R}^3$ as

$$\mathbf{c}(\Sigma_\sigma) = \frac{1}{|\mathbf{x}(\Sigma_\sigma)|} \iint_{\mathbf{x}(\Sigma_\sigma)} \mathbf{x}\, dA_\delta, \tag{44}$$

where $|\mathbf{x}(\Sigma_\sigma)|$ denotes the surface area of $\mathbf{x}(\Sigma_\sigma)$ in $\mathbb{R}^3$ with respect to the Euclidean metric δ, see Fig. 3. The center $\mathbf{C}_{\text{CMC}}$ then arises as the limit

$$\mathbf{C}_{\text{CMC}} = \lim_{\sigma \to 0} \mathbf{c}(\Sigma_\sigma) \tag{45}$$

outward along this foliation, provided this limit exists. We refer the interested reader to [5] for more information on this construction and its dependence on the choice of asymptotic coordinates.

6.2.1 Spacetime Equivariance

It was observed by Cederbaum and Sakovich in [6] that the divergence issue of both notions of center of mass for $(M_{T_1}, g^{T_1}, K^{T_1})$—i.e., the one defined via a Hamiltonian systems approached by Beig and Ó Murchadha in [3] and the one defined via foliations—is rooted in the lack of dependence on K in both approaches. Generalizing the Constant Mean Curvature foliation approach, they construct asymptotic "Spacetime Constant Mean Curvature (STCMC)" foliations in asymptotically Euclidean relativistic initial data sets, see below. These allow for the definition of a generally covariant center of mass $\mathbf{C}_{\text{STCMC}}$ as well as a correction term $\mathbf{Z}$ for $\mathbf{C}_{\text{BÓRT}}$ such that

$$\mathbf{C}_{\text{STCMC}} = \mathbf{C}_{\text{BÓRT}} + \mathbf{Z} \tag{46}$$

holds under the weak Regge–Teitelboim conditions in the sense that either both sides of the equation diverge or both converge to the same limit. The definition of $\mathbf{C}_{\text{STCMC}}$ mimicks the definition of $\mathbf{C}_{\text{CMC}}$, see Fig. 3, based on the Spacetime Constant Mean Curvature instead of on the (spatial) Constant Mean Curvature foliation. As expected from the spacetime symmetry, one finds $\mathbf{C}_{\text{STCMC}} = \mathbf{0}$ in the graphical example for the graph function T_1 in (41).

It is also proved in [6] that $\mathbf{C}_{\text{STCMC}}$ evolves in time such that

$$\frac{d}{dt}\mathbf{C}_{\text{STCMC}} = \frac{\mathbf{P}_{\text{ADM}}}{E_{\text{ADM}}} \tag{47}$$

under the Einstein Evolution Equations, just as a freely falling point particle in Special Relativity. This applies even when $\mathbf{C}_{\text{STCMC}}$ does not converge.

Before we end this section, let us briefly address what it means that a surface Σ has "constant spacetime mean curvature": If one considers a 2-surface Σ not only as sitting inside the relativistic initial data set (M, g, K) but also as sitting inside the spacetime generated from this relativistic initial data set via the Einstein Evolution Equations then it can be viewed as a co-dimension 2 surface in this spacetime. As such, it has co-dimension 2 extrinsic curvature, taking the form of a normal vector valued symmetric $(0, 2)$-tensor field. The trace (or average) of this normal vector valued symmetric $(0, 2)$-tensor field is called the *spacetime mean curvature vector (field)* $\vec{\mathcal{H}}$ of Σ. The Lorentzian length of $\vec{\mathcal{H}}$, $\mathcal{H}$, is called the *spacetime mean curvature of* Σ. Then, a *spacetime constant mean curvature surface* is a surface with $\mathcal{H} = \text{const}$. It turns out that one can compute $\mathcal{H}$ from the initial data alone, without any reference to the ambient spacetime, and one finds

$$\mathcal{H} = \sqrt{H^2 - (\text{tr}_\Sigma K)^2}, \tag{48}$$

where $\text{tr}_\Sigma K$ is the (partial) trace of K over Σ and H denotes the (spatial) mean curvature of Σ within the initial data set already considered in the Constant Mean Curvature foliation suggested by Huisken and Yau. On the other hand, taking a more physical perspective, one finds that

$$\mathcal{H}^2 = \theta_+\theta_-, \tag{49}$$

where $\theta_\pm$ denote the null expansions of Σ in the ambient spacetime.

7 Lessons Learned and Current Research

We have seen that coordinates are messy in the following sense: In Newtonian Gravity, when ahistorically considering general asymptotically Euclidean coordinate charts on $\mathbb{R}^3 \setminus C$ outside some compact set C, convergence of the center of mass depends not only on suitable decay of the matter density but also on the choice of coordinate system. Accordingly, in General Relativity, where we generally do not have preferred systems of coordinates, one cannot hope to have convergence of any notion of center of mass in *all* asymptotically Euclidean coordinate charts. It was suggested to remedy such divergence issues by resorting to asymptotic parity conditions, however, as we showed, not all asymptotically Euclidean relativistic initial data sets have asymptotic parity.

It hence remains an open question whether, instead of asking for asymptotic parity, one can find a condition on asymptotic coordinate charts which is geometric (i.e., coordinate independent) just as Bartnik's harmonic coordinates, compatible with translations (and reflections), and implies convergence of $\mathbf{C}_{\mathrm{STCMC}}$, and of course such that every asymptotically Euclidean relativistic initial data set carries such a coordinate system. Coordinate charts satisfying such a condition could then legitimately be considered a natural analog of Cartesian coordinates in General Relativity. This question is currently studied by the authors and our coauthor Jan Metzger.

Acknowledgements This work was supported by the focus program on Geometry at Infinity (Deutsche Forschungsgemeinschaft, SPP 2026). MG also acknowledges support by the Deutsche Forschungsgemeinschaft (DFG, German Research Foundation) under Germany's Excellence Strategy–EXC 2121 *Quantum Universe*–390833306. The authors would like to thank Axel Fehrenbach, Felix Salfelder, Oliver Schoen, Olivia Vičánek Martínez, and Giorgos Vretinaris for help with the graphics.

References

1. Arnowitt, The dynamics of general relativity, Gravitation: an introduction to current research. 227–265 (1962)
2. R. Bartnik, Commun. Pure Appl. Math. **39**, 661–693 (1986)
3. R. Beig, N.Ó. Murchadha, Ann. Phys. **174**, 463–498 (1987)
4. C. Cederbaum, M. Graf, J. Metzger, Initial data sets that do not satisfy the Regge–Teitelboim conditions (2023). Work in Progress
5. C. Cederbaum, C. Nerz, Ann. Henri Poincaré. **16**, 1609–1631 (2015)
6. C. Cederbaum, A. Sakovich, Calc. Var. Part. Differ. Equ. **60**, Paper No. 214, 57 (2021)
7. P. Chruściel, Commun. Math. Phys. **120**, 233–248 (1988)
8. V. Denissov, V. Solovyev, Theor. Math. Phys. **56**, 832–841 (1983)
9. G. Huisken, S. Yau, Inven. Math. **124**, 281–311 (1996)
10. J. Lee, T. Parker, Bull. Am. Math. Soc. (N.S.) **17**, 37–91 (1987)
11. C. Nerz, Calc. Var. Part. Differ. Equ. **54**, 1911–1946 (2015)
12. T. Regge, C. Teitelboim, Ann. Phys. **88**, 286–318 (1974)
13. R. Schoen, S. Yau, Commun. Math. Phys. **65**, 45–76 (1979)
14. E. Witten, Commun. Math. Phys. **80**, 381–402 (1981)

Primordial Black Holes and Higgs Vacuum Decay

Ruth Gregory

Abstract

Phase transitions are part of everyday life, yet are also believed to be part of the history of our universe, where the nature of particle interactions change as the universe settles into its vacuum state. The discovery of the Higgs [1,20], and measurement of its mass suggests that our vacuum may not be entirely stable, and that a further phase transition could take place. This article is based on a talk in the Oldenberg Series, and reviews how we find the probability of these phase transitions, discussing work I have performed on how black holes can dramatically change the result! Apart from a brief update at the end, this article mostly follows the content of the talk.

1 Executive Summary

This work was motivated by the observation that phase transitions typically are seeded by impurities, yet the techniques used by theorists [18,23,24,52] to compute the probability of decay are extremely idealised–with huge mathematical simplifications assumed in order to make computations tractable. In the case of the Higgs vacuum, one intriguing possibility is that the self coupling of the Higgs could become negative at large Higgs values [7,9,28,31], leading to the conclusion that we live in a metastable vacuum, so the question of just how accurate these idealised computations are becomes of very direct relevance! In the work that I review [15–17,40], we take the simplest possible impurity for seeding vacuum decay: a black hole. This breaks the symmetry of the standard theoretical description, yet maintains sufficient theoretical control that the computations can be done largely analytically. The punchline of this article is that black holes change the computation....*enormously*!

R. Gregory (✉)
Department of Physics, King's College London, The Strand, London WC2R 2LS, UK
e-mail: ruth.gregory@kcl.ac.uk

B. Hartmann and J. Kunz (eds.), *Gravity, Cosmology, and Astrophysics*,
Lecture Notes in Physics 1022, https://doi.org/10.1007/978-3-031-42096-2_12

2 The Coleman Computation

In this introductory section, I motivate and review the Euclidean method used to compute tunnelling amplitudes in field theory. This discussion is based largely on the series of papers by Coleman [18,23,24].

2.1 Motivation of the Euclidean Method

The phenomenon of tunnelling is a uniquely quantum mechanical one—classically, if a particle does not have enough energy to scale a barrier, it will remain forever on one side; quantum mechanically however there is always a small probability of it being found across this seemingly unscalable hurdle. Calculating the probability of this process is one of the first computations we meet in Introductory Quantum Mechanics—we study the time independent Schrödinger equation in one spatial dimension

$$E\Psi = -\frac{\hbar^2}{2m}\frac{d^2\Psi}{dx^2} + V(x)\Psi \tag{1}$$

for a potential that has a simple, square, barrier:

$$V(x) = \begin{cases} 0 & x < 0 \ \ \& \ \ x > d \\ V_0 & 0 < x < d \end{cases} \tag{2}$$

where $V_0 > E$. The solution for Ψ is oscillatory outside the barrier, and exponential underneath it (see Fig. 1).

$$\Psi(x) = \begin{cases} Ie^{i\omega x} + Re^{-i\omega x} & x < 0 \\ Ae^{\Omega x} + Be^{-\Omega x} & 0 < x < d \\ Te^{i\omega x} & x > d \end{cases} \tag{3}$$

where $\omega^2 = 2mE/\hbar^2$, and $\Omega^2 = 2m(V_0 - E)/\hbar^2$. Continuity of Ψ and it's derivative at each side of the barrier gives 4 boundary conditions, allowing the wave function to be solved completely. The probability that a particle will tunnel through the barrier is therefore

$$\frac{|T|^2}{|I|^2} = \left[1 + \frac{V_0^2 \sinh^2 \Omega d}{4E(V_0 - E)}\right]^{-1} \sim e^{-2\Omega d} \tag{4}$$

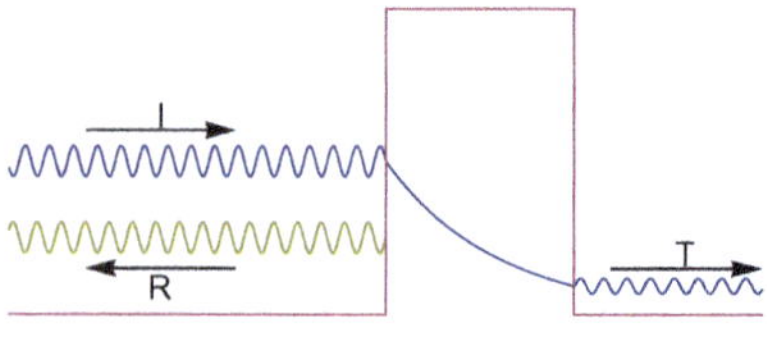

Fig. 1 A sketch of the 1+1-dimensional Schrödinger tunnelling calculation

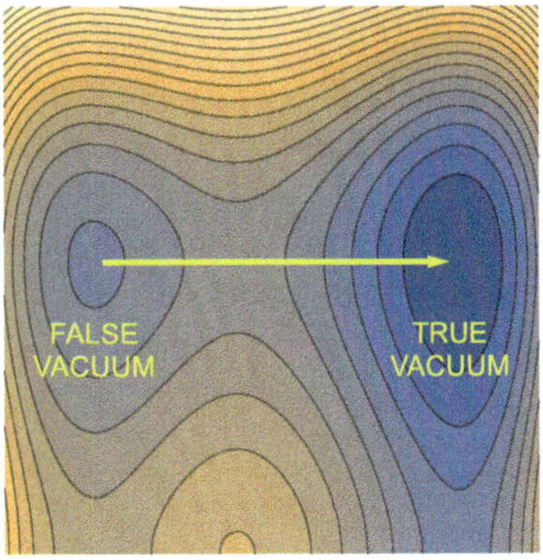

Fig. 2 The "escape" path from the local false vacuum is a saddle point in the inverted potential space

which is strongly dominated by an exponential factor representing the strength of the barrier the particle has to tunnel through:

$$\Omega d = \frac{1}{\hbar}\int_0^d \sqrt{2m(V_0 - E)}dx \tag{5}$$

Computing this leading order exponential suppression is the aim of the Euclidean method.

Now consider the following problem. A classical particle is at the tip of a square well of depth ΔV, it falls in, transferring the potential to kinetic energy, transits the well and goes up the other side. Using $\frac{1}{2}m\dot{x}^2 = \Delta V$, we have

$$\int \sqrt{2m\Delta V}dx = \int \sqrt{2m\Delta V}\dot{x}d\tau = \int 2\Delta V d\tau = \int \left(\Delta V + \frac{1}{2}m\dot{x}^2\right)d\tau = S_E \tag{6}$$

but this is just the Euclidean action for the motion of said point particle! For more general potentials, this gives an intuitive visualisation of the tunnelling amplitude calculation. We invert the potential and consider the classical motion of a "particle" from the (now unstable) local maximum to an exit point, and back again. This motion was called the *bounce* [23], and the one-dimensional tunnelling can readily be extended to multiple dimensions, with the bounce becoming a saddle trajectory between the local maximum (of the inverted Euclidean potential) to an exit point. This *most probable escape path* picture, developed by Banks, Bender, and Wu [3] (see Fig. 2) led naturally to the Euclidean field theory approach of Coleman and others [23,52].

2.2 Tunnelling in Field Theory

In Quantum Field Theory (QFT) fields typically have an interaction potential, with the minimum of the potential representing the vacuum of the theory. It might be, however, that this minimum is not unique, either due to nonlinearly realised symmetry, or

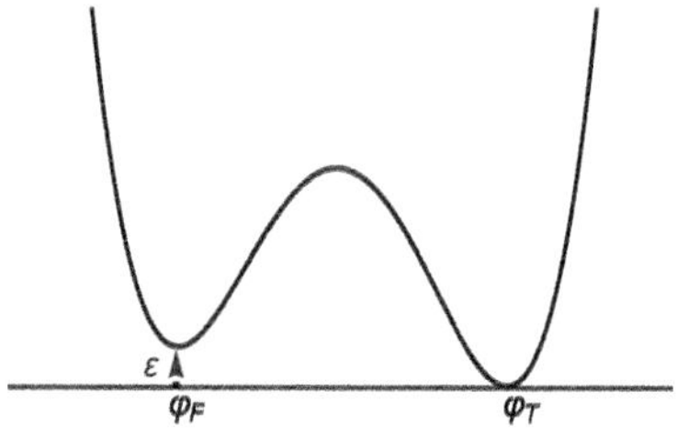

Fig. 3 A sketch of the type of potential relevant for the instanton computation

because there is more than one minimum in the potential separated in field space. Since the vacua are at distinct values of the field, the process of tunnelling from one minimum to another with lower overall energy will give rise to a *first order* phase transition, which we expect to proceed via bubble nucleation.

To find the probability of nucleation, we can follow the Euclidean prescription, taking the equations of motion and analytically continuing to Euclidean time, $t \to i\tau$, finding a solution known as an *instanton* which describes the formation of a bubble of true vacuum inside the false. For simplicity, I will discuss this process for a simple scalar field

$$\mathcal{L}_\varphi = \frac{1}{2}(\partial\varphi)^2 - V(\varphi) \tag{7}$$

where φ is a real scalar field, and $V(\varphi)$ is a potential with a false (local minimum) and true (global minimum) vacuum (see Fig. 3), where the difference in energy (ε) is assumed small relative to the potential barrier to be traversed. As is conventional, I have set $c = \hbar = 1$ in the ensuing discussion.

Before outlining the full procedure, it is instructive to consider the physically motivated simplification of the problem: the *Goldilocks Bubble*. Consider a bubble of radius R that fluctuates into existence. The bubble will cost energy to form, as between the vacua the field has to transit from φ_F to φ_T, passing through a region with high potential and gradient energy; this gives rise to a *wall* which will carry energy momentum, hence "cost" to form. On the other hand, the interior volume has dropped from energy ε to zero energy. If the bubble is small, the surface area to volume ratio means the gain from energy will be too small to sustain the bubble and it will recollapse. If the bubble is large, then the energy budget will allow the bubble to grow, but it will be "expensive" to form in the first place—there is therefore a "just right" bubble size where the bubble is initially at rest at formation. Putting the details in, and writing σ for the energy per unit area of the bubble wall we see that the energy budget is

$$\delta E = 2\pi^2 R^3\sigma - \frac{\pi^2}{2}\varepsilon R^4 \tag{8}$$

recalling that we are in 4 Euclidean dimensions so that the bubble is a 3-sphere. δE is stationary at $R = 3\sigma/\varepsilon$, for which $\delta E = 27\pi^2\sigma^4/2\varepsilon^3$, hence we expect that the bubble will nucleate at this radius with a probability

$$\mathcal{P} \sim e^{-27\pi^2\sigma^4/2\varepsilon^3\hbar} \tag{9}$$

It turns out that this *thin wall* intuition of Coleman is extremely efficient at extracting the essence of the full computation.

To outline the full computation, one must solve the Euclidean field equations

$$\frac{1}{\rho^3}\frac{d}{d\rho}\left[\rho^3\frac{d\varphi}{d\rho}\right]+\frac{\partial V}{\partial\varphi}=0 \tag{10}$$

where $\rho^2=\tau^2+\mathbf{x}^2$ is the distance from the origin in Euclidean $\mathbb{R}^4$. Taking the potential to have the approximate form

$$V(\varphi)\approx\frac{\lambda}{4}(\varphi^2-\eta^2)^2-\frac{\varepsilon}{2\eta}(\varphi-\eta) \tag{11}$$

where $\varepsilon \ll \lambda\eta^4$, then V has two local minima, approximately at $\pm\eta$, with the vacua energies separated by ε. To leading order, the equation of motion for φ is

$$\varphi''+\frac{3\varphi'}{\rho}+\lambda\varphi(\varphi^2-\eta^2)=0 \tag{12}$$

which can be solved numerically with the boundary conditions $\varphi\to-\eta$ as $\rho\to\infty$, $\varphi(0)\approx\eta$. Performing this integration, the solution is well approximated by

$$\varphi\simeq\eta\tanh\left[\sqrt{\lambda/2}\,\eta(R-\rho)\right] \tag{13}$$

where $R\approx 3\sigma/\varepsilon$, and the energy per unit area of the wall is given by

$$\begin{aligned}\sigma&=\frac{1}{R^3}\int_0^\infty\rho^3\left[\frac{1}{2}\varphi'^2+V(\varphi)\right]d\rho=\frac{\lambda\eta^4}{2R^3}\int_0^\infty\rho^3\mathrm{sech}^4\left[\sqrt{\frac{\lambda}{2}}\,\eta(R-\rho)\right]d\rho\\&\approx\sqrt{\frac{\lambda}{2}}\frac{4\eta^3}{3}\end{aligned} \tag{14}$$

This analytic approximation is excellent for $\sqrt{\lambda/2}\,\eta R\gg 1$, and even for potentials that have far thicker bubble walls it gives a very good ballpark estimate for the tunnelling probability.

2.3 Tunnelling with Gravity

In the previous discussion, a key feature was that the vacuum energy was different for the true and false vacua, but we know that energy gravitates, hence the false vacuum will have a different gravitational behaviour than the true vacuum, and the bubble wall will also have a gravitational signature. The impact of gravity was first worked out in the paper of Coleman and de Luccia [24] using the thin wall approximation discussed above. This is based on the seminal work of Israel [51], describing the gravitational effects of thin shells (in this case, the bubble wall).

First, it is worth making a few remarks about the Euclidean approach with gravity. Below the Planck scale, we expect that spacetime is essentially classical, but that gravity can contribute to quantum effects through the impact of spacetime curvature, and the back-reaction of quantum fields on the spacetime. Usually, we take some fixed classical background and quantise around this, an approach used in black hole thermodynamics and cosmological perturbation theory. There is a broader sense in which we can use gravity semi-quantum mechanically however, and that is by using the partition function approach of Gibbons and Hawking [36]. While the philosophy of this requires some finessing, the basic methodology is clear: we extend the partition function to include the Einstein-Hilbert action,

$$S_{GH} = -\frac{1}{16\pi G}\int_{\mathcal{M}} d^4x\sqrt{g}(R-2\Lambda) + \int d^4x\mathcal{L}_{\text{matter}}\left(+\frac{1}{8\pi G}\int_{\partial\mathcal{M}} d^3x\sqrt{h}K\right) \tag{15}$$

where the Gibbons-Hawking boundary term has been included (in brackets) for completeness. At finite temperature we have a finite periodicity of Euclidean time, and typically we integrate over geometries with the same boundary conditions at $\partial\mathcal{M}$. Whereas there is no clear method for dealing with fluctuations, the saddle points of the path integral are unambiguous—these are solutions of the classical Euclidean field equations.

First I will give a qualitative picture of the instanton before describing the CDL calculation in more detail. For pictorial simplicity, consider the case of tunnelling from a finite vacuum energy ε to zero vacuum energy. A finite positive vacuum energy is a positive cosmological constant, which we know to be de Sitter spacetime. This has a Lorentzian description in terms of the surface of a hyperboloid embedded in a five-dimensional spacetime:

$$w^2 + x^2 + y^2 + z^2 - t^2 = \ell^2 = \frac{3}{8\pi G\varepsilon} \tag{16}$$

here, ℓ is the scale of curvature of the de Sitter spacetime, and is related as shown to the false vacuum energy. On rotation to Euclidean time, this becomes a 4-sphere embedded in $\mathbb{R}^5$ (see Fig. 4). Zero vacuum energy on the other hand corresponds to Minkowski spacetime, and is just flat, or planar. Our instanton must therefore cut a cap off the de Sitter 4-sphere and replace it with a flat surface as illustrated in Fig. 4.

We can play the same "Goldilocks bubble" game as before, but noting from (15) that there is potentially a gravitational contribution to energy. First, the cost of the wall, coming from the matter contribution in (15), is the same: $\mathcal{E}_{\text{wall}} = 2\pi^2R^3\sigma$. Next, the naive energy gain from false to true vacuum depends on the volume of the cap excised from the sphere, and this is captured by the bulk contribution to (15):

$$\mathcal{E}_{\text{cap}}(R) = \frac{2}{3}\pi^2\varepsilon\ell^4\left[2 - 2\left(1-\frac{R^2}{\ell^2}\right)^{\frac{3}{2}} + 3\frac{R^2}{\ell^2}\right] \tag{17}$$

Using just these two terms gives an approximate answer that includes the curved geometry of de Sitter space, but it neglects the impact of the (negative) gravitational

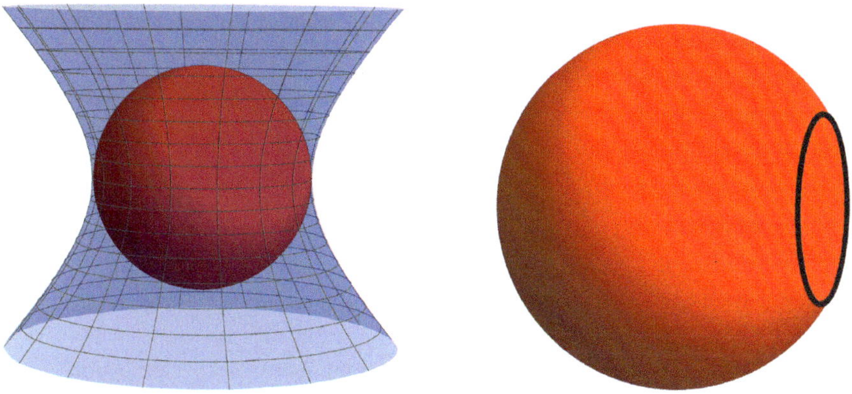

Fig. 4 Left: Lorentzian and Euclidean de Sitter space. Right: the instanton replaces the cap of the sphere with a flat surface. The kink at the interface represents the energy momentum of the bubble wall

potential energy due to the gravitational field of the wall. To capture this, we need to look more closely at the extrinsic curvature of the wall, as this effect is understood as the contribution from the Gibbons Hawking boundary term in the action.

In the Coleman de Luccia (CDL) approach [24], they describe the instanton in the thin wall limit, using the Israel conditions [51]. The wall separates the false vacuum, de Sitter space:

$$ds^2 = d\rho^2 + \ell^2 \sin^2(\rho/\ell) d\Omega_{III}^2 \tag{18}$$

from the true, Minkowski, vacuum

$$ds^2 = dr^2 + r^2 d\Omega_{III}^2 \tag{19}$$

The Israel equations relate the jump in the extrinsic curvature across the wall to the energy in the wall. The wall sits at $r_0 = \ell \sin(\rho_0/\ell) = R$, with normal $d\rho\ /\ dr$. The wall is characterised by an energy-momentum tensor that is proportional to the induced metric on the wall,

$$h_{ab} = g_{ab} - n_a n_b \tag{20}$$

and the wall extrinsic curvature is given by:

$$K_{ab} = -\Gamma^c_{ab} n_c = \begin{cases} \frac{\cos(\rho_0/\ell)}{\ell \sin(\rho_0/\ell)} & \text{FV} \\ \frac{1}{r_0} & \text{TV} \end{cases} \tag{21}$$

Substituting into the Israel equations relevant for the Euclidean wall gives

$$K^+_{\alpha\beta} - K^-_{\alpha\beta} = -\frac{1}{R}\left(1 - \sqrt{1 - \frac{R^2}{\ell^2}}\right) h_{\alpha\beta} = -4\pi G \sigma h_{\alpha\beta} \tag{22}$$

Writing $\bar{\sigma} = 2\pi G\sigma$ for compactness of notation, this is easily seen to be solved by

$$R_0 = \frac{4\bar{\sigma}\ell^2}{1 + 4\bar{\sigma}^2\ell^2} \tag{23}$$

We can then substitute this solution into the full Euclidean action

$$\begin{aligned} \mathcal{B} &= \frac{1}{16\pi G}\int_{\text{cap}}(R - 2\Lambda)\sqrt{g}d^4x + \int_{\text{wall}}\left[\frac{(K_+ - K_-)}{8\pi G} + \sigma\right]\sqrt{h}d^3x \\ &= \mathcal{E}_{\text{cap}}(R) + \mathcal{E}_{\text{wall}}(R) + \mathcal{E}_{\text{grav}}(R) \end{aligned} \tag{24}$$

where we have written the integrals for general R to make connection to the Goldilocks argument, and identify the gravitational potential energy of the wall curvature for general R as:

$$\mathcal{E}_{\text{grav}}(R) = \int_{\text{wall}}\frac{(K_+ - K_-)}{8\pi G}\sqrt{h}d^3x = -\frac{2\pi R^2}{4G}\left[1 - \left(1 - \frac{R^2}{\ell^2}\right)^{\frac{1}{2}}\right] \tag{25}$$

Making this action stationary with respect to R recovers the Coleman de Luccia R_0 as expected, and gives the tunnelling exponent

$$\mathcal{B}_{CDL}(R_0) = \frac{\pi\ell^2}{G}\frac{16\bar{\sigma}^4\ell^4}{(1 + 4\bar{\sigma}^2\ell^2)^2} \tag{26}$$

Why is this computation important for the Higgs vacuum? In particle physics, we describe fundamental interactions via the Standard Model (SM) Lagrangian, which encodes the bosons, fermions, and their interactions.

$$\begin{aligned} \mathcal{L}_{SM} = &-\frac{1}{2}\text{Tr}\,\mathbf{G}_{\mu\nu}\mathbf{G}^{\mu\nu} - \frac{1}{2}\text{Tr}\,\mathbf{W}_{\mu\nu}\mathbf{W}^{\mu\nu} - \frac{1}{4}F_{\mu\nu}F^{\mu\nu} \\ &+ \left(\mathbf{D}_\mu\Phi\right)^\dagger\mathbf{D}^\mu\Phi + \mu^2\Phi^\dagger\Phi - \frac{\lambda}{2}\left(\Phi^\dagger\Phi\right)^2 + \ldots.. \end{aligned} \tag{27}$$

A key feature of this is the Higgs scalar field and its self coupling, or the Higgs potential. The self coupling of the Higgs, λ acquires radiative corrections, and changes with energy scale $V(\Phi) \sim \lambda(\Phi)|\Phi|^4/4$. As was realised some years ago [54,56,62,68], this could have implications for the stability of the Higgs vacuum. The calculation depends on the masses of other fundamental particles, particularly the top quark (see [2,66] for more recent results) and the values of the mass of the Higgs and the top quark put us in a region where the self-coupling could potentially become negative at very high energies [7,9,28,30,31,38,50].

Whether or not we should be concerned at the metastability of the Higgs vacuum then becomes an issue of computing the probability of decay. If we use this Coleman de Luccia result, the half life of decay is around 10^{138} years, well in excess of the age of the universe! We might therefore think this metastability is not an issue, but this would be to ignore the fact that the calculation I have just described is incredibly idealised—very much a "spherical cow"! It is therefore time to revisit our understanding of vacuum decay, and to explore whether we can take a step towards a more realistic set-up, including the effect of an impurity.

3 Thin Wall Bubbles with Black Holes

In this section, I review how including a black hole as a seed for false vacuum decay changes the above picture, and the key results from [15–17,40] (see also [21,37,59,67]). I will stick with the thin wall approximation, as this allows a largely analytic approach, and review the numerical work in the next section.

3.1 Bubbles with Black Holes

The Coleman de Luccia bubbles discussed in the last section are incredibly ideal—the universe is completely smooth and featureless, akin to a supercooled fluid. Since we are discussing the gravitational effect of the vacuum, we should also think about what happens if the universe is not featureless and isotropic. The simplest possible impurity we can think of is a black hole. This is an exact solution to Einstein's gravitational equations, and can be added to the picture without adding any further matter content. We also know how to treat a black hole in Euclidean space—indeed, this was one of the most dramatic conceptual discoveries of the Gibbons-Hawking paper [36]: that a Euclidean black hole naturally has a periodic time, and demanding non-singularity of the Euclidean geometry mandates a periodicity of Euclidean time $\Delta\tau = 8\pi GM$ (See Fig. 5). Given that the partition function for a field at temperature T is described by a period (Euclidean) time coordinate with periodicity $\beta = 1/T$, this naturally leads to the conclusion that a black hole has temperature $T = 1/8\pi GM$—consistent with Hawking's breakthrough result just two years earlier [44]. Of course, there are niggles with this interpretation, not least that a black hole is not in thermal equilibrium with its surroundings, but the correspondence between this Euclidean approach and the considerably more complex Lorentzian computation is compelling, and to this day we compute the temperature of a black hole by looking at its Euclidean continuation.

Now consider the geometrical set-up. The idea is that we have a black hole in the false vacuum that acts as a seed for vacuum decay. A bubble will therefore nucleate around the black hole, leaving the same exterior solution, but replacing the interior with the true vacuum and, potentially, a remnant black hole that will in general have a different mass to the original seed. We are therefore looking for solutions for a thin bubble wall that match two different vacuum energies (i.e., cosmological constants) and two different masses of black hole:

$$ds^2 = f_\pm(r)d\tau_\pm^2 + \frac{dr^2}{f_\pm(r)} + r^2 d\Omega_{II}^2 \tag{28}$$

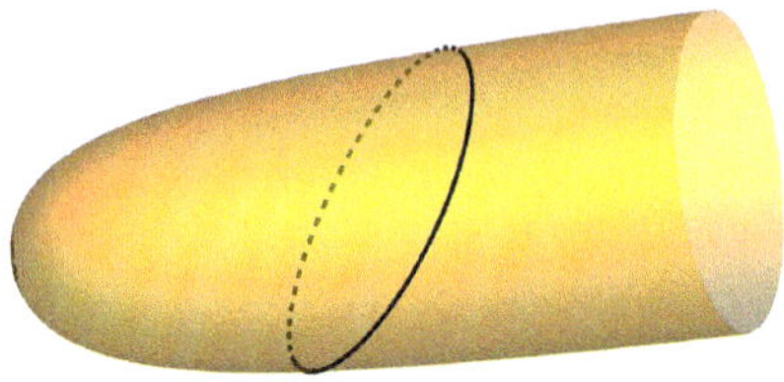

Fig. 5 A sketch of a wall wrapping the periodic Euclidean time direction. Note that this is a very special solution where the periodicity of the wall motion precisely matches that of the black hole

where

$$f_\pm(r) = 1 - \frac{2GM_\pm}{r} - \frac{\Lambda_\pm}{3}r^2 \tag{29}$$

is the metric potential for a black hole with a cosmological constant. One point of confusion can be around the meaning of M_+ in the instanton space, where the only *black hole* is the one at the centre of the solution, with horizon determined by M_-. However, recall that gravity in four dimensions propagates—we experience the impact of the moon's gravity here on Earth via the tides, and indeed the sun's gravity via the neap and spring differences in tide. In other words, we can deduce locally the masses of these celestial objects by looking at local tidal forces. It is exactly the same with the exterior M_+ mass–we detect the original seed mass because the tidal forces outside the bubble are still those of the original seed.

The next remark to make when looking for an exact solution in GR is whether we are imposing an Ansatz on the geometry, or whether our solution is general. Mathematically, introducing a black hole alters the symmetry of the CDL set-up, with $SO(4)$ symmetry, by breaking the equivalence between spatial and timelike directions, leaving an $SO(3) \times U(1)$ symmetry: the spherical symmetry of the black hole we see in space, and the $U(1)$ of the periodic Euclidean time direction. Fortunately, the problem of the exact solution was solved some time ago in the context of braneworld models, where a general Birkhoff theorem was proven for braneworlds (a more general case of the thin wall) [10], meaning that the solution described above in (28) is indeed a fully general solution to the equations of motion of a wall with the desired symmetry.

Returning to the Goldilocks bubble concept, many of the components are similar, yet there is a crucial difference. The general bubble will have a trajectory $R(\tau)$ around the Euclidean time circle which now must be determined from the Israel equations. There is also another key difference around background subtraction, described presently, meaning that there is no alternative but to roll up our sleeves and proceed with the calculation.

The following discussion paraphrases the analysis in [16], picking out the key calculational steps. The bubble is described by a trajectory in both the plus and minus Euclidean black hole spacetimes (28)

$$X^a = (\tau_\pm(\lambda), R(\lambda), \theta, \phi) \tag{30}$$

where we parametrise the trajectory by the local time, λ on the bubble wall, i.e.,

$$f_\pm \dot{\tau}_\pm^2 + \frac{\dot{R}^2}{f_\pm} = 1 \tag{31}$$

where dots denote derivatives with respect to λ. Taking normals that point towards increasing r,

$$n_\pm = \dot{\tau}_\pm \, dr_\pm - \dot{r} \, d\tau_\pm, \tag{32}$$

(noting $\dot{\tau}_\pm \geq 0$ for orientability), the Israel junction conditions are

$$f_+ \dot{\tau}_+ - f_- \dot{\tau}_- = -2\bar{\sigma} R. \tag{33}$$

We can use (31) to rewrite this as a Friedman-like equation for $\dot{R}$:

$$\dot{R}^2 = 1 - \left(\bar{\sigma}^2 + \frac{\bar{\Lambda}}{3} + \frac{(\Delta\Lambda)^2}{144\bar{\sigma}^2}\right) R^2 - \frac{2G}{R}\left(\bar{M} + \frac{\Delta M \Delta\Lambda}{24\bar{\sigma}^2}\right) - \frac{(G\Delta M)^2}{4R^4\bar{\sigma}^2}, \tag{34}$$

where $\Delta M = M_+ - M_-$ is the black hole mass difference, and $\bar{M} = (M_+ + M_-)/2$ the average, with similar expressions for Λ. There are also two accompanying equations for the time coordinates on each side of the wall,

$$f_\pm \dot{\tau}_\pm = \frac{\Delta\Lambda}{12\bar{\sigma}} R + \frac{\Delta M}{2\bar{\sigma} R^2} \mp \bar{\sigma} R \tag{35}$$

which can be integrated once $R(\lambda)$ is found.

This gives an effective potential for the motion of the brane that allows us to quickly characterise the motion, which is easily found numerically. The generic solution is periodic in λ, therefore we are now in the interesting position of having to fix the periodicity of Euclidean time not by the black hole mass, but by the wall trajectory.

3.2 Computing the Bounce Action

Having found this wall trajectory, we then need to compute the Euclidean action of the thin wall instanton:

$$\begin{aligned} I_E = &- \frac{1}{16\pi G}\int_{\mathcal{M}_+} \sqrt{g}(\mathcal{R}_+ - 2\Lambda_+) - \frac{1}{16\pi G}\int_{\mathcal{M}_-} \sqrt{g}(\mathcal{R}_- - 2\Lambda_-) \\ &+ \frac{1}{8\pi G}\int_{\partial\mathcal{M}_+} \sqrt{h} K_+ - \frac{1}{8\pi G}\int_{\partial\mathcal{M}_-} \sqrt{h} K_- + \int_{\mathcal{W}} \sigma\sqrt{h} \end{aligned} \tag{36}$$

and subtract the action of the background

$$I_{\rm FV} = -\frac{1}{16\pi G}\int_{\mathcal{M}} \sqrt{g}(\mathcal{R}_+ - 2\Lambda_+) \tag{37}$$

to find the bounce action for the decay amplitude.

In these expressions, $\mathcal{M}$ is the geometry of the background false vacuum seed space, $\mathcal{W}$ is the location of the wall in the instanton geometry, which is a submanifold of $\mathcal{M}$. $\mathcal{M}_-$ is the true vacuum part of the instanton geometry inside the bubble wall, and $\mathcal{M}_+$ the part exterior to the bubble, with $\partial\mathcal{M}_\pm$ refers to the boundaries of each of $\mathcal{M}_\pm$ induced by the wall, each of which have a normal pointing outwards to larger r, in agreement with the Israel prescription. In general, as described in [16], there may also be additional boundary or bulk terms required for renormalisation of the action.

Having noted these expressions, a very important observation is that in order to perform the background subtraction of the false vacuum geometry, we must have the *same* geometries at large r (or outside the wall). This means that $\mathcal{M}_+$ must be *identical* to the portion of $\mathcal{M}$ that lies outside the wall. We can therefore write

$$\begin{aligned}\mathcal{B} =& \frac{1}{16\pi G}\int_{\mathcal{M},r<R} \sqrt{g}(\mathcal{R}_+ - 2\Lambda_+) - \frac{1}{16\pi G}\int_{\mathcal{M}_-} \sqrt{g}(\mathcal{R}_- - 2\Lambda_-)\\ &+ \frac{1}{8\pi G}\int_{\partial\mathcal{M}_+} \sqrt{h}K_+ - \frac{1}{8\pi G}\int_{\partial\mathcal{M}_-} \sqrt{h}K_- + \int_{\mathcal{W}} \sigma\sqrt{h}\end{aligned} \tag{38}$$

where hopefully the notation is self explanatory.

Now let us consider the implication of the periodicity of the general solution $R(\lambda)$, β, versus the natural periodicity of a black hole spacetime

$$\beta_0 = \frac{4\pi}{f'(r_h)} = \frac{4\pi r_h}{1 - \Lambda_\pm r_h^2} \tag{39}$$

having used $f(r_h) = 0$ at the horizon to simplify the expression. If $\beta \neq \beta_0$, then we will have a conical singularity at the centre of the Euclidean space $r = r_h$. In early work on tunnelling with black holes [4,49], the implication of this mild singularity was neglected, as it was believed to be unimportant, however it is reasonably straightforward to compute, as explained in Appendix A of [40]. In essence, one smooths out the conical singularity with a family of regular metrics with curvature confined to proper distance $\rho < \epsilon$ from the 'horizon', the degree of the singularity in the Ricci scalar is ρ^{-1}, whereas the volume element scales as ρ. The nett result is that the integral

$$\int_{\rho<\epsilon} d^4x \mathcal{R}\sqrt{g} = 2\mathcal{A}\delta \tag{40}$$

is finite, and proportional to the deficit angle δ, and the area of the horizon $\mathcal{A} = 4\pi r_h^2$. To compute the deficit angle, note that in the natural black hole geometry, the angular coordinate $\theta_0 = 2\pi\tau/\beta_0$ has a range $[0, 2\pi]$. With periodicity β however, this changes to $[0, 2\pi\beta/\beta_0]$, hence the deficit angle is $\delta = 2\pi - 2\pi\beta/\beta_0$. We therefore see that the contribution to the action coming purely from the conical deficit is

$$I_{\text{conical}} = -\frac{1}{16\pi G}\int_{\rho<\epsilon} d^4x\mathcal{R}\sqrt{g} = -\frac{(\beta_0-\beta)}{\beta_0}\frac{\mathcal{A}}{4G} \tag{41}$$

Returning to the general expression (38), the first line, which is the bulk contribution, can now be computed as

$$\begin{aligned}\mathcal{B}_{\text{bulk}} =& \frac{\Lambda_+}{6G}\int_0^\beta d\tau_+(R^3 - r_{h,+}^3) - \frac{\Lambda_-}{6G}\int_0^\beta d\tau_-(R^3 - r_{h,-}^3)\\ &+ \frac{(\beta_+-\beta)}{\beta_+}\frac{\mathcal{A}_+}{4G} - \frac{(\beta_--\beta)}{\beta_-}\frac{\mathcal{A}_-}{4G}\\ =& \frac{(\mathcal{A}_+-\mathcal{A}_-)}{4G} - \frac{1}{4G}\int d\lambda R^2\left(f_+'\dot{\tau}_+ - f_-'\dot{\tau}_-\right)\\ &- \frac{\beta}{4G}\left[\frac{\mathcal{A}_+}{\beta_+} - \frac{\mathcal{A}_-}{\beta_-} - 2G(M_+-M_-) + \frac{2\Lambda_+}{3}r_{h,+}^3 - \frac{2\Lambda_-}{3}r_{h,-}^3\right]\end{aligned} \tag{42}$$

However, noting that $\beta_\pm = 4\pi/f_\pm'(r_{h,\pm})$, and using the horizon relation $f_\pm(r_{h,\pm}) = 0$, the term in square brackets on the final line vanishes.

We are now left with the boundary and wall terms, which can be combined into an expression dependent on the wall trajectory:

$$\mathcal{B}_{\text{wall}} = \int_{\mathcal{W}}\sqrt{h}\left[\frac{(K_+-K_-)}{8\pi G} + \sigma\right] = -\frac{\sigma}{2}\int_{\mathcal{W}}\sqrt{h} = \frac{1}{2G}\int d\lambda\, R\left(f_+\dot{\tau}_+ - f_-\dot{\tau}_-\right) \tag{43}$$

having used $f_+\dot{\tau}_+ - f_-\dot{\tau}_- = -2\bar{\sigma}R$.

Putting all of these results together, the action of the bounce is:

$$\mathcal{B} = \frac{\mathcal{A}_+}{4G} - \frac{\mathcal{A}_-}{4G} + \frac{1}{4G}\oint d\lambda\left\{\left(2Rf_+ - R^2f_+'\right)\dot{\tau}_+ - \left(2Rf_- - R^2f_-'\right)\dot{\tau}_-\right\} \tag{44}$$

This computation was the core result of [15–17,40]. In general, for a given seed mass, (34) will have bubble solutions for a range of remnant masses M_-. However, evaluating (44) shows that these have different bounce actions, hence different decay rates. For a given seed mass there will be a unique remnant mass that has the lowest action. Since the probability of decay is the exponent of the negative of this action, the decay is dominated by this minimum action value. Figure 6 (Fig. 5 from [40]) shows how the bounce action depends on the seed mass for tunnelling from a positive to zero vacuum energy for a value of wall tension $\bar{\sigma}\ell = 0.2$. For each seed mass (expressed as a ratio of M_+ to the maximum allowed mass in de Sitter, the Nariai mass M_N) there is a range of allowed remnant masses. The minimal action configuration is shown as a dashed red line.

Explicit computation shows that for small seed masses, the minimal action bounce removes the seed black hole is altogether but it still τ−dependent. For larger seed masses, the minimal action bounce is static, leaving behind a remnant black hole, with increasing remnant mass as the seed mass is increased. These two branches meet at a critical mass with the lowest action bounce which is both static, and without a black hole remnant. We can read off from the parameter range discussion in the appendix of [16] the value of M_+ at which this occurs. For example, if $\Lambda_- = 0$, we get:

$$GM_+ = \frac{128\bar{\sigma}^3}{3(\Lambda_+ + 12\bar{\sigma}^2)^2} \tag{45}$$

or,

$$\frac{M_+}{M_p} = \frac{128\pi}{27} \frac{(\bar{\sigma}\ell/2)^3}{(1+4\bar{\sigma}^2\ell^2)^2} \frac{\ell}{L_p} \tag{46}$$

where M_p and L_p are the Planck mass and length scale respectively. In order to remain well below the Quantum Gravity scale, we require $M_+/M_p \gg 1$, and we have the bound $\bar{\sigma}\ell < 1/2$, hence we require large ℓ/L_p with $\bar{\sigma}\ell$ not too small. However, both $\bar{\sigma}$ and ℓ are determined by the properties of the scalar potential, and these conditions are not particularly realistic, and for the Higgs potential, not relevant at all. Hence, the most likely bounce is one which has seed mass much larger than this critical value, and is therefore static and with a remnant black hole.

The most important take home message of this section is that adding a black hole decreases the bounce action, hence *enhances* decay. The parameters, ($M_\pm$, $\Lambda_\pm$, $\bar{\sigma}$) used to produce Fig. 6 are all approximately of a similar order, close to unity, therefore the suppression of the bounce action is not as marked as it will be for more realistic parameter values. We will see this when exploring the Higgs vacuum in the

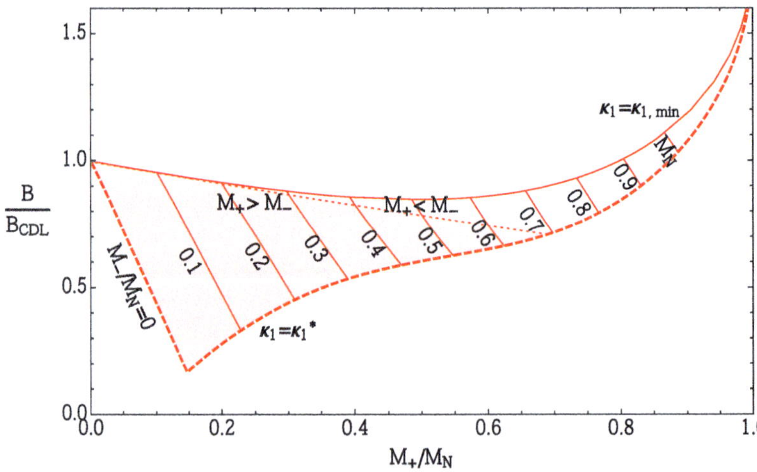

Fig. 6 A comparison of the CDL bounce action and the black hole seeded action for tunnelling from positive to zero vacuum energy. (Fig. 5 from [40].) The seed mass is expressed as a ratio of M_+ to the maximum allowed mass in de Sitter, the Nariai mass M_N, and the values of the remnant mass are labelled. The κ annotations refer to combinations of the seed and remnant parameters and are given in [40]

next section. However, before discussing the implications for the Higgs vacuum, I will present two analytic examples of bubbles, first re-deriving the CDL result from the perspective of the static patch, which relies crucially on the computation of the conical singularity integral, then the result for static instantons.

3.2.1 CDL Revisited

The CDL instanton has $M_+ = M_- = 0$, and let $\Lambda_- = 0$ so that we are tunnelling from a positive false vacuum energy $\varepsilon = \Lambda_+/8\pi G = 3/8\pi G\ell^2$ to the Minkowski true vacuum as before. With these parameter values, (34), simplifies considerably,

$$\dot{R}^2 = 1 - \left(\bar{\sigma} + \frac{1}{4\bar{\sigma}\ell^2}\right)^2 R^2 = 1 - \omega^2 R^2 \tag{47}$$

where ω is defined above, together with the equations for $\tau_\pm$

$$\left(1 - \frac{R^2}{\ell^2}\right)\dot{\tau}_+ = (\omega - 2\bar{\sigma})\, R \quad ; \qquad \dot{\tau}_- = \omega R \tag{48}$$

From which we read off the solution as

$$\begin{aligned} R &= \omega^{-1}\cos\omega\lambda \\ \tau_- &= \omega^{-1}\sin\omega\lambda \\ \tau_+ &= \ell \arctan\left[\frac{\sin\omega\lambda}{\ell(\omega - 2\bar{\sigma})}\right] \end{aligned} \tag{49}$$

Note that this solution has the size determined by ω^{-1}, which is identical to the CDL bubble side, (23), found by analytically continuing global de Sitter. In contrast however, the periodicity of this solution, $\lambda \in [-\frac{\pi}{2\omega}, \frac{\pi}{2\omega}]$, translates to $\Delta\tau_- = 2/\omega$ and $\Delta\tau_+ = 2\ell \operatorname{arccot}[\ell(\omega - 2\bar{\sigma})]$. However, the de Sitter false vacuum in the static patch has a natural periodicity $2\pi\ell$. Thus, the instanton computation will take into account the regularization of the conical deficit integral implicitly, and can therefore be regarded as a check of this procedure.

Substituting this solution into (44), noting that with $M_\pm = 0$, $f_+ = 1 - r^2/\ell^2$, $f_- = 1$, and the entropy terms vanish, gives

$$\mathcal{B} = \oint d\lambda \frac{R}{2G}(\dot{\tau}_+ - \dot{\tau}_-) = \frac{\pi}{4G}\left(\ell - \sqrt{\ell^2 - 1/\omega^2}\right)^2 = \frac{\pi\ell^2}{G}\frac{16\bar{\sigma}^4\ell^4}{(1 + 4\bar{\sigma}^2\ell^2)^2} \tag{50}$$

i.e., the CDL result (26). This result would not have been obtained without the correct evaluation of the Einstein-Hilbert integral over the conical singularity. Clearly, the result for a physical process should not depend on the choice of coordinates for the analytic continuation. Here, we have analytically continued the static time patch, whereas the canonical CDL computation analytically continues the global time coordinate. These are different analytic continuations, but with the right calculational method, we get the same result.

3.2.2 Static Bubbles

Another example where we can find the tunnelling amplitude analytically is when the instanton is independent of Euclidean time, $R(\lambda) = R_0$. In this case (31) implies that $f_\pm \dot{\tau}_\pm^2 = 1$. Then the individual components of the Israel equations

$$\frac{f_+\dot{\tau}_+ - f_-\dot{\tau}_-}{R} = \sigma = \frac{f_+' - 2\ddot{R}}{2f_+\dot{\tau}_+} - \frac{f_-' - 2\ddot{R}}{2f_-\dot{\tau}_-} \tag{51}$$

can be seen to imply that the wall integral in (44) vanishes, hence we are led to the remarkably simple result

$$\mathcal{B} = \frac{\mathcal{A}_+}{4G} - \frac{\mathcal{A}_-}{4G} \tag{52}$$

i.e., the probability of tunnelling from a seed to remnant black hole via a static instanton is simply the Boltzmann suppression due to the entropy difference:

$$\mathcal{P} \sim e^{-\delta S/\hbar} \tag{53}$$

This turns out to be the key result of this work—that black holes can give rise to static instantons where the tunnelling rate is given by the the Boltzmann factor associated to the drop in entropy of the configuration.

4 The Fate of the Higgs Vacuum

The previous discussion of bubbles used the thin wall approximation. This was a physically intuitive argument developed by Coleman to get analytic results quickly in an era of minimal computing power. In reality, the Higgs potential does not have a form in which thin walls are even a reasonable approximation to a bubble. In this section I review the thick wall bubbles relevant for Higgs vacuum decay. I then discuss the ultimate fate of the Higgs vacuum, first comparing the decay rate to other quantum processes with a black hole, then remarking on the mutual constraint between vacuum metastability and cosmological primordial black holes.

4.1 From Thin to Thick: Higgs Vacuum Bubbles

For shorthand, the Higgs potential is written as a function of the magnitude of the Higgs, $\phi = |\Phi|$, and the potential is expressed in terms of an effective quartic coupling constant $\lambda_{\rm eff}$,

$$V(\phi) = \frac{1}{4}\lambda_{\rm eff}(\phi)\phi^4 \tag{54}$$

Radiative corrections have been computed [6,22,28,32,35], and in [16,17], were approximated by the analytic three-parameter fit

$$\lambda_{\rm eff}(\phi) = \lambda_* + b\left(\ln\frac{\phi}{M_p}\right)^2 + c\left(\ln\frac{\phi}{M_p}\right)^4. \tag{55}$$

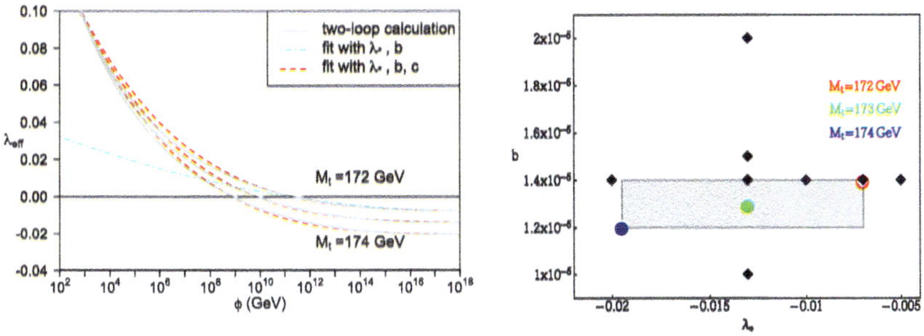

Fig. 7 Left: The three parameter fit of the high-energy effective coupling used for vacuum decay results in [17]. All three parameters can be fixed by matching to the SM calculation for a given Higgs and Top quark mass. The plots show Higgs mass $M_{\mathrm{H}} = 125\,\mathrm{GeV}$ and top quark masses 172 GeV ($\lambda_* = -0.007$), 173 GeV ($\lambda_* = -0.013$) and 174 GeV ($\lambda_* = -0.00195$). A two parameter fit is shown for comparison. Right: The values of λ_* and b panned over in the numerical integrations presented in the branching ratios in Sect. 4.2

which provides a good approximation to the exact results (which have to be computed individually for given Top and Higgs masses) over a wide range of energy, and allows efficient numerical evaluation of the instanton equations of motion. (See Fig. 7.)

This potential does not admit thin wall solutions, so to link Higgs bubbles to the thin wall bubbles of the previous section, we add a quantum gravity motivated correction to the scalar potential [5, 13, 29, 39, 55, 57]:

$$V(\phi) = \frac{1}{4}\lambda_{\mathrm{eff}}(\phi)\phi^4 + \frac{1}{6}\lambda_6\frac{\phi^6}{M_p^2} + \dots \tag{56}$$

The effect of λ_6 is to add a definite minimum to the potential at large values of ϕ with sufficient barrier to give an approximate thin wall. Using the observation that realistic bubbles have static instanton solutions, the Ansatz used in [17] was

$$ds^2 = f(r)e^{2\delta(r)}d\tau^2 + \frac{dr^2}{f(r)} + r^2(d\theta^2 + \sin^2\theta d\varphi^2), \tag{57}$$

where f contains contributions from the black hole, the bubble, and any vacuum energy

$$f = 1 - \frac{2G\mu(r)}{r} \tag{58}$$

in the function $\mu(r)$. The redshift function $\delta(r)$ on the other hand responds purely to the energy momentum of the bubble. The equations of motion are

$$f\phi'' + f'\phi' + \frac{2}{r}f\phi' + \delta' f\phi' = V_\phi, \tag{59}$$

$$\mu' = 4\pi r^2\left(\frac{1}{2}f\phi'^2 + V\right), \tag{60}$$

$$\delta' = 4\pi G r\phi'^2 \tag{61}$$

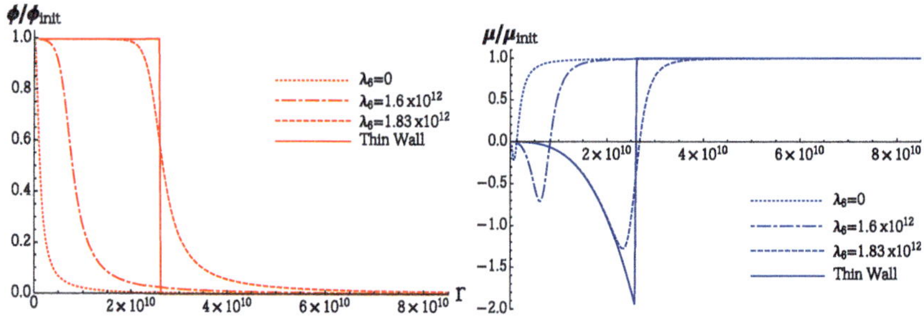

Fig. 8 An illustration of the effect of switching on the λ_6 parameter. The thin wall solution is shown as a solid line, with intermediate values of λ_6 shown. Note however the large values of the coupling that are necessary to produce anything approaching a thin wall

The equations for μ and ϕ can be decoupled from the δ equation, which can be found from substituting the solution for ϕ in (61).

Figure 8 shows the solutions for the bounce, changing λ_6 to tune the bubble wall solutions from the thin wall of the previous section to the thick wall of the SM Higgs with $\lambda_6 = 0$. The full details of the process are given in [17].

Having found the solution, the exponent in the bounce amplitude is given by the entropy shift between the seed and remnant black hole. In this case, considering the decay of the Higgs vacuum, the false vacuum energy is taken to be zero, therefore we are tunnelling to a negative vacuum energy. Writing $\mu(r_-) = \mu_-$ as the boundary value of μ at the horizon of the remnant black hole, and $\mu_\infty = M_+$ as the asymptotic value of μ we have

$$\mathcal{B} = \frac{\mathcal{A}_+}{4G} - \frac{\mathcal{A}_-}{4G} = \frac{\mu_\infty^2 - \mu_-^2}{2M_p^2} \tag{62}$$

Thus, having found the numerical bubble solution, the bounce action is easily extracted in terms of the values of μ at each of the boundaries.

The main output of the numerical work in [17] is that the bubble is diffuse, with the Higgs gradually varying from the SM value to large ϕ at the remnant black hole horizon. Due to the weak coupling of gravity, this profile has a relatively small effect on the mass function, so $(\mu_\infty - \mu_-)/M_p \ll 1$. The vacuum decay rate has the form $\mathcal{A}e^{-\mathcal{B}}$, where $\mathcal{A}$ is a pre-factor determined in field theory by the determinant of fluctuations around the saddle point together with a factor of $\sqrt{\mathcal{B}/2\pi}$ for each translational zero mode [18]. Here, we have a single zero mode from the time-translation symmetry, and estimate the determinant factor (the evaluation of the determinant is problematic in Euclidean Quantum Gravity) by horizon timescale $(GM_+)^{-1}$. Putting this together, our estimate for the vacuum decay rate is:

$$\Gamma_D \approx \left(\frac{\mathcal{B}}{2\pi}\right)^{\frac{1}{2}} (GM_+)^{-1} e^{-\mathcal{B}} \tag{63}$$

We now turn to the consequences of this result.

4.2 A Comparison with Hawking Evaporation

Having computed the vacuum decay rate in the presence of a black hole, we note that, as is characteristic for a tunnelling process, it is exponentially suppressed. However, no matter how strongly suppressed, if the half life is significantly less than the age of the universe this is problematic! On the other hand, we also have Hawking's result [44], that black holes evaporate, and taking Page's result for the evaporation rate [60]

$$\Gamma_H = \frac{\dot{M}}{M} \approx 3.6 \times 10^{-4} (G^2 M_+^3)^{-1} \tag{64}$$

we find the branching ratio of the tunnelling rate to the evaporation rate is

$$\frac{\Gamma_D}{\Gamma_H} \approx 44 \frac{M_+^2}{M_p^2} \mathcal{B}^{1/2} e^{-\mathcal{B}} \tag{65}$$

Whether this branching ratio is larger than unity for semi-classical black holes centres around the comparison between the power law pre-factors of M_+/M_p, which must be large, and the size of the exponent $\mathcal{B}$ for those values. Having already noted that the difference in the mass function is small for the thick Higgs bubbles, a rough estimate of the bounce action is $\mathcal{B} \sim M_+ \delta M / M_p^2$, hence we can get a quick estimate of the behaviour of this branching ratio

$$\frac{\Gamma_D}{\Gamma_H} \approx 44 \left(\frac{M_+}{M_p} \right)^{5/2} \sqrt{\frac{\delta M}{M_p}} e^{-M_+ \delta M / M_p^2} \tag{66}$$

Figure 9 shows a plot of this function, and how it depends on this mass difference, which itself in turn depends on the energy contained in the diffuse Higgs bubble. While M_+ itself is large relative to the Planck mass (so that the spacetime curvature remains safely below the Planck scale), δM will be much smaller, as it is dependent on the scale of the Higgs relative to the Planck scale, integrated over the bubble.

As Fig. 9 shows eloquently, as δM drops, the tunnelling rate is not just faster than evaporation, but becomes strongly dominant (by factors of 10^{10} or so) for seed masses around a gram. The lifetime of a black hole of this mass is around 10^{-25}s, hence if the tunnelling process is strongly dominant, the decay will be essentially instantaneous.

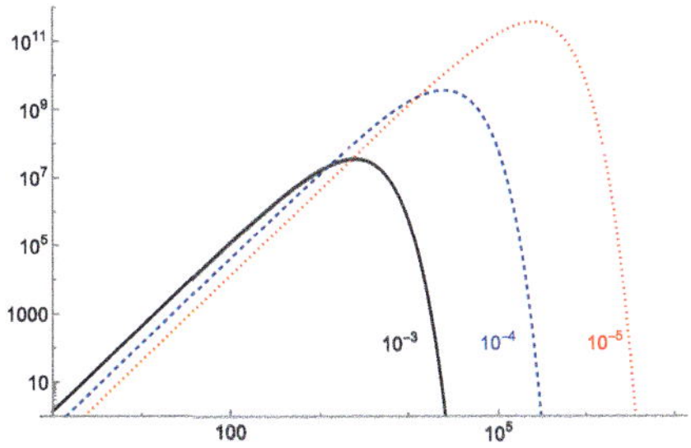

Fig. 9 The branching ratio strongly depends on the difference in the mass function generated by the diffuse bubble

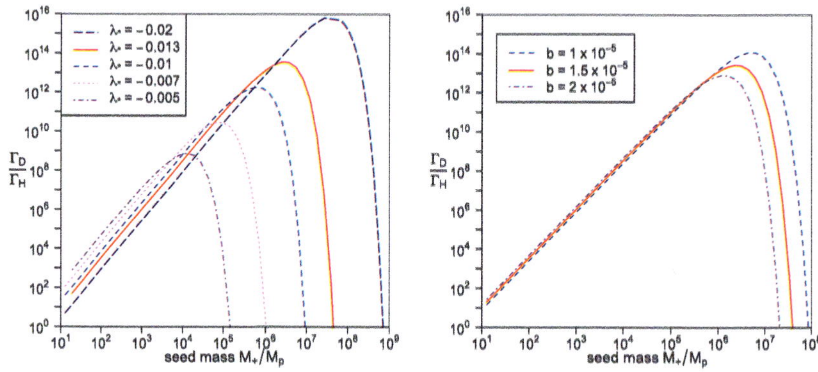

Fig. 10 The branching ratio of vacuum decay to evaporation shown as a function of seed mass for different values in the 3 parameter fit to the SM potential. On the Left, the impact of varying λ_* with $b = 1.4 \times 10^{-5}$. On the Right, the impact of varying b with $\lambda_* = -0.013$. (Note that the value of the coupling at the SM scale then fixes c in terms of b and λ_*)

Having acquired some intuition on how important the vacuum decay process is likely to be, we now show the actual results of [17] for the vacuum decay branching ratios scanning through various fits to the SM potential.

As Fig. 10 shows, there is a wide range of black hole masses for which vacuum decay is the dominant process. In terms of the three parameter fit to the running Higgs coupling, the bare quartic coupling λ_* has the biggest impact on the branching ratio, however it is clear that for a wide range of potentials, vacuum decay is inevitable if a black hole has small enough mass.

From Fig. 10, we see that in order to catalyse vacuum decay a black hole has to be very light–of order a Kg or less! The only possibility for this type of black hole is that it is a *Primordial Black Hole* (PBH), i.e., one that is not formed by gravitational mass of a star, but rather formed from strong perturbation fluctuations in the very early universe [19]. Black holes formed at or below a mass of $\sim 4.5 \times 10^{14}$g would have already lost sufficient mass by evaporation to lie within this range of catalysis. Given that we are in the (presumed) metastable vacuum currently, we can deduce that there was no PBH with mass lighter than $\sim 4.5 \times 10^{14}$g in our past light cone, however the question of whether black holes lighter than this are incompatible with a metastable vacuum is a little more subtle, as described in [27], as the black hole not only has to seed decay, but the true vacuum would have to have percolated to encompass our current Hubble volume. Figure 11 gives a sketch of how the bubbles would expand and percolate if formed at a redshift $Z = 10$. The main outcome of the cosmological study is that if PBH's are formed strongly peaked around a mass of 5×10^{14} or less, then this is incompatible with a metastable Higgs vacuum. (See also [25,58] for constraints from high energy particle collisions and extra dimensions).

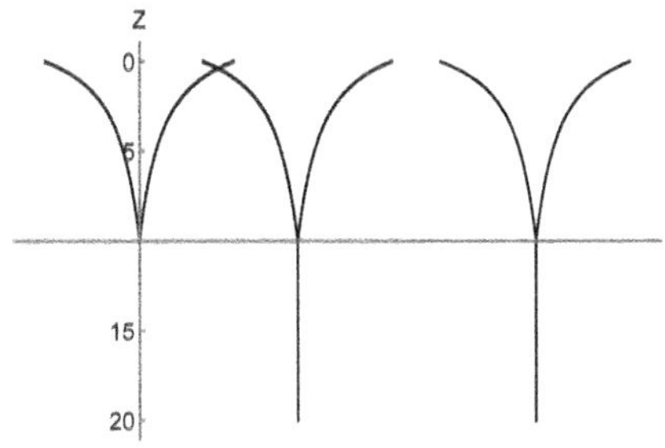

Fig. 11 A sketch of the bubble percolation process for black holes reaching the critical mass range for vacuum decay at a redshift of 10

5 Closing Thoughts

In this presentation, I reviewed my work on vacuum decay catalysis by black holes. In these examples, decay proceeds via bubble nucleation and is described by a semi-classical Euclidean approach. One can have concerns that this is not a well-defined approach for including gravity in quantum processes, but then to be consistent one should never use the Euclidean method for computing the temperature of a black hole. My opinion is that one either accepts *all* the usual compromises for quantum processes with gravity, or seeks an alternative. Given the success of the inflationary paradigm in explaining the perturbation spectrum of the microwave background, this would be a rather extreme point of view! Nonetheless, the result that having an impurity enhances and catalyses decay is very much in keeping with the way first order phase transitions manifest in nature.

Other explorations of the impact of black holes on tunnelling processes include the Hawking Moss instanton [45]—an even more perplexing quantum process in which the entire universe up-tunnels to a higher cosmological constant before relaxing to a lower vacuum energy than the initial state. Here, the black hole also enhances the tunnelling probability [41,42] while at the same time raising interesting questions around constraints on allowed vacuum transitions. Tunnelling in extended theories of gravity (e.g., [26,43,61]) both with and without seeds has also been studied, with broadly similar outcomes.

The use of the regularisation of singular instantons raises questions around analytic continuation and whether the thermal nature of the environment has fully been taken into account (see e.g., [14,46,48,53,64,65]). What these results generally indicate is that the Euclidean method is somehow dissatisfying as a description of quantum tunnelling, and that perhaps the time is ripe to consider alternative approaches to describing tunnelling in both field theory and gravity (see, e.g., [12,33,47]) or indeed to testing tunnelling, and the implications of seeds, in the lab [8,11,34,69].

Ultimately, whether or not the Standard Model is metastable, the challenge to critique our set of tools and methodology in semi-classical gravity and field theory gives a timely and compelling impetus to re-examine our description of nonperturbative processes, and hopefully will lead to new and deeper understanding.

Acknowledgements It is a pleasure to thank all my collaborators on this program of work: Tom Billam, Philipp Burda, Leo Cuspinera, De Chang Dai, ShiQian Hu, Katie Marshall, Florent Michel, Naritaka Oshita, Sam Patrick, Dejan Stojkovic, Ben Withers, and a special thanks to my primary collaborator, Ian Moss. This work has been supported by the STFC grants ST/P000371/, ST/J000426/1. I also acknowledge the Aspen Center for Physics, supported by National Science Foundation grant PHY-2210452, and the Perimeter Institute for Theoretical Physics. Research at Perimeter Institute is supported by the Government of Canada through the Department of Innovation, Science and Economic Development Canada and by the Province of Ontario through the Ministry of Colleges and Universities.

References

1. G. Aad et al., ATLAS collaboration. Phys. Lett. B **710**, 49 (2012)
2. G. Aad et al., ATLAS. JHEP **11**, 150 (2019)
3. T. Banks, C.M. Bender, T.T. Wu, Phys. Rev. D **8**, 3346–3378 (1973)
4. V. Berezin, V. Kuzmin, I. Tkachev, Phys. Lett. B **207**, 397 (1988)
5. B. Bergerhoff, M. Lindner, M. Weiser, Phys. Lett. B **469**, 61 (1999)
6. F. Bezrukov, M.Y. Kalmykov, B.A. Kniehl, M. Shaposhnikov, JHEP **1210**, 140 (2012)
7. F. Bezrukov, M. Shaposhnikov, Zh. Eksp, Teor. Fiz. **147**, 389 (2015)
8. T.P. Billam, R. Gregory, F. Michel, I.G. Moss, Phys. Rev. D **100**(6), 065016 (2019)
9. K. Blum, R.T. D'Agnolo, J. Fan, JHEP **03**, 166 (2015)
10. P. Bowcock, C. Charmousis, R. Gregory, Class. Quant. Grav. **17**, 4745 (2000)
11. J. Braden, M.C. Johnson, H.V. Peiris, S. Weinfurtner, JHEP **07**, 014 (2018)
12. J. Braden, M.C. Johnson, H.V. Peiris, A. Pontzen, S. Weinfurtner, Phys. Rev. Lett. **123**(3), 031601 (2019)
13. V. Branchina, E. Messina, M. Sher, Phys. Rev. D **91**, 013003 (2015)
14. V. Briaud, A. Shkerin, S. Sibiryakov, Phys. Rev. D **106**(12), 125001 (2022)
15. P. Burda, R. Gregory, I. Moss, Phys. Rev. Lett. **115**, 071303 (2015)
16. P. Burda, R. Gregory, I. Moss, JHEP **1508**, 114 (2015)
17. P. Burda, R. Gregory, I. Moss, JHEP **1606**, 025 (2016)
18. C.G. Callan, S. Coleman, Phys. Rev. D **16**, 1762 (1977)
19. B.J. Carr, S.W. Hawking, Mon. Not. Roy. Astron. Soc. **168**, 399 (1974)
20. S. Chatrchyan et al., CMS collaboration. Phys. Lett. B **710**, 26 (2012)
21. P. Chen, G. Domènech, M. Sasaki, D.h. Yeom, JHEP **07**, 134 (2017)
22. K.G. Chetyrkin, M.F. Zoller, JHEP **1206**, 033 (2012)
23. S. Coleman, Phys. Rev. D **15**, 2929 (1977)
24. S. Coleman, F. De Luccia, Phys. Rev. D **21**, 3305 (1980)
25. L. Cuspinera, R. Gregory, K. Marshall, I.G. Moss, Phys. Rev. D **99**(2), 024046 (2019)
26. L. Cuspinera, R. Gregory, K.M. Marshall, I.G. Moss, Int. J. Mod. Phys. D **29**(01), 2050005 (2020)
27. D.C. Dai, R. Gregory, D. Stojkovic, Phys. Rev. D **101**(12), 125012 (2020)
28. G. Degrassi, S. Di Vita, J. Elias-Miro, J.R. Espinosa, G.F. Giudice, G. Isidori, A. Strumia, JHEP **1208**, 098 (2012)
29. A. Eichhorn, H. Gies, J. Jaeckel, T. Plehn, M. Scherer, R. Sondenheimer, JHEP **04**, 022 (2015)
30. J. Elias-Miro, J.R. Espinosa, G.F. Giudice, G. Isidori, A. Riotto, A. Strumia, Phys. Lett. B **709**, 222 (2012)
31. J. Ellis, J. Phys. Conf. Ser. **631**(1), 012001 (2015)
32. J.R. Espinosa, G.F. Giudice, A. Riotto, JCAP **0805**, 002 (2008)
33. J. Feldbrugge, N. Turok, Ann. Phys. **454**, 169315 (2023)
34. O. Fialko, B. Opanchuk, A.I. Sidorov, P.D. Drummond, J. Brand, EPL **110**(5), 56001 (2015)
35. C. Ford, D.R.T. Jones, P.W. Stephenson, M.B. Einhorn, Nucl. Phys. B **395**, 17 (1993)
36. G.W. Gibbons, S.W. Hawking, Phys. Rev. D **15**, 2752 (1977)

37. D. Gorbunov, D. Levkov, A. Panin, JCAP **10**, 016 (2017)
38. A. Gorsky, A. Mironov, A. Morozov, T.N. Tomaras, JETP **120**, 399 (2015), [Zh. Eksp. Teor. Fiz. **147**, 399 (2015)]
39. E. Greenwood, E. Halstead, R. Poltis, D. Stojkovic, Phys. Rev. D **79**, 103003 (2009)
40. R. Gregory, I.G. Moss, B. Withers, JHEP **1403**, 081 (2014)
41. R. Gregory, I.G. Moss, N. Oshita, JHEP **07**, 024 (2020)
42. R. Gregory, I.G. Moss, N. Oshita, S. Patrick, JHEP **09**, 135 (2020)
43. R. Gregory, S.Q. Hu, *Seeded Vacuum Decay with Gauss-Bonnet*, arXiv:2305.03006 [hep-th]
44. S.W. Hawking, Commun. Math. Phys. **43**, 199 (1975)
45. S.W. Hawking, I.G. Moss, Phys. Lett. B **110**, 35 (1982), [Adv. Ser. Astrophys. Cosmol. **3**, 154 (1987)]
46. T. Hayashi, K. Kamada, N. Oshita, J. Yokoyama, JHEP **08**, 088 (2020)
47. T. Hayashi, K. Kamada, N. Oshita, J. Yokoyama, JCAP **05**(05), 041 (2022)
48. M. He, K. Kohri, K. Mukaida, M. Yamada, JCAP **01**, 027 (2023)
49. W.A. Hiscock, Phys. Rev. D **35**, 1161 (1987)
50. G. Isidori, G. Ridolfi, A. Strumia, Nucl. Phys. B **609**, 387 (2001)
51. W. Israel, Nuovo Cimento Soc. Ital. Phys. **B44**, 4349 (1966)
52. I.Y. Kobzarev, L.B. Okun, M.B. Voloshin, Sov. J. Nucl. Phys. **20**, 644 (1975), [Yad. Fiz. **20**, 1229 (1974)]
53. K. Kohri, H. Matsui, Phys. Rev. D **98**(12), 123509 (2018)
54. I.V. Krive, A.D. Linde, Nucl. Phys. B **432**, 265 (1976)
55. Z. Lalak, M. Lewicki, P. Olszewski, JHEP **1405**, 119 (2014)
56. M. Lindner, M. Sher, H.W. Zaglauer, Phys. Lett. B **228**, 139 (1989)
57. F. Loebbert, J. Plefka, *Quantum Gravitational Contributions to the Standard Model Effective Potential, Vacuum Stability*, arXiv:1502.03093 [hep-ph]
58. K.J. Mack, R. McNees, Phys. Rev. D **99**(6), 063001 (2019)
59. K. Mukaida, M. Yamada, Phys. Rev. D **96**(10), 103514 (2017)
60. D.N. Page, Phys. Rev. D **13**, 198 (1976)
61. A. Rajantie, S. Stopyra, Phys. Rev. D **95**(2), 025008 (2017)
62. M. Sher, Phys. Rept. **179**, 273 (1989)
63. A. Shkerin, S. Sibiryakov, Phys. Lett. B **746**, 257 (2015)
64. A. Shkerin, S. Sibiryakov, JHEP **08**, 161 (2022)
65. A. Shkerin, S. Sibiryakov, JHEP **11**, 197 (2021)
66. A.M. Sirunyan et al., CMS. Eur. Phys. J. C **80**(7), 658 (2020)
67. N. Tetradis, JCAP **1609**(09), 036 (2016)
68. M.S. Turner, F. Wilczek, Nature **D79**, 633 (1982)
69. A. Zenesini, A. Berti, R. Cominotti, C. Rogora, I.G. Moss, T.P. Billam, I. Carusotto, G. Lamporesi, A. Recati, G. Ferrari, *Observation of False Vacuum Decay via Bubble Formation in Ferromagnetic Superfluids* arXiv:2305.05225 [hep-ph]

Correction to: Gravity, Cosmology, and Astrophysics

Betti Hartmann and Jutta Kunz

Correction to: B. Hartmann and J. Kunz (eds.), *Gravity, Cosmology, and Astrophysics*, Lecture Notes in Physics 1022, https://doi.org/10.1007/978-3-031-42096-2

The book was originally published electronically on the publisher's internet portal non open access. With the editor(s)' decision to opt for Open Choice the copyright of the book changed on 17-May-2024 to "© The Authors 2024" and the chapters are forthwith distributed under a Creative Commons Attribution under a CC BY 4.0 license and the copyright holder updated to 'The Author(s)'. The funder credit line in the Front matter has been updated. The book has also been updated with these changes.

The updated version of the book can be found at https://doi.org/10.1007/978-3-031-42096-2

B. Hartmann and J. Kunz (eds.), *Gravity, Cosmology, and Astrophysics*,
Lecture Notes in Physics 1022, https://doi.org/10.1007/978-3-031-42096-2_13

Glossary

Accretion disc: a disc/torus of matter in orbital motion around a compact object

AdS/CFT correspondence: conjectured duality between a gravitational theory in a d-dimensional Anti-de-Sitter space-time (AdS) and a $(d-1)$-dimensional conformal field theory (CFT)

Asymptotically flat manifold: a Riemannian manifold such that the complement of a compact set is diffeomorphic to R^n without a ball and such that in these coordinates the metric and its partial derivatives up to second order decay to the Euclidean metric respectively zero sufficiently fast at infinity. The correspoding coordinates are called asymptotically flat coordinates

Asymptotically flat initial data set: an asymptotically flat manifold with a symmetric (0,2)-tensor K decaying to zero sufficiently fast in the asymptotically flat coordinates. K plays the role of the second fundamental form if the initial data set is embedded into spacetime.

(Asymptotic) parity conditions: conditions sometimes imposed on asymptotically flat coordinates demanding a stronger decay on the odd respectively even parts of g respectively K ensuring convergence of the center of mass integrals. Also called Regge-Teitelboim conditions.

Axial perturbation: Stellar oscillation of odd-parity according to the spherical harmonics

Black hole: a solution to the equations of gravity models that possesses a space-time curvature singularity shielded from observation by an external observer by an event horizon

Brown dwarfs: substellar objects that are not massive enough to start the nuclear fusion from Hydrogen to Helium that is typical for ordinary stars

Chameleon scalar field: a scalar field that couples to a matter fluid (i.e. dark energy candidate)

Center of mass: the center of mass of an asymptotically flat manifold (or initial data set) is given via the limit of a flux integral through large coordinate spheres, if this limit exists. An alternative description can be given using the limit of the coordinate centers of large constant (spacetime) mean curvature surfaces

B. Hartmann and J. Kunz (eds.), *Gravity, Cosmology, and Astrophysics*,
Lecture Notes in Physics 1022, https://doi.org/10.1007/978-3-031-42096-2

Conformal field theory: a quantum field theory with conformal invariance; typically used to describe systems with scale invariance, e.g. critical points in statistical physics or condensed matter physics

Cosmic string: a line-like topological defect that could have formed during one of the phase transitions in the early universe whenever a U(1) symmetry remains unbroken; can intersect and form cosmic string loops

Ellis-Bronnikov wormholes: Specific types of wormholes described by static spherically symmetric solutions of general relativity coupled to a phantom scalar field.

Entanglement: a quantum information theoretic concept that is also used in the context of quantum field theories and captures the degree of inability to factor a quantum state into a product of states

Equation of state: A relation between the energy density, pressure and temperature of the dense matter in neutron stars

Event horizon: boundary separating the interior and exterior of a black hole such that the interior cannot influence the exterior; surface of infinite redshift

Grand Unified Theory: a suggested model that unifies the three fundamental interactions that can be described by a gauge theory, i.e. the electromagnetic, the weak nuclear and the strong nuclear interaction

Gravitational lensing: phenomenon similar to what happens when a light ray crosses an optical lens: the trajectory changes, but instead of an optical lens, a massive body (e.g. a galaxy or a black hole) curves the space-time leading to a beding of the photons' path

Harmonic coordinates: coordinates on a Riemannian manifold such that the coordinate functions are harmonic, i.e., satisfy $\Delta x^i = 0$

Higgs boson: quantum excitation of the Higgs field, a scalar field that plays a crucial role in the Higgs mechanism which—through spontaneous symmetry breaking—gives mass to the elementary gauge bosons in the Standard Model of Particle Physics

Holography: collection of conjectured dualities between gravitational theories and lower-dimensional quantum field theories. It is used as an approach to quantum gravity. The best studied example of holography is the AdS/CFT correspondence

Hydrodynamics: describes the mechanics of fluids, in particular the dynamics of fluids

Inflation: a cosmic epoch of exponential expansion of space in the very early universe

Isospectrality: the property of having identical or degenerate spectra; here, it refers to the observation that the quasinormal modes of the massless Ellis-Bronnikov wormhole exhibit a threefold degeneracy, indicating isospectrality among the axial and polar modes

Kerr space-time: analytically given solution of the vacuum Einstein equation that is stationary and axisymmetric; describes the exterior of a black hole that possesses mass and angular momentum

Killing vector: a vector that generates isometries, i.e. moving along the flow generated by the Killing vector field does not change the metric

Naked singularity: a space-time singularity that is not shielded from view by an event horizon (in contrast to the space-time singularity at the centre of black holes)

Neutron star: a compact object with around one to two solar masses; a product of the collapse of a massive star

Polar perturbation: stellar oscillation of even-parity according to the spherical harmonics

Quasinormal mode: characteristic frequency and damping time of the gravitational waves emitted from the ringing object; characteristic features of compact objects that describe their response to perturbations

Ray tracing: method that traces backward the geodesic equation from our "screen" to the massive body, recording the number of turns around it with different impact parameters

Scalar field: associates a scalar value to every point in a space(-time)

Scalar-tensor theory: an alternative gravity theory which extends General Relativity by introducing scalar field(s)

Schwarzschild space-time: analytically given solution of the vacuum Einstein equation that is static and spherically symmetric; describes the exterior of a black hole that possesses mass

Shadow: dark central area of the direct imaging of a black hole, from which we have not received photons; the *inner shadow* is the smallest possible shadow we can observe

Stochastic gravitational wave background: a background of random gravitational waves that is suggested to exist in the universe; sources can be cosmological (early universe) or astrophysical

Transfer functions: obtained using ray tracing procedure; correspond to the radius of the m-th intersection between the light ray and the accretion disk

Vacuum decay: if a model possesses several local minima (of the energy) it is possible that the system decays from the false vacuum to the true vacuum; our universe seems to be currently in such a false vacuum

White dwarfs: Compact objects and end-state of stars in which the gravitational collapse is prevented by the pressure of degenerate electrons; cannot be more massive than the so-called Chandrasekhar bound of 1.4 solar masses

Wormhole: a solution to the equations of gravity models that connects one asymptotically flat space-time to another one; bridges that connect different regions of space-time

The manufacturer's authorised representative in the EU is Springer Nature Customer Service Centre GmbH, Europaplatz 3, 69115 Heidelberg, Germany. If you have any concerns regarding our products, please contact ProductSafety@springernature.com

Printed and bound by CPI Group (UK) Ltd, Croydon, CR0 4YY
28/07/2026
02179195-0003